The True and perfect
Seat upon Horse-back

The Gentleman's DICTIONARY

In Three Parts:

I. *The Art of Riding the Great Horse*

II. *The Military Art*

III. *The Art of Navigation*

By Georges Guillet de St. Georges

FACSIMILE

SHERWIN & FREUTEL *Publishers*

Los Angeles : 1970

SBN 8404-0006-3

LIBRARY OF CONGRESS CATALOG CARD NUMBER 77-103247

THE GENTLEMAN's DICTIONARY.

In THREE PARTS.

VIZ.

I. The *Art of Riding the Great Horſe*: Containing the Terms and Phraſes us'd in the *Manage*, and the Diſeaſes and Accidents of Horſes.

II. The *Military Art*; Explaining the Terms and Phraſes us'd in Field or Gariſon; The Terms relating to *Artillery*; The Works and Motions of *Attack* and *Defence*; and the Poſt and Duty of all the Officers of the Army: Illuſtrated with Hiſtorical Inſtances, taken from the Actions of our Armies.

III. The *Art of Navigation*; Explaining the Terms of Naval Affairs; as Building, Rigging, Working, and Fighting of Ships; the Poſt and Duty of Sea-Officers, *&c.* With Hiſtorical Examples taken from the Actions of our Fleet.

Each Part done Alphabetically, from the Sixteenth Edition of the Original *French*, Publiſhed by the Sieur *GUILLET*, and Dedicated to the *Dauphin*. With Large Additions, Alterations and Improvements, adapted to the Cuſtoms and Actions of the *Engliſh*; and above Forty Curious Cuts, that were not in the Original.

LONDON:

Printed for *H. Bonwicke* in St. *Paul's* Church-Yard; *T. Goodwin*, *M. Wotton*, *B. Tooke*, in *Fleetſtreet*; and *S. Manſhip* in *Cornhill*. 1705.

THE

Publishers PREFACE.

THE Sieur Guillet *having published in* French, Les Arts de l'Homme d' Epee, *in the Form of a Dictionary, in Three distinct Alphabets; which has now born Sixteen Impressions in a few Years; we thought it would be acceptable to the Publick, to make it speak* English. *But finding 'twas capable of many Improvements, and was altogether accommodated to the* French *Customs and Actions; we chose, instead of a bare Translation, to take in such Improvements and Corrections, as were handed to us by Persons thoroughly vers'd in the respective Arts; and to make such Alterations, as would better suit the Customs and History of* England.

The General Design of this Work, is to serve not only those who are bred to the Sword, but all

that pretend to a Gentlemanlike Education; and at the ſame time to gratify the Curioſity of others, who will, doubtleſs, be fond of knowing the true Import of ſuch Terms and Phraſes, as are daily met with in Common Diſcourſe, and are uſually tack'd to the Buſieſt Actions of Life; Now that England *is retriev'd from Luxury and Softneſs, and once more fam'd for Manly Exerciſes and Actions.*

The Alphabet which leads the Van in this Performance, explains all that relates either to a Horſe, or a Horſeman. It includes not only the Definitions of Things and Phraſes uſed in the Manage, or Academies for Riding; but likewiſe the Terms relating to the Condition, Uſe, Feeding, and Treatment of a Horſe; to the Diſeaſes of Horſes, and the uſual Remedies, with the manner of giving 'em; and, in fine, to all the Appurtenances of that Uſeful Animal. In compiling this Part, the Sieur Guillet *made uſe of the Aſſiſtance of Meſſieurs* de Bournonville, *and* du Pleſſis, *(both Riding-Maſters to the* Dauphin *of* France) *and above all, of Mr.* Solleyſel, *the Excellent Author of the* Compleat Horſeman; *to which our Reader is frequently referr'd. In tranſlating this Part, we have taken care to do Juſtice to the* French, *and at the ſame time to bring it as near to our Jockey Terms, as the Nature of the Thing would allow.*

The

The Second Part ranges the Military Terms in an Alphabetical Order; most of which are Illustrated with particular Phrases, pointing to the Famous Instances of English *Bravery and Conduct; especially those which late Years have produced. The Sieur* Guillet's *Illustrations were all taken from the Actions of the* French, *which we chose to Exchange for* English: *And at the same time we thought it not improper to give new Definitions of many Terms, pursuant to the present Practice of our own Armies and Engineers; as well as to add a great many useful Things that were either invented out of* France, *or brought to Light since the Sieur* Guillet *writ. To render this Part yet more Instructive, we have Inserted Accurate Cuts of the Engines, Instruments, Contrivances,* &c. *us'd in War; and have prefix'd to it a General Draught of all the Parts of Fortification, and the Method of Approaches in Sieges; which we take to be the distinctest and justest that has yet appear'd. In fine, the Reader will find this Part altogether new moulded; which is owing to the Care and Capacity of a Gentleman, who is perfectly well vers'd in all the Arts retaining to the* Military Profession, *and was an Eye-Witness of all the Actions of our Army, in this and the last War.*

As for the Third and Last Part, relating to Navigation, *we have Calculated the Sea-Terms*

and

and Phrases, the Parts and Materials of Ships, and all the other Terms of Art, according to the English *Practice of* Navigation, *and ways of Speaking: And have thrown in Historical Instances of the* English *Naval Exploits; which all the World knows to surpass those of other Nations, both in Glory and Number. To this Part we have prefix'd an exact Draught of a Ship compleatly Rigg'd, with Explicatory References to all its Parts, Ropes,* &c. *And in fine, we reckon we may, without Vanity, affirm, That in this, as well as the Second Part, we have far out-done our Original.*

THE

THE
Gentleman's Dictionary.

PART I.
The Manage: Or the Art of Riding.

In the ensuing Alphabet the Terms are ranged in an Alphabetical Order: And in regard the manner of applying is not less necessary than their Explication, you have particular Phrases tack'd to the most essential Words; and those so contriv'd, that they not only shew in what manner these Terms should be employed in a regular way of speaking, but likewise lay open the Grounds of the Matter in hand.

The Definition of the Art of Riding.

The Art of Riding teaches at once how to form both the Horseman and the Horse: The former it teaches a good Seat upon Horseback, a free, easy, disengaged Posture, and the means of making his Hand accord with his Heels: The Horse it instructs, as much as is possible, how to carry well, to take his Aids gentle and fine, to fear the Corrections that can fix him to a Walk, a Trot, and a Gallop; and then to manage or work upon all sorts of Airs; that thus broken and managed, he may be of Use in the Dangers of War, in the Necessities of Life, and sometimes in the Pomp and Splendor of Festivals and Publick Shews.

ABATE; A Horse is said to abate or take down (*Rabattre*) his Curvets, when working upon Curvets he puts his two Hind-legs to the ground both at once, and observes the same exactness in all the times. Your Horse abates his Curvets very well; for in regard he *harps* or *grapples* with his two Hams, and in working keeps his Legs very low, he abates his Curvets with a good Grace;

 that

that is, his two hinder Feet touch the ground together.

ACTION of the Mouth is the Agitation of the Tongue and the Mandible of a Horſe, that by champing upon the Bridle, keeps his Mouth freſh. You ſee by this white ropy Foam, that this Horſe has the *Action* of the Mouth; which is a ſign of Vigour, Mettle, and Health.

ACULER; a *French* Word us'd in the Academies, importing, that a Horſe in working upon Volts in the Manage, do's not go far enough forwards at every Time or Motion, ſo that his Shoulders embrace or take in too little ground, and his Croupe comes too near to the Center of the Volt. This Horſe has *aculé*, becauſe the Horſeman did not turn his hand, and put him on with the Calf of his inner Leg. Horſes have a natural Inclination to this Fault, in making Demivolts. When the *Italians* work a Horſe upon the Demivolts, call'd *Repolons*, they affect to make 'em *aculer*, or cut ſhort. See *Entabler*, and *Repolon*.

AGE of a Horſe, is the Space of Time elapſed ſince he was foal'd: Which deſerves a very particular regard, upon the account that a Horſe has the advantage of being capable to ſhew and ſet forth the different Progreſs of his firſt Years; partly by the Number of his Teeth, according as he caſts or ſhoots 'em forth one after another, or as theſe change and give place to freſh ones; partly by black Specks or Marks placed by Nature upon ſome of 'em; and partly by ſome other Preſumptions of leſs Certainty, ſuch as thoſe taken from the Knots or Joynts of the Tail. The Youngeſt of theſe Horſes can't be above Two Years and a Half old; for he is now but Caſting his Nippers. The other that's a Caſting his Middling Teeth, is between Three and a Half and Four Years. And the third, that now puts forth his Tuſhes, and whoſe Corner-Teeth are upon the point of appearing, muſt be about Four Years old. But as for that Old *Engliſh* Horſe, his Age is counterfeited; he has had Tricks play'd with his Teeth: You ſee by his Corner-Teeth that are ſo Long, and Clear of Fleſh, that he's certainly countermark'd; And tho' it would ſeem as if he had not yet ras'd, and were not yet Seven Years old, he begins to ſeel. See *Teeth*, *Shell-tooth'd*, *Tail*, *Raze*, and *Seel*.

AID; to Aid, Aſſiſt, or Succour a Horſe, is to ſuſtain and help him to work true, and mark his Times or Motions with a juſt Exactneſs. Aſſiſt your Horſe with the Calves of your Legs; help him with a nice tender Heel; aid him with your Tongue. 'Tis not enough to aid this Horſe with the

the Rod, he must have harsher *Aids*.

Aids are the Helps or Assistance that the Horseman gives from the gentle and moderate Effects of the Bridle, the Spur, the Caveson, the Poinson, the Rod, the Action of the Legs, the Motion of the Thighs, and the Sound of the Tongue. We give these *Aids* to prevent the Correction or Chastisement that is sometimes necessary in Breaking and Managing a Horse. You'll never ride well, unless you be very attentive and active without precipitancy, in not losing or missing your Times, and in giving the Aids seasonably: For without that, you'll accustom your Horse to dose upon it. If this Horse does not obey the Aids of the Calves of his Legs, help him with the Spur, and give him a prick or two. This Sorrel-Horse has his Aids very nice; that is, he takes them with a great deal of Facility and Vigor. This Gentleman gives his Aids very fine; that is, he animates and rouzes up the Horse seasonably, and helps him at just Turns, in order to make him mark his Times or Motions justly. This Barb knows the Aids; he obeys or answers the Aids; he takes them fine. You do not give the Aids of the Caveson with discretion; you make a Correction of 'em, which will baulk your Horse. See *Brouiller*.

Inner Aids, *Outer Aids*. See *Inside*, *Enlarge*, and *Narrow*.

AIR is a Cadence and Liberty of Motion accommodated to the natural Disposition of the Horse, which makes him work in the Manage, and rise with Obedience, Measure, and Justness of Time. Some Riding-Masters take the word *Air* in a strict Sense, as signifying a Manage that's higher, slower, and more artful or design'd than the *terra a terra*. But others give it a larger Signification, including under that sense a *terra a terra*; for if a Horse manages well in a *terra a terra*, they say the Horseman has happily hit the Air of the Horse. In general, the Walk, Trot, and Gallop, are not accounted *Airs*. And yet some very good Riding-Masters would understand by *Air* the Motion of a Horse's Legs upon a Gallop. For instance, they'll say such a Horse has not the natural Air; that is, in galloping he bends his Fore-legs too little. You should give or form an Air to your Horse, for he has no natural Air; and since his Haunches are very good, he's capable of the Manage, if you do but learn him an Air. All your Horses have an Air naturally; that is, they have Motion enough with their Fore-legs to take a Cadence, if they're put to work at *terra a terra*. This Horse always takes his Lesson with his own Air. Fix or confirm that Horse in

the Air he has taken. This Sorrel takes the Air of the Curvets; but that presents himself with an Air for *Caprioles*. This Mare has no Inclination nor Disposition to these Airs. See *Pesate*.

High Airs or High Manage, are the Motions of a Horse that rises higher than *terra a terra*, and works at Curvets, Balotades, Croupades, and Caprioles. In regard that Horse has the Beginning or first Steps of rais'd Airs, and of himself affects a high Manage, you ought to use this his Disposition discreetly, that he may not be dishearten'd or baulk'd; for your high Airs make a Horse angry when he's too much put to't; and you ought to have suppl'd his Shoulders very well before you put him to leap. See *Pesate*, and *Leaping*.

AMBLE, is the Going or Pace of a Horse; the Motion of which is two Legs of a side rais'd and set down together, after which the two Legs of the other side rise and come down in the same manner; each side observing an alternate Course. The first Pace of young Colts or Foals is the Amble; but they quit it as soon as they have strength enough to trot. You have done in your Breed what is commonly practised in *England*; for to put your Horse upon an Amble, you have put Locks upon 'em, and Wisps of Hay round the Pasterns of their hinder Legs: But at the same time you did not consider that there's no such thing as an Amble in the Manage, and that the Riding-Masters allow of nothing but Pace, Trot, and Gallop: And the reason they give for it, is, That they can put a Horse from a Trot to a Gallop without stopping him; but can't pass from an Amble to a Gallop without a stop, which loses time and interrupts the Justness and Cadence of the Manage.

Amble free; A Horse is said to amble free that goes a good Amble when led by the Halter in one's hand.

Amble broken. See *Aubin*.

ANTICOR (or *Avant-Coeur*) is a preternatural Swelling of a round Figure, almost as big as the half of one's fist, which being occasioned by a Sanguine and Bilious Humour, appears in a Horse's Breast, opposite to the Heart. Your Horse has got an *Anticor* that may kill him unless you bring it to Suppuration by good Remedies. But the Anticor that's upon my *English* Horse do's not trouble me at all; for the Swelling lessens upon the Suppuration, which is a good sign. See *Grass*.

ANVIL is a Mass of Iron, sometimes forged with a Hammer; the upper part of which is cover'd with Steel solder'd, in order to forge upon it either cold or hot; the soft and

pliable

pliable Iron, ſuch as that of *Spain* and *Turkey*, being work'd in the former, and the harder Iron in the latter Method. Your forged Anvils are preferable to thoſe that are caſt; and the extremities both of the one and the other terminate in little rais'd Anvils, which ſerve for rounding Horſe-Shoes.

APPUI or *Stay* upon the hand, is the reciprocal Senſe between the Horſe's Mouth and the Bridle-hand; or the Senſe of the Action of the Bridle in the Horſeman's hand. The true and right *Appui* of the Hand is the nice Bearing up or Stay of the Bridle; ſo that the Horſe aw'd by the Senſibility and Tenderneſs of the Parts of his Mouth, dare not reſt too much upon the Bitt-mouth, nor chack or beat upon the hand to withſtand it. Such a Horſe has a dull deaf *Appui*; that is, he has a good Mouth, but his Tongue's ſo thick that the Bitt can't work or bear upon the Barrs: For the Tongue being not ſenſible or tender, as the Bars, is benum'd or harden'd by the Bitt, and ſo the *Appui* is not good. The Bitt do's not preſs the Bars in the quick, by reaſon of the Groſsneſs of the Tongue, or elſe of the Lips. Your Horſe has a Reſt or Stay that forces the hand; which ſhews that he has a bad Mouth. This Horſe has no *Appui*, no Reſt upon the hand; that is, he dreads the Bit-mouth; he's apprehenſive of the hand; and he can't ſuffer the Bitt to preſs or bear tho' never ſo little upon the Parts of the Mouth; and thus it comes to paſs that he do's not eaſily obey the Bridle.

A Horſe who is taught a good *Appui*. If you mean to give that Horſe a good Reſt upon the Hand, it behoves you to gallop him, and put him often back: A Long-ſtretch Gallop is likewiſe very proper for the ſame end, for in galloping, he gives the Horſeman an opportunity of bearing him upon the Hand. Such a Horſe has too much *Appui*, he throws himſelf too much upon the Bitt. A Horſe that has a Fine Stay or Reſt upon the Hand, *i. e.* equal, firm and light, or one that obeys the Bridle. See *Hand*.

A Full *Appui* upon the Hand, is a Firm Stay, without reſting very heavy, and without bearing upon the Hand. Horſes for the Army ought to have a Full *Appui* upon the Hand.

A more than Full Reſt or *Appui* upon the Hand, is ſaid of a Horſe that's ſtop'd with ſome force, but ſtill ſo that he does not force the Hand. This *Appui* is good for ſuch Riders as depend upon the Bridle inſtead of their Thighs.

ARCH'D. A Horſe is ſaid to have Arch'd Legs, when his Knees are bended archwiſe. This Expreſſion relates

to the Fore-quarters, and the Infirmity here ſignified happens to ſuch Horſes as have their Legs ſpoil'd with travelling. The Horſes call'd *Braſſicourts* have likewiſe their Knees bended archwiſe, but this Deformity is natural to them.

ARM of a Horſe. See *Fore-Thigh.*

To *Arm.* A Horſe is ſaid to Arm himſelf when he preſſes down his Head as if he would *Chock*; and bends his Neck ſo as to reſt the Branches of the Bridle upon his Counter, in order to diſobey the *Bitt-mouth*, and guard his Bars and his Mouth, which are reliev'd by over bending the Neck. Since your Horſe Arms himſelf, give him a Knee'd Branch, that will raiſe him, and make him carry his Head well. See to *Carry low.*

Arm with the Lips. A Horſe is ſaid to Arm himſelf with the Lips, when he covers his Bars with his Lips, and makes the preſſure of the Bitt too deaf and firm; this is commonly done by Thick-lip'd Horſes: You muſt order your Bitt-maker to forge you a Bitt-Mouth with a Canon or Scratch-Mouth that's broader near the Bankets than at the place of its preſſure or reſt upon the Barrs; and this will hinder your Horſe to Arm himſelf with his Lips. Sometimes we ſay, the Lip arms the Barr, *i. e.* covers or ſcreens it. See *Diſarm.*

Armand. See *Drench.*

ARRESTS are Mangy Humours upon the Sinews of the Hinder-legs of a Horſe, between the Ham and the Paſtern; they ſeldom appear upon the Shank-Sinew. Their Name is taken from their likeneſs to the Arreſts or Small Bones of a Fiſh. See *Rat-tail.*

ARZEL, A Horſe is ſaid to be Arzel that has a White Mark upon the Far-foot behind. Your Superſtitious *Cavaliers* perſuade themſelves that by an unavoidable Fatality ſuch Horſes are unfortunate in Battels; and ſuch is the ſtrength of this Prejudice, that they do not care to uſe 'em.

ASSIST. See to *Aid.*

ATTAINT, is a Blow or Wound receiv'd by a Horſe in his Hinder-feet, from another Horſe that follows him too cloſe: This Word is likewiſe us'd to ſignifie a Blow that a Horſe's Foot receives from the Fore or Hinder oppoſite Foot; or a Blow given by one of the Hinder-feet ſtriking againſt the Cronet of the Fore-foot. Your Horſe could not have given himſelf a Ruder Attaint, for I find with the Probe, that it penetrates between the Hoof and the Coffin-Bone; which gives reaſon to ſuſpect that the Tendon is affected, and that the Attaint reaches to the Cronet.

Upper-Attaint, is a violent Blow given with the Toe of the Hind-feet, upon the Sinew of the Fore-legs.

AUBIN, is a Broken Going, or Pace of a Horſe, between an Amble and a Gallop, which is not eſteemed.

AVERTI: A *French* Word us'd in the manage, as applied to the Pace or Motion of a Horſe; ſignifying a Motion that's injoin'd, regulated and requir'd in the Leſſons. *Pas Ecouté*, and *Pas d'Ecole* (*i.e.* liſtening Paces, or School Paces) ſigniſie the ſame thing.

B.

BACK, or *Reins* of a Horſe. Such a Horſe has Vigor in his Back; he goes upon his Fillets, he has good Loyns. A Horſe that works in the Manage upon his Loins, is one that lowers his Hips and his Neck, and raiſes his Fillets. A Weak-back'd Horſe is apt to ſtumble. Such a Horſe defends himſelf with his Back, he leaps and plays with his Fillets, and doubles his Reins to incommode the Rider.

Back of a Horſe. To Mount a Horſe *à dos* (in *French*) is to Mount him Bare-back, or without a Saddle.

BALOTADES, are the Leaps of a Horſe between two Pillars, or upon a Straight Line, made with Juſtneſs of Time, with the Aids of the Hand, and the Calves of the Legs; and that in ſuch a manner, that when his Fore-feet are in the Air, he ſhews nothing but the Shooes of his Hinder-feet, without yerking out. Thus 'tis that the Air or Manage of *Balotades* differs from *Caprioles*, for a Horſe that works at *Caprioles*, yerks or ſtrikes out his Hinder-Legs with all his force, keeping them near and even. *Balotades* differ likewiſe from *Croupades* in this; that in the former the Horſe ſhews his Shooes, when he lifts or raiſes his Croup, but in *Croupades* he draws his Hinder-feet under him. Your Horſe will take to *Balotades*, when you let him go upon *Caprioles*; for when the great Fire and Mettle of the *Caprioles* is paſt, Horſes take to *Balotades* of themſelves, and then to *Croupades*, unleſs a *Poinſon* in a Hard Hand makes 'em yerk out, and continue the Air of *Caprioles*. See *Yerk*.

To make a Croſs upon *Balotades*. See *Croſs*.

BALZANE. See *White-foot*.

BANDS of a Saddle, are Two Pieces of Iron, Flat, and Three-Fingers Broad, nail'd upon the Bows of the Saddle, one on each ſide, contriv'd to hold the Bows in the Situation that makes the Form of the Saddle. To put a Bow in the Band, is to Nail down the Two Ends of each Band to each ſide of the Bow. Beſides theſe Two Great Bands, the Fore-bow has a Small one, call'd the Wither-Band, and a Creſcent to keep

up the Wither-Arch. The Hinder-Bow has likewiſe a ſmall Band to ſtrengthen it.

BANQUET, is that ſmall part of the Branch of a Bridle that's under the Eye, which being rounded like a Small Rod, gathers and joins the Extremities of the Bitt to the Branch, and that in ſuch a manner, that the Banquet is not ſeen, but cover'd by the Cap, or that part of the Bitt that's next the Branch.

Banquet-Line, is an Imaginary Line, drawn by the Bitt-makers along the *Banquet*, in forging a Bitt, and prolong'd upwards and downwards, to adjuſt the deſign'd Force or Weakneſs of the Branch, in order to make it ſtiff or eaſie; for the Branch will be hard and ſtrong, if the Sevil-hole is on the outſide of the Banquet-line, with reſpect to the Neck; and the Branch will be weak and eaſie, if the Sevil-hole is on the inſide of the Line, taking the Center from the Neck. See *Branch* and *Shoulder*.

BARB, is a Horſe brought from *Barbary*: Such Horſes are commonly of a Slender Light Size, and very Clean and Thin in the Legs. Your *Spaniſh* and *Engliſh* Horſes are much better Bodied, and have larger Legs than the *Barbs*. All the Colts of our Breed, are come of a *Barbary* Stallion. We always choſe *Barbs* for our Stallions, becauſe they are always Mettled, and have good Wind and Heels; and 'tis commonly ſaid, *Barbs* may die, but they never grow old: The contrary of which is obſerv'd in your *Friezland* large *Dutch* Horſes (*Rouſſins*) for they grow ſoon old, and are long a dying, ſo that they prove troubleſome to their Maſters. The Duke of *Newcaſtle* has well obſerv'd, that the Vigor and Mettle of *Barbs* never ceaſes but with their Life.

BARBLES are Knots of Superfluous Fleſh that grow up in the Channels of a Horſe's Mouth; that is, in the Interval that ſeparates the Barrs, and lies under the Tongue.

BARDELLE, is a Saddle made in the Form of a Great Saddle, but only of Cloath ſtuff'd with Straw, and tied tight down with Packthread, without either Leather, Wood, or Iron: They are not us'd in *France*, but in *Italy* they trot their Colts with ſuch Saddles, and thoſe who ride them are call'd *Cavalcadours*, or *Scozone*.

BARNACLES; the Word commonly us'd for what the *French* call *Mouraille*, which ſee.

To BAR a Vein, or Strike it, is an Operation perform'd by the Farrier upon the Veins of a Horſe's Legs, and the other Parts of his Body, with intent

intent to ſtop the Courſe, and leſſen the Quantity of the Malignant Humours that prevail there. Your Horſes have got Traverſe-Mules or Kib'd-Heels, and Rat-Tails or Arreſts in the Hinder-Legs, you muſt Barr a Vein. In order to Barr a Vein, the Farrier opens the Skin above it, and after diſengaging it, and tying it both above and below, he ſtrikes between the Two Ligatures.

BARS of a Horſe's Mouth are the Ridge, or the higheſt Parts of that Place of the Gum that never bears any Teeth, and is ſituated between the Grinders and the Tuſhes on each ſide of the Mouth; So that that part of the Gum which lies under, and at the ſide of the Barrs, retains the Name of Gum. The Barrs are that part of the Mouth upon which the Bitt ſhould reſt, or have its *Appui*; for though only a ſingle Canon bears upon the Tongue, the Barrs are ſo tender and ſenſible, that they feel the Effect of it even through the Thickneſs of the Tongue. Since your Horſe's Barrs are ſo very ſenſible, he has a Fine Light Mouth; and in a whole Days Riding, you have him all along upon the Hand, with an Even, Firm and Light *Appui*, or Reſt. This Horſe has Round Hard Barrs, that are not very ſenſible, you muſt make him a Bitt that will Rouze him, and make him Feel it, ſuch as a Bitt that pulls all along; that is, one that does not bend, to give room to the Tongue in the Middle. Theſe are deſperate Bars, which have been ſo broke and cicatris'd, that now they are inſenſible. Your Horſe ſeems to have a very Fine Tender Mouth, for his Barrs are Sharp and Edged like thoſe of a *Barbary*-Horſe. Your Horſe's Lip guards, that is, covers the Barr.

A Horſe is ſaid to fall foul of the Barr, (in *French*, *Embarrer*) when in the Stables he entangles his Legs upon the Partition-Barr that's put to ſeparate two Horſes, and keep 'em from falling upon one another. Your *Barbs*, and your Vigorous Tickliſh Horſes are apt to fall foul of the Barr, and when they do, they ſtruggle, and fling, and wound themſelves in the Hocks, the Thighs, and the Legs, and are in danger of laming themſelves, unleſs you ſpeedily cut the Cord that keeps up one End of the Barr, and ſo ſuffer that End to fall to the Ground.

BAY Colour. A *Bay*-Horſe is what we commonly call Red inclining to Cheſnut. This Colour varies ſeveral ways; 'tis a Dark-Bay, or a Light-Bay, according as 'tis more or leſs deep: And we have likewiſe *Dappled-Bays*. See *Mirouette*.

ette. All Bay-Horses have black Manes, which distinguishes them from the Sorrel that have red or white Manes.

BEAN or Lampas. See *Lampas*.

Bean, or Eye of a Bean. See *Eye*.

BEARD, or *Under-beard*, or *Chuck* of a Horse; is that part underneath the lower Mandible on the out-side, and above the Chin, which bears the Curb of the Bridle, See *Curb* and *Genette*.

BEAT upon the hand. See *Chack*.

Beat; to beat the Dust or the Powder, is said of a Horse that at each Time or Motion does not take in Ground or Way enough with his Fore-legs. A Horse beats the Dust at *terra a terra*, when he do's not imbrace or take in Ground enough with his Shoulders, and makes all his Times or Motions too short, as if he made 'em in one place. He beats the dust at Curvets, when he do's 'em too precipitantly, and too low. He beats upon a Walk, when he walks too short, and makes but little way, whether in straight Lines, Rounds, or Passaging.

BELLY; a Thick-*belly'd*, a Well-body'd, a Well-flank'd Horse; that is, a Horse that has large, long, and well-made Ribs; or such as are neither too narrow nor too flat. This Horse has Belly enough, for his Flank shews that he feeds well. Such a Horse has no Body, he's thin-flank'd; that is, his Ribs are too narrow, or short, and the Flank turns up: which makes his Body look flankless, like a Grey-hound. A Horse of this nature we commonly call in *French*, an *Estrac*; which generally speaking is a fine sort of tender Horses, not very fit for Travelling or Fatigue, unless they feed very heartily. We reject all Coach-Horses that are not well-bodied, all that are narrow or thin-gutted, and seem to have the Hide or Skin of their Flanks stitch'd upon their Ribs: But a Hunter is not the worse lik'd for being light-belly'd; nay on the contrary, he's preferr'd to a thicker flank'd Horse, provided he's well winded, of good Mettle, light, and a great Eater. Since this Runner has lost it's Belly, if you mean to give it a better Belly, and make it thicker gutted, you must turn it to grass. Instead of the word *Belly* or *Gut* we frequently use *Flank*. See *Handful*, *Estrac*, *Light-belly'd*, and *Flank*.

BIDET. See *Nag*.

BITT, or Horse-Bitt, (in *French*, *Mords*) in general signifies the whole Machine of all the iron Appurtenances of a Bridle; as the Bit-mouth, the Branches, the Curb, the Sevil-holes, the Tranchefil, and the Cross Chains: But oftentimes it signifies only the *Bit-mouth* in particular.

Bit-mouth (in *French*, *Embouchure*) is a piece of Iron forged ſeveral ways, in order to be put into a Horſe's Mouth, and to keep it in ſubjection. Of theſe Bit-mouths, ſome are ſingle Canon-mouths, ſome are Canon-mouths with an Up-ſet or mounting Liberty, ſome Scatch-Mouths, ſome Mouths after the Form of a Barge, ſome with two long turning Olives, and ſeveral other ſorts; all with different Liberties for the Tongue, or without Liberty. But all Bit-mouths ought ſtill to be proportion'd to the Mouth of the Horſe, according as 'tis more or leſs cloven and wide, or more or leſs ſenſible and tender, according as the Tongue and the Lips are higher or flatter, and as the Palate is more or leſs fleſhy: Obſerving withal that if the Horſe be old, the Palate will always have but little Fleſh upon it. Your Horſe has a falſe Mouth; and tho' the Bit-mouth is well ordered, he do's not obey. See *Liberty*.

A *Bit-mouth*, all of a piece, without a Joint in the middle, is call'd by the *French* a Bit that preſſes *de l' entier*. See *Barrs*.

BLACK; More, or Coal-black is the Colour of a Horſe that's of a deep, ſhining, and lively Black.

BLAZE. See *Star*, and *White-Face*.

BLEYNE is an Inflammation occaſion'd by Blood putrified in the inner part of the Coffin towards the Heel, between the Sole and the Coffin-Bone. My Horſe is not lame now, for he's cur'd of his *Bleyne*. See *Hoof-caſt*.

BLOSSOM or Peach-colour'd Horſe, is one that has his Hair white, but intermix'd all over with Sorrel and Bay Hairs. Such Horſes are ſo inſenſible, and hard both in the Mouth and the Flanks, that they are ſcarce valued; beſides, that they are apt to turn blind.

To BLOW upon the Hair or Fruſh. See *Hair*, and *Bouillon*.

BOAR; A Horſe is ſaid to boar, when he ſhoots out his Noſe as high as his Ears, and toſſes his Noſe in the wind. See *Wind*.

BODY; A Horſe well-body'd; *i. e.* thick-flank'd, high-belly'd. See *Belly*, and *Light-belly'd*.

BOLSTERS of a Saddle, are thoſe parts of a great Saddle which are rais'd upon the Bows, both before and behind, to hold the Rider's Thigh, and keep him in a Poſture of withſtanding the Diſorders that the Horſe may occaſion. Commonly Saddles have no Hind-Bolſters. We uſe the Expreſſion of fitting a Bolſter, (in *French*, *Chauſſer une Batte*) when

we

we put the Cork of the Saddle into the Bolſter, to keep it tight. That Part of the Saddle being formerly made of Cork, took firſt that name, tho' now 'tis made of Wood.

BOUILLON is a Lump or Excreſcency of Fleſh that grows either upon or juſt by the Fruſh, inſomuch that the Fruſh ſhoots out like a Lump of Fleſh, and makes the Horſe halt; and this we call the Fleſh blowing upon the Fruſh. Your Manage Horſes, which never wet their foot, are ſubject to theſe Excreſcencies, which make 'em very lame.

BOULETE; A Horſe is call'd *Bouleté*, when the Fetlock or Paſtern-joynt (in *French, Boulet*) bends forward, and out of its natural Situation; whether thro' violent Riding, or by reaſon of being too ſhort-jointed, in which caſe the leaſt Fatigue will bring it.

BOWS of a Saddle, are two Pieces of Wood laid Archwiſe, to receive the upper part of a Horſe's Back, to give the Saddle its due Form, and to keep it tight. The Fore-bow which ſuſtains the Pommel, is compos'd of the Withers, the Breaſts, the Points or Toes, and the Corking. The Withers is the Arch that riſes two or three fingers over the Horſe's Withers. The Breaſts are placed where the Arch or upper part of the Bow ends. The Points or Toes are the lower part of the Bow: and the Corking is pieces of Wood, formerly pieces of Cork, upon which we fit and make faſt the Bolſters. The Hind-bow bears the *Trouſſequin*, or quilted Roll. The Bows are cover'd with Sinews; that is, with Bulls Pizzles beaten, and ſo run all over the Bows to make 'em ſtronger. Then they ſtrengthen 'em with Bands of Iron to keep 'em tight, and on the lower ſide of the Bows, nail on the Saddle-ſtraps, with which they make faſt the Girths.

BOUT; A Horſe is ſaid to be *A Bout*, when he is overdone and quite ſpent with Fatigue.

BOUTE; A Horſe is called *Bouté*, when his Legs are in a ſtraight Line, from the Knee to the Coronet. Short-joynted Horſes are apt to be *Bouté*; and, on the other hand, long-joynted Horſes are not.

BRANCHES of the Bridle, are Two Pieces of Iron bended, which in the interval between one and the other bear the Bitt-Mouth, the Croſs-Chains, and the Curb; ſo that on one end they anſwer to the Head-ſtall, and on the other to the Reins, in order to keep the Horſe's Head in ſubjection. A Hardy, Bold or Strong Branch, is one that brings in the Head. A Weak Branch is a Branch that was formerly us'd for raiſing the Head, but now is diſus'd,

eſpecially ſince the diſcovery of the Error of thoſe who fancy'd that it rais'd after the ſame manner with the Kned-Branches. Mr. *Solleyſel* has publiſh'd an excellent Treatiſe of Bitts and Branches. See *Banquet* and *Shoulder*.

BRASSICOURT, or *Brachicourt*, is a Horſe whoſe Fore-Legs are naturally bended Arch-wiſe; being ſo call'd by way of diſtinction from an Arch'd-Horſe, whoſe Legs are bow'd by Hard Labour.

BRAYE, an obſolete *French* Word, made uſe of by ſome to ſignifie the Entry of a Horſe's Throat, or the Extremity of the Channel towards the Lower Maxillary Bones.

BREAK. To break a Horſe in Trotting, is to make him Light upon the Hand by Trotting, in order to make him fit for a Gallop. To break a Horſe for Hunting, is to Supple him, and make him take the Habit of Running. 'Tis a Furious Fatigue to run Horſes at full ſpeed before they are Broke. When this Horſe is Broke, he'll run well. This Thick well-knit Horſe is ſtrong indeed, but 'twill be a long time before he's Broke.

BREASTS, part of the Bow of a Saddle. See *Bows*.

BREAST of a Horſe. See *Counter*.

BREAST-PLATE, or Tee, is the Strap of Leather that runs from one ſide of the Saddle to the other, over the Horſe's Breaſt, in order to keep the Saddle tight, and hinder it to ſlide backwards when the Horſe goes upon a Riſing-Ground.

BREATH, or Wind. This Word ſignifies ſometimes the eaſie Reſpiration of a Horſe; and ſometimes it implies the Eaſe and Reſt, or Repoſe of a Horſe. As, Give your Horſe Breath, do not ride him down. Give that Leaping Horſe a long Breathing time between the Turns or Repetitions of his Manage. This Horſe is well Winded. Your Hunters ſhould be Long-winded. This *Barb* has always held his Wind equally upon his Manage. This Horſe is Maſter of his Wind, or Breath. This laſt Expreſſion is applied to Horſes that ſnort, and your Jockeys take Snorting for a ſign of a Long-winded Horſe. See *Snort*.

A Thick-winded Horſe (in *French*, *Cheval gros d' haleine*) is a Horſe that without being Purſive, wants Breath, and wheezes very much upon a Trot or a Gallop. This Horſe is Thick-winded, by reaſon that the Reſpiration Paſſages are too narrow.

BREED, (in *French Haras*) is the Place where Mares for Breed, and Stallions are kept in order to raiſe a Stud. To keep a Breed, to govern and manage a Breed. All the Mares in this Breed have

taken;

taken; *i. e.* they are with Foal. To make a Good Breed, you can't chooſe a better Stallion than a *Spaniſh* Horſe; nor better Stud-Mares than your *Naples* Mares. Out of this Breed I have had two *Barbary Echape's*.

BRIDLE of a Horſe is a Contrivance made of Straps or Thongs of Leather, and Pieces of Iron, in order to keep a Horſe's Head in ſubjection and obedience.

Inſtead of ſaying, Pull the Bridle, we ſay, Bear your Hand: And for Slacking the Bridle, we ſay, Slack your Hand: Lower your Hand: Yield with your Hand.

To cleave to, or hold by the *Bridle*, implies the Fault of an Ill Horſeman, who inſtead of ſlacking his Hand, when a Horſe is diſorderly, clings to it, as if 'twere to the Mane. Such a Horſe-man does not ſit firm by no means; he wants the Habit, or elſe the Force of clinging cloſe with his Thighs: When the Horſe flies out, and is diſorderly, he holds by the Bridle as if 'twere the Pommel of the Saddle. See *Light-hand* and *Pontlevis*.

Touch, or *Check* of the *Bridle*. See *Ebrillade* and *Saccade*.

To *Swallow the Bridle* (in *French*, *Boire la Bride*, or to Drink the Bridle) is ſaid of a Horſe that has too wide a Mouth, and too ſmall a Bitt-Mouth; inſomuch that the Bitt riſes too high, and Gathers or Furles the Lips, and miſplaces it ſelf above that place of the Barrs where the Preſſure ſhould be; and by this means the Curb is likewiſe diſplac'd, and ſhov'd too high.

Bridle-hand, is the Horſeman's Left-hand; the Right-hand is call'd the Spear or Sword-hand.

BRIDON. See *Snaffle*.

BRILLANT: A Briſk High Mettled Stately Horſe is called *Brillant*, as having a Rais'd Neck, a Fine Motion, Excellent Haunches upon which he riſes, tho' never ſo little put on. Such a Horſe Champs upon his Bridle with a Good Grace. I ſaw but now in the King's Querry one of the moſt *Brillant* and beſt manag'd Horſes that can be. I have juſt bought a Horſe that knows nothing, but he's *Brillant* and if right taken, will ſucceed in the Manage.

BRING *in* a Horſe, is to keep down the Noſe of a Horſe that Bores and Toſſes his Noſe up to the Wind; this we do with a good ſtrong Branch. See *Banquet* and *Wind*.

BROUILLER: A *French* Word us'd in the Academies, importing that a Horſe, when put to any Manage, Plunges, Traverſes, and appears in Diſorder. This Gentleman is not Maſter of his Legs, he makes his Horſe *Brouiller*; *i. e.* he makes him Traverſe and Caſt

down

down his Head, the Spur being too hard for him. Sit very close upon that *Barb*, for he has his Aids so fine; that if you cling with your Thighs never so little beyond the Due, and alter your Legs, you'll make him *Brouiller*, and hinder him to work in the Due Manage. See *Thighs*.

BUTTERIS is an Instrument of Steel fitted to a Wooden-Handle, with which they pare the Foot, or cut the Hoof of a Horse.

BUTTON of the Reins of a Bridle, is a Ring of Leather with the Reins pass'd through it, which runs all along the length of the Reins. To put a Horse under the Button, is done when a Horse is stop'd without a Rider upon his Back, the Reins being laid upon his Neck, and the Button lower'd so far down, that the Reins bring in the Horse's Head, and fix it to the true Posture or Carriage. 'Tis not only the Horses which are manag'd in the Hand that must be put under the Button, for the same Method must be taken with such Horses as are bred between two Pillars, before they are back'd.

C.

CADENCE, is an equal Measure or Proportion observ'd by a Horse in all his Motions, when he's thoroughly manag'd, and works justly at Gallop, *terra a terra*, and the Airs; so that his Times or Motions have an Equal Regard to one another, that one does not imbrace or take in more Ground than the other, and that the Horse observes his Ground regularly. This Horse works always upon the same *Cadence*; he follows the *Cadence*; he does not change his *Cadence*, he remains equally between the two Heels. He is fine and gentle in all his Aids, and when put to the Manage, he never interrupts his *Cadence*. This Horse has so Fine a Mouth, and works with so much liberty in the Shoulders and Haunches, that he keeps his *Cadence* with great Facility: Nay, he takes a very good *Cadence* upon his Airs, without stepping false, without jumbling, and works equally to both Hands. See *Countertime* and *Time*.

CALADE, or *Basse*, is the Descent, or Sloping Declivity of a Rising Manage Ground, being a small Eminence upon which we ride down a Horse several times, putting him to a short Gallop, with his Forehams in the Air, to make him learn to ply or bend his Haunches, and form his Stop upon the Aids of the Calves of the Legs, the Stay of the Bridle, and the Cavesson, seasonably given; for without these

these Aids, he would throw himself too much upon his Shoulders, and not bend his Haunches. Work your Horse in a *Calade* after the *Italian* way; ride him straight, and then you make good use of the *Calade*. These *Calades* will discourage your Horse, and perhaps ruin his Hams, for you have pitch'd upon too steep a Declivity; and besides, you do not make the Aids of the Bridle accord with those of the Calves of the Legs.

CANON-MOUTH of a Bitt, is a round but long piece of Iron, consisting sometimes of two pieces that couple and bend in the middle; and sometimes only of one piece, that does not bend, as in the Canon-Mouth *à Trompe*. *Canon-Mouths* of all sorts are contriv'd to keep the Horse in subjection; and are so contriv'd that they rise gradually towards the middle, and ascend towards the Palate; to the end that the void Space left underneath may give some liberty to the Tongue.

CAPARASSON or Horse-cloth, is a sort of Cover for a Horse. For led Horses 'tis commonly made of Linnen Cloth, border'd round with Woollen, and enrich'd with the Arms of the Master upon the middle which covers the Croupe, and with two Cyphers on the two sides. The Caparassons for the Army are sometimes a great Bear's-skin; and those for Stables are of single Buckram in Summer, and of Cloth in Winter.

CAPRIOLES or Leaps of *firma a firma*, are Leaps that a Horse makes in one and the same place, without advancing forwards, and that in such a manner, that when he's in the Air and at the height of his Leap, he yerks or strikes out with his Hinder-legs even and near. A Capriole is the most difficult of all the high Manage, or rais'd Airs. It differs from a Croupade in this, that in a Croupade the Horse do's not shew his Shoes; and from a Balotade in this, that in a Balotade he do's not yerk out. Your Horse will never work well at Caprioles, unless you put him between two Pillars, and teach him to raise first his Fore Quarters, and then his Hind Quarters, while his Fore are yet in the Air; for which end you must give the Aids of the Whip, and the *Poinson*. If you would teach your Horse to make Caprioles, and yerk out handsomely with his hinder Feet, stay and help him with your hand, and your heels. This Leaping Horse takes to Caprioles of himself, for he makes equal Leaps, and that upon the hand, *i. e.* without forcing the hand, and resting heavy upon the Bridle. See to *Yerk*.

CARACOL is an oblique Piste or Tread traced out in Semi-rounds, changing from one hand to another, without observing a regular ground. When Horse advance to charge in Battel, they sometimes ride up in *Caracols*, to perplex the Enemy, and make 'em doubtful whether they are about to take 'em in the Front, or in the Flank. *Caracol* is a *Spanish* word, and in that Language signifies the Motion that a Squadron of Horse makes, when upon an Engagement the first Rank has no sooner fired their Pistols, than they divide and open into two Half-Ranks, the one wheeling to the right, and the other to the left, along the Wings of the Body to the Rear. Every Rank observes the same Order after firing; and the Turning or Wheeling from the front to the rear is call'd a *Caracol*.

To Caracole, is to go in the form of Half-rounds.

CAREER; this Word signifies both the Ground that's proper for the Manage, and the Course or Race of a Horse that do's not go beyond two Hundred Paces. This Barb makes a very good Career from pacing to stopping. This *English* Horse do's not furnish his Career; that is, he does not finish his Course with the same Swiftness, and does not move so short and swift at the middle and end, as at the beginning. This *Spanish* Horse is fit for the Ring; he has a short and swift Career, and holds it for a Hundred Paces.

CARRY *low*; (in *French Porter bas*;) A Horse is said to Carry Low that has naturally a soft ill-shap'd Neck, and lowers his Head too much. All Horses that arm themselves Carry Low; but a Horse may Carry Low without arming: For when he arms himself, his Neck is too supple, and he wants to evade the subjection of the Bridle; but when he Carries Low, he has his Neck ill-plac'd, and ill-made. Your Horse Carries Low, try whether a *French* Branch or a Gigot will raise him; for tho' they can rectify Nature, I doubt much if they can change it.

To Carry well, or in a becoming Posture, (in *French, Porter en beau lieu*) is said of a Horse whose Neck is rais'd or arch'd, and who holds his Head high, without constraint, firm and well-placed.

CAST Hair or Hoof. A Horse casts or sheds his Hair at least once a year. Every Spring he Casts his Winter-coat, and takes a Summer one; and sometimes in the end of Autumn he puts on his Winter-hair, in case he has been ill-curry'd, or ill-cloath'd, or kept in a cold Stable. Sometimes he Casts likewise his Hoof; and this happens frequently to your Coach-Horses that come from

B Hol-

Holland: Fot theſe being bred in a moiſt marſhy Country, their Hoofs are too ſlabby, and of too weak a Conſiſtence; ſo that coming into a dryer Country, where they're fed with leſs juicy Provender, they gradually caſt their Hoof, by reaſon that their foot grows, and another firmer Hoof is form'd. Since your Horſe caſts his Hoofs, make the Farrier give 'em a good form in Shoeing him, or elſe his Feet will grow flat, and like an Oyſter-ſhell.

CAVALCADOUR is a Word us'd at the Court of *France*, and the Families of the Blood; ſignifying the Querry that's Maſter of the Horſe. Thus we ſay, the Querry Cavalcadour of the Queen's Stables; the Querry Cavalcadour of Monſieur or the Duke of *Orleans*'s Stables. In *Italy* this word ſignifies the Perſon who trots Colts with *Bardelle* Saddles. See *Bardelle*.

CAVALLERISSE is an old *Italian* word, now diſus'd in *France*, ſignifying a Perſon vers'd in the Art of governing and breaking Horſes.

CAVALIER, in a warlike ſenſe, ſignifies all Soldiers that are mounted on Horſeback: But in the Manage it implies one that underſtands Horſes, and is practis'd in the Art of Riding them.

CAVESSE *de More*. See *More's-head*.

CAVESSON, is a ſort of Noſe-band, ſometimes of Iron, ſometimes of Leather or Wood, ſometimes flat, and ſometimes hollow or twiſted, which is clapt upon a Horſe's Noſe to wring it, and ſo forward the ſuppling and breaking of the Horſe. The *Caveſſon* of Leather, and thoſe of Wood, are made uſe of when we put Horſes between two Pillars; and when we ſay a Horſe takes the Ropes, we mean the Ropes or Straps of that ſort of *Caveſſon*. An Iron *Caveſſon* ſaves and ſpares the Mouth of Young Horſes when we break 'em; for by the help of it we accuſtom 'em to obey the Hand, and to bend the Neck and Shoulders, without hurting their Mouth, or ſpoiling their Barrs with the Bitt. Now, an Iron *Caveſſon* is a Semicircle, or a Band of Iron bended to an Arch, conſiſting of two or three Pieces join'd by Hinges, and this we clap upon the Noſe of a Young Horſe. Some *Caveſſons* of Iron are twiſted or wreath'd, and ſome are flat, which bear equally upon the Noſe, and are indeed the beſt. A *Caveſſon, à Siguette*, or a Biting *Caveſſon* is hollow in the middle, and notch'd like a Saw upon the two ſides of its Concavity, in order to pinch the Noſe of a Surly or Stiff-necked Horſe. The *Caveſſon* call'd *Camare* was arm'd with little Teeth, on very Sharp Points

Points of Iron, which tore and abuſed a Horſe; ſo that at preſent 'tis baniſh'd the Academies, and the very Name is ſcarce known among the Bittmakers of *Paris*. The *Siguette* is likewiſe in diſgrace. All Iron *Caveſſons* are mounted with a Head-ſtall, a Throat-band, and two Straps or Reins, with three Rings; one Rein we paſs through the Middle Ring, when we mean to make a Horſe work round a Pillar; or for want of a Pillar, round a Man that ſtands in the Center. Through the two ſide Rings, we paſs the two Reins which the Rider holds in his Hand, or makes faſt to his Saddle, in order to keep a Horſe's Head in ſubjection, and to ſupple his Shoulders. See *Ropes*.

CHACK, or Beat upon the Hand. A Horſe is ſaid to Chack or beat upon the Hand, when his Head is not ſteady, but he toſſes up his Noſe, and ſhakes it all of a ſudden, to avoid the Subjection of the Bridle. You have a *Turkiſh* Horſe that retains the Fault of that Country Horſes; he Beats upon the Hand; and the beſt Bitts that are, nor the beſt Hand, can never fix his Head. This Horſe Chacks upon the Hand like a *Croat* or *Croatian* Horſe; which proceeds from this, that his Barrs are too ſharp and ridged, or edged, ſo that he can't bear the preſſure of a Bitt, though never ſo gentle. If your Horſe had not too ſenſible, or too tender a Mouth, he would not Beat upon the Hand: But in order to fix and ſecure his Head, you need only to put under his Noſe-band a ſmall flat Band of Iron, bended Arch-wiſe, which anſwers to a *Martingale*. This will hinder him to beat upon the Hand, but will not break him of the Habit; for as ſoon as the *Martingale* is taken off, he'll fall into the ſame Vice again. See *Wind*.

CHAIN; the Croſs *Chain* of a Bridle. See *Tranchefile*.

CHANFRIN, is the Fore-part of a Horſe's Head, extending from under the Ears, along the Interval between the Eye-brows, down to his Noſe.

CHANFRAIN-BLANC. See *Star* or *Blaze*.

CHANGE a Horſe, or Change Hand, is to turn or bear the Horſe's Head from one Hand to the other, from the Right to the Left, or from the Left to the Right. You ſhould never change your Horſe without puſhing him forward upon the Turn; and after the Turn, puſh him on ſtraight, in order to a Stop. This Horſe changes from the Right with an ugly Grace. See *Entier*, *Nails*, *Walk* and a *Paſſade* of *Five times*.

CHANGE Hair or Hoof. See *Caſt*.

CHANNEL of the Mouth of a Horſe, is that Concavity in the middle of the Lower Jaw, appointed for a place to the Tongue; which being bounded on each ſide by the Barrs, terminates in the Grinders, or Maxillary-Teeth. The *Barbles* grow in this *Channel*.

CHAPELET, is a Couple of Stirrup-Leathers mounted each of 'em with a Stirrup, and joining at top in a ſort of Leather-Buckle, call'd the Head of the *Chapelet*, by which they are made faſt to the Pommel of the Saddle, after being adjuſted to the Rider's Length and Bore: They are uſed both to avoid the trouble of taking up or letting down the Stirrups every time that the Gentleman mounts on a different Horſe and Saddle, and to ſupply the want of the Academy Saddles, which have no Stirrups to 'em.

CHAPERON of a Bitt-mouth, is a Word only us'd for Scatch-Mouths, and all others that are not Canon-mouths; ſignifying the End of the Bitt that joins to the Branch juſt by the *Banquet*. In Scatch-mouths the *Chaperon* is Round, but in others 'tis Oval: And the ſame part that in Scatch and other Mouths is called *Chaperon*, is in Canon-Mouths called *Fronccan*.

CHARBON, (*i.e. Coal*) is an obſolete *French* Word, ſignifying that little Black Spot or Mark that remains after a Large Spot in the Cavity of the Corner-teeth of a Horſe, about the Seventh or Eighth Year, when the Cavity fills, and the Tooth being ſmooth and equal, is ſaid to be *ras'd*.

CHARGE, is a Preparation or an Ointment of the conſiſtence of a Thick Decoction, applied to the Shoulder-Splaits, Inflamations and Sprains of Horſes. The Parts affected are rubb'd and chafed with this Compoſition, after which you may cover 'em with Sinking-Paper, if you will. *Charges* are made two ways, *viz.* either with *Emmiellures*; *i.e.* a Mixture of Honey, Turpentine, Suet, and the other Drugs; or with a Remolade, which is a Mixture of the Lees of Wine with the Drugs of the *Emmiellure*. Your Farriers confound the Names of *Charge*, *Emmiellure* and *Remolade*, and indifferently uſe one for t'other.

CHASTISEMENTS, or Corrections, are the Severe and Rigorous Effects of the *Aids*; for when the *Aids* are given with Severity, they become Puniſhments.

CHAUSSE *Trop-Haut*; a White-footed Horſe is ſaid to be ſuch, when the White Marks run too high upon the Legs.

CHEST Founder'd. See *Foundered*.

CHEVALER; (a *French* Word;) a Horſe is ſaid to *Chevaler*,

valer, when in passaging upon upon a Walk, or a Trot, his Far Fore-leg crosses or overlaps the other Fore-leg every second time or motion. See *to Passage*.

CHINE, or Spine of the Back; (in the *French Echine*, and among the Ancient *Italian* Masters *Esquine*) is the Backbone, or the Ridge of the Back of a Horse.

CHINK. See *Clift*.

CLAMPONNIER, or *Claponnier*; an obsolete Word, signifying a Long-jointed Horse; that is, one whose Pasterns are long, slender and over pliant. The Word is properly applicable only to Bulls or Cows, for *la Claponniere* (in *French*) is in them what the Pastern is in a Horse.

CLIFT, *Chink*, *Crack* or *Chap*; (in *French Avalure*) is a Deficiency in the New, Soft and Rough Uneven Hoof, that grows in Horses Feet upon the *Hoofcast*.

CLOSE *Behind* (in *French*, *Crochu*; *i. e.* Crooked;) is a Horse whose Hoofs come too close together; commonly such Horses are good. The Country People in *France* call such a Horse *Jarretier*; but that Word is in disuse at Court.

To *Close* (*Fermer*) a Passade justly, is when the Horse ends the Passade with a Demivolt in good order, well narrow'd and rounded, and terminates upon the same Line upon which he parted; so that he's still in a condition to part from the Hand handsomely, at the very last time or Motion of his Demivolt. The *French* call this *Fermer*, or *Serrer la Demivolte*.

COFFIN of a Horse is the whole Hoof of the Foot above the Cronet, including the Coffin-bone, the Sole and the Frush. For want of knowing how to cure such a *Bleyme*, your Horse is Hoof-cast, and his Coffin is fallen.

Coffin-Bone is a small Spongy Bone inclosed in the midst of the Hoof, and possessing the whole Form of the Foot. See *Tendon*.

COLD, or Rheum, is a Flux of Impure Humours, which are evacuated by a Horse's Nostrils, and occasion a Cough less or more, a loathing of Food, and a heaving or beating of the Flanks.

COLT implies both the *He* and *She* of a Mare's casting: They lose this Name when they are about four Years old, for then we begin to back 'em. They are not capable of any great Labour or Fatigue till the Upper-tushes have cut the Skin, which happens about four Years, or four and a half. You should not put this Colt in the Manage, he is not above three Years old; you'll weaken his Back, and quickly ride him down; stay till he's at least five Years old, and then he'll have

more Vigor and Memory. See *Amble.*

To COMMENCE, or initiate a Horſe, is to put him to the Firſt Leſſons, in order to break him. To commence this Horſe, you muſt work him round the Pillar. See *Rope.*

CORD. See *Rope.*

CORK of a Saddle. See *Bolſter.*

CORNERS, or Angles of the Volte, are the Extremities of the Four Lines of the Volte, when you work in a Square. To make the Four Corners; to work upon the Four Corners. See *Four* and *Square.*

Corner-Teeth of a Horſe, are the Four Teeth that are placed between the Middling Teeth and the Tuſhes, being two above, and two below, on each ſide of the Jaw, which ſhoot forth when the Horſe is four Year and a half old.

CORONET, or *Cronet* of a Horſe, is the loweſt part of the Paſtern, which runs round the Coffin, and is diſtinguiſh'd by the Hair which joins and covers the upper part of the Hoof. Look to your Horſe's *Coronet*, he has given himſelf an Attaint. Your Horſe has a Crepance, or Malt-worm in his *Coronet.*

CURVETS are Leaps of an indifferent heighth, which a Horſe makes in raiſing firſt his two Fore-legs in the Air, and making the two hinder Feet follow with an equal Cadence; ſo the Haunches go down together, after the Fore-feet have touch'd the Earth, in continual and regular Reprizes. A Horſe that's put to the Air of Corvets, a Horſe that makes Corvets, that works at Corvets, that takes to Corvets of himſelf. This Horſe beats the Duſt in his Corvets, for he plays 'em too faſt, and too low. You have work'd this Horſe in Corvets with ſo little care, that he has got a Jardon. Since that Horſe has got the Spavin, he'll abate his Corvets from the greateſt heighth, for it being a Dry-Spavin, 'twill make him lift his Legs. See to *Harp* or *Grapple*, to *Beat the Duſt*, *Abate* and *Time.*

To make a Croſs in Corvets. See *Croſs.*

COUCH'D. A Horſe couched upon his Volts. See *Volte Couched.*

COUNTER of a Horſe, is that part of a Horſe's Fore-hand which lies between the Shoulder, and under the Neck. I ſee an Anticor in your Horſe's Counter, you'll be obliged to put *Ellebor-Roots* into it. See *Herber.*

Counter-Time, (in *French*, *Contre-temps*) is the defence or reſiſtance of a Horſe that interrupts his Cadence, and the Meaſure of his Manage: This is occaſioned either by a bad Horſe-man, or by the Malice of the Horſe. This Leaping Horſe has made two or three Counter-

times,

times, and instead of raising his Fore-quarters, has continued to yerk behind. This Horse has broke the Justness of his Manage by his Counter-times, and the Rider has but sorrily seconded the Aids of the Bridle with the Aids of the Heels.

Counter-mark'd. A Horse is said to be Counter-mark'd, when his Teeth are artificially made hollow by a Farrier's Graver; and a False Mark is made in the Hollow Place, in imitation of the Eye of a Bean; with intent to make people believe that the Horse is not above six years old, and so conceal his Age, which in effect is far beyond what the Mark would seem to intimate.

Counterpoise, or Ballance of the Body, is the liberty of the Action and Seat of a Horseman, acquired by practising in the Manage: So that in all the Motions made by the Horse, the Horse-man does not incline his Body more to one side than to the other, but continues in the middle of the Saddle, bearing equally on his Stirrups, in order to give the Horse the seasonable and proper Aids. This Gentleman keeps his Counterpoise so well, that he is always prepar'd against the Surprizes and Disorderly Motions of the Horse. See *Seat*.

COURSE, or Race. This Word, which is not received in the Manage, signifies upon other occasions, a Gallop at full speed. This *Barb* is a good Courser, and well winded. This *English* Racer has won the Course.

CRACK, or Cleft, is a Chop in the Pasterns or Fetlocks of a Horse, which throws out Red Stinking Water. See *Cleft*.

CRAMP is a Stiffness in the Leg of a Horse, who oftentimes when he's drawn out of the Stable, drags his Leg for fifty or sixty Paces, as if he had no Motion in his Hough; but after that the Cramp disappears.

CRATCHES is a Swelling on the Pastern, under the Fetlock, and sometimes under the Hoof; upon which Score 'tis distinguish'd into the Sinewy Cratches which affect the Sinew, and those upon the *Cronet*, call'd *Quitter-bones*. Since your Horse has a *Quitter-bone*, you must cut the Tendon.

CREAT is an Usher to a Riding-Master, or a Gentleman bred in the Academy, with intent to make himself capable of teaching the Art of Riding the Great Horse.

CREPANCE is a Chap or Cratch in a Horse's Legs, given by the Spunges of the Shooes of one of the Hinder-feet crossing and striking against the other Hinder-foot. This Cratch degenerates into an Ulcer.

CREVICE; *i. e.* Chop, Clift, or Chink.

CROATS, or *Cravats*, are Horses brought from *Croatia*, a Frontier of *Hungary*; which for the most part beat upon the Hand, and bear up to the Wind; that is, bear their Neck high, and thrust out their Nose, shaking their Head. The *Croats* are subject to be hollow or shell-tooth'd.

CRONET. See *Coronet*.

CROSS: To make a *Cross* in *Corvets*, to make a *Cross* in *Balotades*, is to make a sort of Leaps or Airs with one Breath forwards, backwards, and sideways, as in the Figure of a *Cross*. Some talk of making a *Cross* in *Caprioles*, but that can't be; for the Horses that should make *Caprioles* backwards, would appear resty, and such as we call *Ramingue*, which would not work according to the just exactness of the Manage: Not to mention that the most vigorous Horse that is, can't with one Breath mark the whole *Cross* in *Caprioles*.

CROWN'D: A Horse is said to be crown'd, when by a Fall, or any other Accident, he is so hurt or wounded in the Knee, that the Hair sheds and falls off, without growing again. This Horse is Crown'd, you'll find it a hard matter to put him off; for Crown'd Horses are not very saleable, because they are suspected to be apt to fall upon their Knees.

CROUPADES are higher Leaps than those of *Corvets*, which keep the Fore and Hindquarters of the Horse in an equal height, so that he trusses his Hind-legs under his Belly, without yerking or shewing his Shooes. *Groupades* differ from *Caprioles* and *Balotades* in this, that in *Croupades* the Horse does not yerk as he does in the other two Airs.

High-Croupades are *Croupades* rais'd above the ordinary height. I'm about to put this Horse to the Air of *Croupades*. This Horse presents right for *Croupades*. He works in *Croupades*; he makes *Croupades*. See *Yerk*.

CROUPE of a Horse is the Extremity of the Reins, above the Hips. In making the Volts, a Horse's Shoulders should be opposite to his *Croupe*. Those who us'd this Expression, mean'd, that a Horse walking sideways, and upon two *Pistes*, his Shoulders make one Tract while the *Croupe* makes another. But after all, this is not a just way of speaking; for at that rate the Shoulders are not opposite to the *Croupe* in a Straight Line, by reason that half the Shoulders marches before the *Croupe*, and the Horse bending his Neck a little, looks into the Volt. See *Head in*, and *Truss'd*.

To *Gain the Croupe*, is when one Horse-man makes a Demitour

tour upon another, in order to take him upon the *Croupe*. If in a Combat you are hard put to it by your Enemy, make a Demi-Pyroet at the end of the Passade, and gain his *Croupe*.

Without slipping the Croupe (*Sans que la Croupe echape*). This Expression is used for Volts and a Gallop, and signifies, without traversing, without letting the *Croupe* go out of the Volte or the Tread of the Gallop.

Your Horse has a *Tettar* in his *Croupe*. See *Tettar*.

CURB is a Chain of Iron made fast to the upper part of the Branches of the Bridle, in a Hole called the *Eye*, and running over the Beard of the Horse. Your Horse will never be right upon the Hand unless you give him a longer Curb, for the uneasiness of his Beard, occasion'd by this Curb, makes him toss his Nose up to the Wind, and chack and beat upon the Hand. Your *English* Watering Bitts have no Curbs; and the *Turkish* Bitts, call'd *Genettes*, have a Ring that serves in the room of a Curb. See *Genette*.

To give a Leap upon the Curb, is to shorten the Curb, by laying one of the Mails or S like Joints of the Chain over the rest.

Curb is a Hard and Callous Tumour, that runs along the inside of a Horse's Hoof; that is to say, on that part of the Hoof that is opposite to the Leg of the same side.

CUT. See *Interfere*.

To *Cut* or *Geld* a Horse, is to render him impotent; after which he is called a Gelding (in *French*, *Hongre*) by way of distinction from a Stone-horse. Commonly your *Roussins*, (*i. e.* your Strong Thick-bodied *Dutch* Horses) are *Stone-horses* and not *Geldings*. The best way to cure a Horse of Biting and Kicking is to Geld him.

To *Cut* the Round, or *Cut* the Volte, is to change the Hand when a Horse works upon Volts of one Tread; so that dividing the Volt in two, he turns and parts upon a Right Line to recommence another Volt. In this sort of Manage, the Riding-Masters are wont to cry, *Coupez, Coupez le Rond*; Cut, cut the Round.

D.

DAPPLE-BLACK, is a Black-horse, that in his Black Skin or Hair has Spots or Marks, which are yet Blacker and more shining than the rest of the Skin. When *Bay-Horses* have Marks of a Dark-bay, we call 'em *Dappled-Bays*, or *Bays a Miroir*.

DECEIVE. A Horse is said to be deceived (*Trompe*) upon a *Demivolt* of one or two Treads, when working (for instance)

inſtance) to the Right, and not having yet furniſh'd above half the *Demivolt*, he's preſs'd one time or motion forwards, with the Inner-Leg, and then is put to a *Repriſe* upon the left, in the ſame *Cadence* with which he begun; and thus he regains the Place where the *Demivolt* had been begun to the Right, and works to the Left. Thus you may deceive a Horſe upon any Hand.

DEVUIDER: A Term in the Academies, applied to a Horſe that in working upon *Volts*, makes his Shoulders go too faſt for the *Croupe* to follow, ſo that inſtead of going upon two Treads, as he ought, he endeavours to go only upon one: Which comes from the reſiſtance he makes in defending againſt the Heels; or from the, fault of the Horſe-man that's too haſty with his Hand. See *Haſten*.

DISARM: To Diſarm the Lips of a Horſe, is to keep 'em ſubject and out from above the Barrs, when they are ſo large as to cover the Barrs, and prevent the true Preſſure or *Appui* of the Mouth, by bearing up the Bitt, and ſo hindring the Horſe to feel the effects of it upon the Barrs. Give your Horſe a Bitt with a *Canon Coupé*, or *Cut*, which will Diſarm his Lips; or elſe put the *Olives* upon him, which will have the ſame Effect.

DIS-ERGOT. See *Ergot*.

To *Diſgorge*, (in *French*, *Degorger*) is to Diſcuſs or Diſperſe an Inflamation or Swelling: Your Horſe's Legs are Gorged or Swelled, you muſt walk him out to Diſgorge 'em.

DISUNITE: A Horſe is ſaid to Diſunite, that drags his Haunches, that gallops falſe, or upon an ill Foot. See *Gallop falſe*.

DOCK, (or *Trouſſequeue*) is a large Caſe of Leather as long as the *Dock* of a Horſe's Tail, which ſerves for a Cover to the Tails of Leaping Horſes, and is made faſt by Straps to the Crupper, and has Leathern Thongs that paſs between his Thighs, and along the Flanks to the Saddle Straps, in order to keep the Tail tight, and to hinder it to whisk about, or make the Horſe appear broader at the *Croupe*.

DOUBLE: To Double the Reins: A Horſe Doubles his Reins when he leaps ſeveral times together to throw his Rider. This *Ramingue* doubles his Reins, and makes *Pontlevis*'s. See *Pontlevis*.

DRENCH; (in *French*, *Armand*) is a ſort of Decoction prepared for a Sick Horſe, and compoſed of ſeveral Drugs, mentioned in Mr. *Solleyſel*'s *Compleat Horſe-man*. They put the Drench or *Armand* upon the end of a Bull's Pizzle, and thruſt

thrust it down his Throat, in order to recover his Appetite and Strength.

DRY: To put a Horse to Dry-Meat, is to feed him with Corn and Hay, after taking him from the Grass, or housing him.

DULL: The Marks of a Dull Stupid Horse (in *French, Marques de ladre*) are White Spots round the Eye, and on the Tip of the Nose, upon any general Colour whatsoever: These Marks are hard to be distinguished in a White-horse: Though the Vulgar take these Spots for Signs of Stupidity, 'tis certain they are a great sign of the Goodness of a Horse; and the Horses that have them, are very sensible and quick upon the Spur.

DUST: To beat the Dust. See *Beat*.

E.

EAR; *Wide-ear'd*; (in *French Oreillard*): A Horse is said to be such, if the Root or lower part of the Ear is placed too low, and the Ear it self is too large. Take care of these Two Wide-Ear'd Nags, their Ears are ill placed, and flap too far down; do but observe that with their Great Broad Ears, they mark all the Times or Points of a Walk or Trot, as if they were Hogs.

EBRILLADE, is a Check of the Bridle, which the Horse-man gives to the Horse by a Jerk of one Rein, when he refuses to turn. An *Ebrillade* differs from a *Saccade* in this, that a *Saccade* is a Jerk made with both Reins at once. Most People confound these two Words, under the general Name of a Check, or Jerk of the Bridle; (in *French, un Coup de Bride*): But let that be as it will, 'tis always a Chastisement, and no Aid, and the Use of it is banished the Academies.

ECAVESSADE, is a Jerk of the Cavesson.

ECHAPER; *Laisser Echaper de la Main*: To suffer a Horse to escape, or slip from the hand; A Gallicism us'd in the Academies, implying, to give him head, or put on at full speed. When you give your Horse head, guide him streight. See *Nails*.

ECHAPE; An *Echapé* (so call'd in *French*) is a Horse got between a Stallion and a Mare, of a different Breed, and different Countries. In this Breed I have had two *Barbary Echapés*, and two of *Spanish* Horses.

ECOUTE: A Pace or Motion of a Horse is said to be *Ecouté*, or Listening, when a Horse rides well upon the hand and the heels, compactly put upon his Haunches; and hears

hears or liſtens to the Heels or Spurs, and continues duly balanced between the Heels, without throwing to either ſide. This happens, when a Horſe has a fine Senſe of the Aids of the Hand and Heel.

ECURIE is a Covert-place for the Lodging or Houſing of Horſes. Our Horſes are but ill ſtabled in this *Ecurie.* The *French* uſe the word (*Etabler*) to *ſtable*, as a Verb. But in the room of the Noun they imploy *Ecurie.* This Horſe's Hair ſtares, and is planted upright, becauſe he has been too coldly ſtabled in this *Ecurie.*

ECUYER; A *French* word (in *Engliſh*, Querry) has different Significations in *France.* In the Academy, or Manage, the Riding-Maſter goes by the name of *Ecuyer.* In the King of *France*'s Houſhold there are Querries, or *Ecuyers de quartier*, who help the King in Mounting his Horſe, and Alighting; and follow his Majeſty upon Horſe-back, and carry his Sword. The Queen's Gentlemen Uſhers, and the Maſters of the Horſe to Princes and Perſons of Quality, are called *Ecuyers.* Beſides all which, there are the *Ecuyers Cavalcadours*; for which ſee *Cavalcadour.*

EFFECTS of the Hand are taken for the Aids, *i. e.* the Motions of the Hand, that ſerve to conduct the Horſe. There are four Effects of the Hand, or four ways of making uſe of the Bridle; namely, to puſh a Horſe forwards, or give him head; to hold him in; and to turn the hand either to the right or left. See *Nails.*

EGUILLETTE; *Nouer l' Eguillette.* See *Yerk.*

EMBARRER. See *Barr.*

EMBRACE *the Volt.* A Horſe is ſaid to Embrace a Volt, when in working upon Volts he makes a good way every time with his Fore-legs. Such a Horſe has Embraced a good deal of Ground; for from the place where his Fore-feet ſtood, to where they now ſtand, he has Embraced or gone over almoſt a Foot and a half. If he does not Embrace a good deal of ground, he'll only beat the duſt; that is, he'll put his Fore-feet juſt by the place from whence he lifted them. Thus the oppoſite Term to Embracing a Volt, is, *Beating the duſt.* A Horſe cannot take in too much ground, provided his Croupe does not throw out, that is, provided it does not go out of the Volt. See *Beat.*

EMIELLURE. See *Charge.*

ENCRAINE; an old obſolete and improper Word, ſignifying a Horſe wither-wrung, or ſpoil'd in the Withers.

ENLARGE a Horſe, or make him go large, is to make him embrace more Ground than he

be cover'd. This is done, when a Horse works upon a round, or upon Volts, and approaches too near the Center, so that 'tis desir'd he should gain more ground, or take a greater Compass. To Enlarge your Horse, you should prick him with both Heels, or aid him with the Calves of your Legs, and bear your Hand outwards. Your Horse narrows, enlarge him, and prick him with the inner Heel, sustaining him with the outer Leg in order to press him forwards, and make his Shoulders go. Upon such occasions the Riding-Masters cry only, *Large, Large.* See *In.*

ENSEMBLE; *Bien ensemble.* See *Haunches.*

ENTABLER; A word us'd in the Academies, as applied to a Horse whose Croupe goes before his Shoulders in working upon Volts: For in regular Manage, one half of the Shoulders ought to go before the Croupe. Your Horse *Entables*; for in working to the right, he has inclination to throw himself upon the right Heel; and that fault you may prevent by taking hold of the right Rein, keeping your right Leg near, and removing your left Leg as far as the Horse's Shoulder. A Horse can't commit this fault without committing that call'd in the Academies *Aculer*; which see. But *Aculer* may be without *Entabler.* See *Aculer*, and *Embrace.*

ENTIER, the *French* Word for a Stone-Horse. *Entier* is a sort of Resty Horse that refuses to turn; and is so far from following or observing the Hand, that he resists it. Such a Horse is *Entier* on the right Hand; he puts himself upon his right Heel, and will not turn to the right. If your Horse is *Entier*, and refuses to turn to what hand you will, provided he flies or parts for the two Heels, you have a Remedy for him; for you have nothing to do but to put the *Newcastle* upon him, *i. e.* supple him with a Cavesson made after the Duke of *Newcastle*'s way.

Entier; a Bit that presses *de l' entier.* See *Bit-mouth.*

ENTORSES. See *Pastern.*

ENTRAVES, and *Entravons.* See *Locks.*

ENTREPAS is a broken Pace or Going, and indeed properly a broken Amble, that is neither walk, nor trot, but has somewhat of an Amble. This is the Pace or Gate of such Horses as have no Reins or Back, and go upon their Shoulders, or of such as are spoil'd in their Limbs.

EPARER; a Word us'd in the Manage, to signify the Flinging of a Horse, or his Yerking and striking out with his Hind-legs. See *Nouer l' Eguillete*, and *Yerk.* In Caprioles, a Horse must yerk out behind with all his force; but in Balot-

Balottades he ſtrikes but half out; and in Croupades he does not ſtrike out his Hind-legs at all. All ſuch Yerking-Horſes are reckon'd rude Horſes.

ERGOT, is a Stub like a piece of ſoft Horn about the bigneſs of a Cheſnut, placed behind and below the Paſtern-Joynt, and commonly hid under the Tuft of the Fetlock. To diſ-ergot, or take it out, is to cleave it to the quick with an Inciſion-knife, in order to pull up a Bladder full of Water that lyes cover'd with the *Ergot*. This Operation is ſcarce practiſed at *Paris*; but in *Holland* 'tis frequently performed upon all four Legs, with intent to prevent Watery Sores, and other foul Ulcers.

ESCLAME is an obſolete *French* Word, ſignifying a light-belly'd Horſe.

ESQUIAVINE, an old *French* Word, ſignifying a long and ſevere Chaſtiſement of a Horſe in the Manage.

ESTRAPADE is the Defence of a Horſe that will not obey; who to get rid of his Rider, riſes mightily before; and while his Fore-hand is yet in the Air, yerks furiouſly with his Hind-legs, ſtriking higher than his Head was before; and during this Counter-time, goes back rather than advances.

ESTRAC is the *French* Word for a Horſe that is light-body'd, lank-belly'd, thin-flank'd, and narrow-cheſted. See *Belly*, *Light-belly'd*, *Flank*, *Jointee*, and *Green*.

ESTRAPASSER. See *Over-do*.

EXTEND a Horſe: Some make uſe of this Expreſſion, as importing, to make a Horſe go large.

EYE of the Branch of a Bridle, is the uppermoſt part of the Branch, which is flat with a hole in it, for joining the Branch to the Head-ſtall, and for keeping the Curb faſt.

Eye; a Horſe unſhod of one Eye. A rallying Expreſſion, importing, that he's blind of an Eye.

Eye of a Bean is a black Speck or Mark in the Cavity of the Corner-Teeth, which is form'd there about the Age of Five and a half, and continues till Seven or Eight. And 'tis from thence that we uſually ſay, ſuch a Horſe Marks ſtill; and ſuch a one has no Mark. See *Teeth*.

F.

FALCADE. A Horſe makes Falcades, when he throws himſelf upon his Haunches two or three times, as in very quick Corvets; which is done in forming a Stop, and Half-ſtop. A Falcade, therefore, is this Action of the Haunches, and of the Legs

Legs which bend very low, as in Corvets, when you make a Stop or a Half-ſtop. This Horſe ſtops well, for he makes two or three Falcades, and finiſhes his Stop, with a Peſate. This Horſe has no Haunches; he'll make noFalcades. The Falcades of that Horſe are ſo much the prettier, that in making them his Haunches are low. Stop your Horſe upon the Haunches, in making him ply 'em well; ſo that after forming his Falcades, he may reſume his Gallop without making a Peſate; that is, without ſtopping, or marking one Time: And thus he'll make a Half-ſtop. See *Stop*, *Half-ſtop*, *Haunches*, and *Time*.

FALSE. This Horſe is falſe. He Gallops falſe. See *Gallop falſe*.

FARCY is a Poyſon or Corruption that infects the Blood of Horſes, and appears in Swellings like Strings along the Veins, in Knots, and even in Ulcers. See *Fire*.

FARRIER'*s Pouch*; A Leathern Bag, in which they carry Drivers, Nippers, Shoes for all Sizes of Feet, good ſharp Nails, and all that is proper for new Shoeing a Horſe that has loſt his Shoe upon a Road. If you have no Farrier with you, you muſt always have in your Equipage a Farrier's Pouch well provided, and a Groom that knows how to drive Nails.

FEATHER upon a Horſe, is a ſort of natural Frizling of the Hair, which in ſome places riſes above the lying Hair, and there caſts a Figure reſembling the Tip of an Ear of Corn. There are Feathers in ſeveral places of a Horſe's Body, and particularly between the Eyes. Many believe that when the Feather is lower than the Eyes, 'tis a Sign of a weak Eye-ſight: But this Remark is not always certain.

A Roman Feather, (in *French*, *Epee Romaine*) is a Feather upon a Horſe's Neck, being a Row of Hair turn'd back and rais'd, which forms a Mark like a Sword-blade juſt by the Mane.

FEEL; to Feel a Horſe in the hand, is, to obſerve that the Will of the Horſe is in their hand; that he taſtes the Bridle, and has a good *Appui* in obeying the Bit.

To feel a Horſe upon the Haunches, is to obſerve that he plies or bends them; which is contrary to leaning, or throwing upon the ſhoulders.

FERME *a ferme*; A Word peculiar to the Manage-Schools, ſignifying, in the ſame place, without ſtirring or parting. You muſt raiſe that Horſe *ferme a ferme*. This Horſe leaps upon *firma a firma*, and works well at Caprioles. When a young Gentleman comes firſt to the School, the reſt, to play upon him, will aſk him to Gallop upon *firma a firma*.

FETLOCK; (in *French*, *Fanon*) is a Tuft of Hair as big as the Hair of the Mane, that grows behind the Pastern Joint of many Horses: Horses of a Low Size have scarce any such Tuft. This Coach-horse has a large Fetlock; and the other has so much Hair upon his Legs, that if the Coach-man does not take care to keep them clean and tight, he'll be subject to the Watery Sores, call'd the Waters (*Les Eaux* in *French.*) There's an Ergot; (*i. e.* a Round Hard Stub) behind that Fetlock.

FIG is a sort of Wart on the Frush, and sometimes all over the Body of a Horse. The *Figs* that appear in the Frush or Sole, make an Evacuation of Malignant Stinking Humours, that are very hard to cure. For the Cure, see Mr. *Solleysel*'s Book.

FINGART, an absolete *French* Word, signifying a Horse that kicks against the Spurs.

FIRE: To give the Fire to a Horse, is to apply the Firing-Iron red hot to some Preternatural Swelling, in order to discuss it; which is oftentimes done by clapping the Firing-Iron upon the Skin, without piercing through. We give the Fire to Farcy-knots, by running a Pointed Burning-Iron into the Ulcers. We likewise give the Fire for Wrenches of the Pasterns.

FIRING-IRON is a Piece of Copper or Iron, about a Foot long, one end of which is made flat, and forged like a Knife, the Back of it being half an Inch thick, and the Fore-edge about five or six times thinner. When the Farrier has made his Firing-Iron red hot in his Forge, he applies the thinnest part to the Horse's Skin, and so gives the Fire to the Hams, or such places as stand in need of it. The Fire has been well given upon that Horse's Legs; in all appearance the Farrier had a Light-hand. See *Fire*.

FINITEUR, an old Word in the *Italian* Academies, signifying the end of the Career or Course.

FLAME, is a small Instrument of Fine Steel, compos'd of two or three moveable Lancets for blooding a Horse; and sometimes for making Incisions upon occasion, and so supplying the room of an Incision-Knife.

FLANK: A Horse is said to have Little Flanks, to be sorrily Bodied, to be gaunt belly'd, and thin Gutted, (in *French*, *Estrac*) when his flank turns up like a Greyhound, and his Ribs are flat, narrow, and short. A well Flank'd Horse, is one that has wide and well-made Ribs, and a good Body. In this case the Word Flank is used in the room of Gutt. See *Belly*, *Light-belly'd*, *Jointee* and *Green*.

FLESHY

FLESHY *Lump*, or Excrescency. See *Bouillon*.

FLING, is the Fiery and Obstinate Action of an Unruly Horse.

To *Fling* like a Cow, is to raise only one Leg, and give a Blow with it.

To *Fling*, or Kick with the Hind-legs. See *Yerk*.

FLY *the Heels*: A Horse is said to Fly the Heels, when he obeys the Spurs. See *Spur*, and *Heels*.

FONCEAU, is the Bottom or End of a Canon-Bitt-mouth; that is, the part of the Bitt that joyns it to the *Banquet*. See *Chaperon*.

FOOT of a Horse is the Extremity of the Leg, from the *Cronet* to the lower part of the Hoof. The Four Feet are distinguish'd by four different Names: The two Fore-feet are by some call'd the Hands of a Horse; but that Term is in disuse, the common Expression being, the Far Fore-foot, to denote the Right-foot before; and the Near Fore-foot, the Stirrup-foot and the Bridle-hand-foot, to denote the Left-foot before.

Of the two Hinder-feet, the Right is call'd the Far-hind-foot; and when Spears were us'd, 'twas call'd the Spear-foot, because in resting the Spear, the Socket of it answer'd the Right-foot. The Left-hind-foot is call'd the Near-foot behind.

Foot Fat: A Horse is said to have a *Fat Foot*, when the Hoof is so thin and weak, that unless the Nails be drove very short, he runs the risque of being prick'd in shooing. The *English* Horses are very subject to this Disorder. A Horse's *Foot* is said to be *Derobé* (in *French*) *i. e.* rob'd or stolen, when 'tis worn and wasted by going without Shooes, so that for want of Hoof, 'tis a hard matter to shooe him. I would not buy this Horse, because his Feet were *Derobé*, *&c.*

A Horse's Foot is said to be worn or wasted (*usé*) when he has but little Hoof, and not enough for shooeing.

To gallop upon a good *Foot*, or put a Horse upon a good *Foot* (in *French*, *Sur le bon pied*) See *Gallop false*.

FOUNDERING is an Universal Rheumatism, or defluxion of Humours upon the Sinews of a Horse's Legs, which causes so great a Stiffness in the Legs, that they lose their wonted Motion. Your Horse appears very tired, though he has not rid or work'd hard; his Legs are stiff, he can't rise upon his Hinder-legs no more than if his Back were broke: Now all this speaks him foundered. If this Foundering does not make him Chest-founder'd, 'twill be the easier cured; but if he is both Foundered and Chest-founder-ed, and has the Molten-grease,

as it happens sometimes, he will not get over it.

Chest-foundering is the agitation and heaving of a Horse's Flanks, with a Motion almost like that occasioned by a Fever. This may happen when a Horse is rid out of his Wind, and beyond his Strength, and is so over-heated with the fatigue, that he can't breathe.

A Foundered Horse that has been over rid, and has not a free Respiration, is distinguished from a Pursive Horse in this, that the Lungs of a Pursive Horse are altered and much affected by the violent heaving of the Flanks. A Horse is foundered without over riding, when his Inward Parts, or his Blood are over heated and stuffed with Foreing Humours.

To *Founder*, or over ride a Horse, is riding him out of his Breath, and beyond his Strength.

FOUR-CORNERS; to work upon the *Four-Corners*, is to divide (in imagination) the Volt or Round into Four Quarters; so that upon each of these Quarters the Horse makes a Round or two at Trot or Gallop; and when he has done so upon each Quarter, he has made the *Four-Corners*.

FROTH, or *Foam*, is a Moist White Matter that ouzes from a Horse's Mouth. Your Horse champs upon his Bridle, and squeezes out of his Mouth a White Ropy Foam; which argues that he is a Horse of Mettle, Health, and a Cool Fresh Mouth.

FRUSH, or *Frug* of a Horse, is a sort of Tender Horn which arises in the middle of the Sole, and at some distance from the Toe, divides into two Branches running towards the Heel, in the form of a Fork. Look after this Horse, for the Flesh is run in upon the *Frush*; I see an Excrescence or sprouting of Flesh in that part. There's a *Fig* in that Sorrel's *Frush*: And this Roan has a Scabbed *Frush*: And here is another that has a Fat *Frush*; *i. e.* a *Frush* that is too thick and too large.

FUZEE is two dangerous Splents, joining from above downwards. Commonly a *Fuzie* rises to the knee and lames the Horse. *Fuzies* differ from *Screws*, or through-*Splents* in this, that the latter are placed on the two opposite sides of the Leg.

G.

GALLOP is the Motion of a Horse that runs at full speed, in which making a kind of Leap forwards, he lifts both his Fore-legs almost at the same time, and when these are in the Air, and just upon the point of touching the Ground, he lifts both his Hind-Legs almost at

once.

once. A Horſe that has an eaſie light *Gallop*; a Horſe that takes the *Gallop*, that puts himſelf to the *Gallop*. Such a Horſe *gallops* fine; *i. e.* he *gallops* upon his Haunches, he does not preſs heavy upon the Bridle, he bends his Fore-legs well, he has a good Motion with him, he is well coupled, and keeps his Legs united. This Horſe *gallops* like an *Engliſh* Horſe, he grazes or glances upon the Ground, that is, he *gallops* cloſe to the Ground, and does not raiſe his Legs as he ought to do. Such a Horſe goes equally well at a Step, Trot, and Gallop. The Great *Gallop*, or the Hunting *Gallop*, or the *Gallop* with a Long Stretch. *Gallop* with all the Heels; *i. e.* at full ſpeed. A ſhort light *Gallop*; *i.e.* a Slow *Gallop*. This Horſe that wheezes upon a *Gallop* is infirm; but the other that ſnorts, ſhews that he is Long-winded. This Horſe *gallops* well upon two Treads. See *Snort*.

GALOPADE. The Fine *Galopade*, the Short *Gallop*, the Liſtening *Gallop*, the *Gallop* of the School: 'Tis a *Hand-gallop*, or *Gallop* upon the Hand, in which a Horſe galloping upon one or two Treads, is well united, and well *raccourci*, or knit together, well coupled, and well ſet under him. This Horſe makes a *Galopade*, and works with one Haunch in; *i. e.* inſtead of going upon One Tread, whether right out, or in a Circle, he has one Haunch kept in ſubjection, let the turn or change of the Hand be what it will; ſo that the inner Haunch which looks to the Center of the Ground, is more narrow'd, and comes nearer to that Center than the Shoulder does: And thus the Horſe does not go altogether to that ſide, and his way of working is a little more than One Tread, and ſomewhat leſs than Two. The difference between working with one Haunch in, and galloping upon Volts, and managing upon *terra à terra*, is that in galloping upon Volts, and working *terra à terra*, the two Haunches are kept ſubject, and the two Haunches are in, that is, within the Volt; but in galloping a Haunch in, only one is kept ſubject.

To Gallop united, to Gallop upon the good or right foot, is, when a Horſe that gallops right out, having cut the way, or led with either of his Fore-legs, continues to lift that ſame Leg always firſt; ſo that the hinder Leg of a ſide with the leading Fore-leg, muſt likewiſe be rais'd ſooner than the other Hind-leg. For inſtance, if the right Fore-leg leads before the left, then the right Hind-leg muſt likewiſe move ſooner than the left Hind-leg; and in this Order muſt the Horſe continue to go on.

To Gallop falſe, to diſ-unite, to drag the Haunches, to change

feet; to go or run upon false feet, to gallop upon the false foot; is, when the Galloper, having led with one of the Fore-legs, whether right or left, do's not continue to make that Leg always set out first, nor to make the Hind-leg of a side with the leading Leg to move before its opposite Hind-leg; that is to say, the orderly Going is interrupted. A Horse that gallops false, gallops with an unbecoming Air, and incommodes his Rider. My Horse did gallop false, but I have now taught him the right Foot. If your Horse gallops false, or dis-unites, and if you have a mind to put him upon keeping the right foot, and uniting well his Haunches; you must bring to with the Calves of your Legs, and then with the Out-spur, that is, the Spur that's contrary and opposite to the side upon which he dis-unites; so that if he dis-unites to the right, you must prick him with the left Heel.

GANACHES (so called in *French*) are the two Bones, one on each side of the hinder part of the Head, opposite to the Neck or Onset of the Head, which form the Lower Jaw, and give it motion. 'Tis in this place that the *Glands* or Kernels of the *Strangles* and the *Glanders* are placed.

GASKOIN. See *Thigh*.

GATE (in *French*, *Train*) is the Going or Pace of a Horse. This Horse has a good Gate, but the other has a broken Gate. This Horse goes well, but t'other does not.

GELDING, (in *French*, *Hongre*) is a Horse whose Testicles are cut out, so that he is not fit for a Stallion.

To GELD. See *Cut*.

GENET, is a small-siz'd well-proportion'd *Spanish* Horse. Mr. *de la Broue* gives the Name of *Genet* to such *Italian* Horses as are well-made and proportion'd.

GENETTE, or *Turkish* Bit; is a Bit, the Curb of which is all of one piece, and made like a large Ring, and plac'd above the Liberty of the Tongue. When they bridle a Horse, they make his Chin pass thro' this Curb, which surrounds his Beard. This sort of Bits are at present much us'd at the Court of *France*.

Genette; to ride with the Legs *à la Genette*, *i. e.* in the *Genette* or *Spanish* Fashion, is to ride so short, that the Spurs bear upon the Horse's Flank. This would be reckon'd an Indecency in *France* and *England*; but among the *Spaniards* it passes for a piece of Gallantry and handsome Carriage, when they ride upon their *Genettes* in going to Court before the Ladies.

GIGOT, a Branch after the form of a Gigot or Leg, is a Branch, the lower part of which (call'd the *Gargouille*) is round.

GLANDERS, is a thick Slimy and Bloody Humour, proceeding from a Defect in the Lungs, and voided by the Nostrils. A Horse that has the true Glanders is gone, and good for nothing; and whoever buys such a Horse, (in *France*) can oblige the Seller to take him again, any time within Nine Days after Delivery. You must warrant this Horse clear of the Glanders, and Pursyness; and sound, hot or cold. See *Sound*.

GOAD. See *Valet*.

GOING, (in *French*, *Alleure*) is the Pace or Gate of a Horse. This Barb has all his Goings, or Paces, very fine; contrary to what we commonly observe of Barbs; for they are apt to stumble upon a Bowling-Green, unless they be animated, bore up, and put on. This Horse has a cold Gate with him; that is, he does not bend his Knee as he ought to do; and he raises his Legs so little, that he grazes, or sweeps the Ground.

GORGED, *i. e.* swell'd. This Horse's Pastern-Joynt is gorged, and the other has his Legs gorged: You must walk him out to dis-gorge 'em, or take down the Swelling.

GRAPES; a Word sometimes us'd to signify the *Arrests*, or Mangy Tumors that happen in a Horse's Legs. See *Arrests*.

To GRAPPLE (in *French*, *Harper*.) A Horse is said to Grapple, either one, or both Legs; the Expression being peculiar to the Hinder-legs. He grapples both Legs, when he lifts 'em both at once; and raises 'em with precipitation, as if he were a Curvetting. He grapples one Leg, when he raises it precipitantly higher than the other without bending the Ham. Your Horse harps, or grapples, so that he must have the Stringhalt in his Hough.

GRASS; to put a Horse to Grass. To turn him out to Grass to recover him. To take a Horse from the Grass, and keep him at Dry Meat. See *Dry*, and *Green*.

GREEN; (*Vert*) to give a Horse Greens, is to put him to Grass. If in the Spring-time your Horse is over-rid, lean, and gaunt-belly'd, give him *Green* Barley. If he's young, and his Flank not altered, that may recover him.

GROUND. See *Terrain*.

GROUPADES, a Corruption of *Croupades*.

H.

HAIR. In ſpeaking of Horſes, the *French* uſe the Word *Poil*, (*i. e. Hair*) to ſignifie their Colour: And ſometimes 'tis us'd to ſignifie that part of the Flank that receives the Prick of the Spur. After you have prick'd the Horſe, do not leave your Spur in his Hair. Remove your Spur from the Hair, and preſently recover your Legs to their place, for you always have your Spurs in his *Poil*.

To Rub a Horſe *à Poil*, is to rub him down with the Grain, obſerving the natural lying of the Hair, and not going againſt the Hair.

Pale-Hair, or *Poil-Lavé*, is thoſe parts of the Skin that approach more to White than the reſt, as being not of ſo high a Tinge.

Staring Hair, (*Poil Planté*, or Planted Coat) is ſaid of a Horſe whoſe Hair briſtles up, or riſes upright; which Diſorder is owing to his being ill curry'd, not well covered, or too coldly houſed.

HALBERT (in *French*, *Bec de Corbin*) is a ſmall Piece of Iron, one Inch broad, and three or four Inches long, ſolder'd to the Toe of a Horſe's Shooe, which jets out before to hinder a Lame Horſe to reſt or tread upon his Toe. Theſe *Halbert-Shooes* do of neceſſity conſtrain a Lame Horſe, when he goes at a moderate Pace, to tread or reſt on his Heel; which lengthens and draws out the back Sinew, that was before in ſome meaſure ſhrunk.

HALTER for a Horſe, is a Head-ſtall of *Hungary* Leather, mounted with one, and ſometimes two Straps, with a ſecond Throat-band, if the Horſe is apt to unhalter himſelf. If you put a *Halter* upon this Horſe, tye it very ſhort, for if he can but lie down, 'tis enough: Take care of his Head-ſtall or Collar. Such a Horſe has the Tick, he ticks upon his *Halter*.

Halter-caſt, is an Excoriation of the Paſtern, occaſioned by the *Halter* its being entangled about the Foot, upon the Horſe's endeavouring to rub his Neck with his Hinder-feet.

Unhalter. A Horſe is ſaid to Unhalter himſelf, that turns off his *Halter*. Since your Horſe is ſo apt to unhalter himſelf, you muſt get him a *Halter* with a Throat-band.

Strap or String of the *Halter*, (*longe*) is a Cord, or long Strap of Leather made faſt to the *Head-ſtall*, and to the Manger, to tye the Horſe. Do not bridle your Horſe till you ſee if he is *Halter-caſt*. See *Tick*.

HAND, is the Meaſure of a Fiſt clinched, by which we compute the heighth of a Horſe:

The *French* call it *Paume*, and had this Expreſſion and Meaſure firſt imparted to them from *Liege*. A Horſe for War ſhould be 16 or 18 *Hands* high.

Handfull. Two *Handfulls* (in *French, une Jointee*) is as much Grain or Bran as the two *Hands* will hold when joined together. You have a Horſe that's very Lank-belly'd, if you have a mind to make him well Bodied, put a *Jointeè* of Wheat every Morning into his Manger.

Hand: Spear-hand, or *Sword-hand*, is the Horſe-man's *Right-hand*.

Bridle-hand is the *Left-hand* of the Horſe-man. There are ſeveral Expreſſions which relate to the *Bridle-hand*, becauſe that *Hand* gives motion to the Bitt-mouth, and ſerves to guide the Horſe much more than the other helps. A Horſe-man ought to hold his *Bridle-hand* two or three Fingers above the Pommel of the Saddle. This Horſe-man knows how to keep time between his *Hands* and his *Heels*. This Horſe-man has no *Hand*; that is, he does not make uſe of the Bridle but unſeaſonably, and does not know to give the aids or helps of the *Hand* with due nicety.

To keep a Horſe upon the *Hand*, is to feel him in the Stay upon the *Hand*, and to be always prepared to avoid any Surpriſal or Diſappointment from the Horſe.

A Horſe is ſaid to be, or reſt well upon the *Hand*, that never refuſes, but always obeys and anſwers the Effects of the *Hand*, and knows the *Hand*. A Horſeman ſhould make it his buſineſs to make the Horſe know and obey the *Heels*. To make a Horſe right upon the *Hand*, and free in the *Stay*, or *Reſt*, he muſt be taught to know the *Hand* by degrees, and gentle Methods; the Horſe-man muſt turn him, or change *Hands*, ſtop him, and manage with dexterity the *Appui* or *Preſſure* of his Mouth, ſo as to make him ſuffer cheerfully and freely the effect of the Bitt-mouth, without reſiſting or reſting heavy upon the *Hand*. This Horſe has no Stay, he beats upon the *Hand*. The Short, or *Hand-gallop* teaches Horſes to be right upon the *Hand*.

A Light *Hand*. A good Horſeman ought to have a Light *Hand*; that is, he ought only to feel the Horſe upon his *Hand*, in order to reſiſt him when he attempts to ſlip from it; and he ought, inſtead of cleaving to the Bridle, to lower it, as ſoon as he has made his reſiſtance. If a Horſe through an overbearing eagerneſs to go forward, preſſes too much upon the *Hand*, you ought to ſlack your *Hand* at certain times, and keep a hard *Hand* at other times, and ſo diſappoint the Horſe of preſſing continually upon the Bitt. Now this facility or liberty in the

the Horſe-man, of ſlacking and ſtiffening the *Hand*, is what we call a Good *Hand*. Your Horſe works well, but you ſtick too much to the Bridle; inſtead of holding thus by the Bridle, you ought to clap your Thighs cloſe to the Horſe, and keep a Light *Hand*, and then you'll manage with exactneſs. Do not you know, that to have a Light *Hand*, and to manage a Horſe with a Swinging Bridle, is one of the greateſt marks of a good Horſe-man?

To ſlack or yield the *Hand*, is to ſlacken the Bridle.

To hold up, or ſuſtain the *Hand*, is to pull the Bridle in.

To guide a Horſe by the *Hand*, is to turn or change *Hands* upon one Tread.

Part from the *Hand*. See *Part*.

Appui, or Stay of the *Hand*. See *Appui*,

Beat upon the *Hand*. See *Beat*.

To be heavy upon the *Hand*. See *Heavy*.

To preſs upon the *Hand*. See *Preſs*.

To change the *Hand* or turn. See *Change*.

Effects of the *Hand*. See *Effects*.

Haſten the *Hand*. See *Haſte*.

To feel a Horſe upon the *Hand*. See *Feel*.

Full *Hand*, or Full Reſt upon the *Hand*. See *Appui* and *Mouth*.

To have Mares cover'd in the *Hand*. See *Stallion*.

A Horſe is ſaid to force the *Hand*, when he does not fear the Bridle, but runs away in ſpite of the Horſe-man.

To make a Horſe part from the *Hand*, or ſuffer him to ſlip from the *Hand*, is to put on at full ſpeed. To make a Horſe part right from the *Hand*, he ſhould not put himſelf upon his Back or Reins, but bring down his Hips.

All Hands. A Horſe that turns upon *All Hands*, upon a Walk, Trot or Gallop. A Horſe that is not *entier* or reſty for one *Hand*. See *Entier*.

To work a Horſe upon the *Hand*, is to manage him by the effect of the Bridle, without interpoſing any other helps, excepting that of the Calves of the Legs, upon occaſion.

To lead a Horſe in your *Hand*, to walk him in your *Hand*, to trot him in your *Hand*; *i. e.* without mounting him. If you would diſcover whether a Horſe is Lame, trot him in your *Hand* upon a Pav'd place.

Fore-hand and *Hindhand* of a Horſe, is an Expreſſion diſtinguiſhing the Parts of a Horſe, as divided into the fore and hinder Parts, by the ſituation of the Horſe-man's *Hand*. The Parts of the *Fore-hand* are the Head, the Neck, and the Fore-quarters.

quarters. Those of the *Hind-hand* include all the other Parts of the Body. Tho' this Horse looks well in the *Fore-hand*, he's yet better made in the *Hind-hand*. Such a Horse is ill shaped in the *Hind-hand*, he has a Flagging *Croupe*. See *Brillant* and *Movement* or *Motion*.

HAQUENEE, an obsolete *French* Word for an Ambling Horse.

To HARP. See *Grapple*.

HARD *Horse*, is one that is insensible of Whip or Spur. This Horse is so hard, he's good for nothing but a Cart.

HART, or *Stag-Evil*, is a sort of Rheum or Defluxion that falls upon the Jaws, and the other Parts of the *Fore-hand* of a Horse, which hinders him to eat. Sometimes this Distemper affects likewise the Parts of the Hinder-quarters.

HASTE or quicken your *Hand*, (*Hatez la Main, hatez, hatez*) is an Expression frequently used by the Riding-Master, when a Scholar works a Horse upon Volts, and the Master has a mind he should turn his hand quicker to the side on which the Horse works; so that if the Horse works to the right, he turns quicker with his Shoulders to the right. And the like is observ'd if he works to the left.

HAUNCH, or *Hip* of a Horse, is that part of the hind-quarter that extends from the Reins, or Back to the *Hough* or *Ham*. The Art of riding the Great Horse has not a more necessary Lesson than that of putting a Horse upon his *Haunches*; which in other Terms is call'd coupling him well, or putting him well together, or compact; (in *French*, *bien ensemble*, and *sous lui*.) A Horse that can't bend and low'r his *Hips*, throws himself too much upon his Shoulders, and lies heavy upon the Bridle. A Horse is said to be throughly managed (in *French*, *Achevee*) when he bears well upon the hand, knows the *Heels*, and sits well upon his *Hips*. This Horse has his *Haunches* in subjection, and *falques* very well, for in making his *Falquades*, he holds his *Haunches* very low, and bends admirably well. To make a Horse bend his *Hips*, you must frequently make him go backwards, and make use of the Aids of the Hands, and of the Calves of your Legs, in giving him good Stops; and if that does not succeed, try him upon a *Calade*, or Sloping Ground, after the *Italian* Fashion. Your Horse makes his *Hips* accompany his *Shoulders* so well, that he is perfectly right set, See *put upon the Haunches*, *Calade*, *Cavesson*, *Falquade* and *Feel*.

To *Drag the Haunches*, is to change the Leading Foot in galloping. See *Gallop false*.

Head in, and *Hips in*. See *Head*.

To

To Gallop with the *Haunch in.* See *Galopade.*

HEAD of a Horſe, imports the Action of his Neck, and the effect of the Bridle and the Wriſt. This Horſe plants his *Head* well, and obeys the hand. Such a Horſe refuſes to place his *Head*, he ſhoots out his Noſe, and never reſts right upon the hand; he ſtays too much, or too little upon the hand. Such a Horſe appears in a good Poſture, he carries his *Head* well.

To give a Horſe *Head.* See *Partir* and *Echaper.*

Head in, and likewiſe the Hips. You muſt paſſage your Horſe, *Head* and *Croupe in*; *i.e.* work him ſide-ways upon two Parallel Lines, at Step or Trot; ſo that when the Horſe makes a Volt, his Shoulders mark a *Piſte*, or tread, at the ſame time that his *Haunches* give the Tract of another, and the Horſe plying or bending his Neck, turns his *Head* a little within the Volt, and ſo looks upon the Ground he is to go over.

HEARTS: A Horſe of Two *Hearts* (*de deux coeurs*) *i.e.* a Horſe that works in the Manage with conſtraint and irreſolution, and can't be brought to conſent to it. Such Horſes are much of a piece with your *Ramingues*, or Kickers againſt the Spurs.

HEAVY: To reſt heavy, upon the hand, is ſaid of a Horſe who through the Softneſs of his Neck, the Weakneſs of his Back, the Weight of his Fore-quarters, or through Wearineſs, throws himſelf upon the Bridle, but withal, without making any reſiſtance, or any effort to force the Horſe-man's hand. Your Horſe has too great an *Appui* or Reſt upon the Bridle, he's heavy upon the hand, trot him upon his *Haunches*, and ſuſtain or bear up with the Bridle. By ſtopping him, and making him go back frequently, you may make him light upon the hand, and ſo correct that Fault if it comes only from Lazineſs and Stiffneſs; but if it proceeds from a Defect in the Back, or Limbs, there's no remedy for it. My Horſe is heavy upon the hand, but that is not ſo great a Fault as if he preſs'd and reſiſted upon the hand. See *Preſs.*

HEEL of a Horſe, is the lower hinder-part of the Foot, comprehended between the Quarters, and oppoſite to the Toe. Your Horſe is hoof-bound, and to recover his *Heels* you ought to take out his Sole, and keep his *Heels* very wide; by which the *Heels* will be reſtor'd in a Month. This Horſe has Narrow-heels, ſo give him Panton-ſhooes.

Heel of a Horſe-man. This being the part that's arm'd with the Spur, the Word *Heel* is taken form the Spur it ſelf. This Horſe underſtands the *Heels* well; he knows the *Heels*; he

he obeys the *Heels*; he anſwers the *Heels*; he's very well upon the *Heels*: The meaning of all which, is, that the Horſe obeys the Spurs; which, in effect, is flying from 'em. Put your *Heel* to your Horſe; ſtay him; bring to; prick with the right; prick with the left; clap both to his Sides. This Horſe knows the *Heels* in *Curvets*, in *Caprioles*. Make him fly from the Right-heel; make him fly from the Left: Such a Horſe reſiſts the Spurs: he's a *Ramingue*; *i. e*, a Kicker againſt the Spur. To ride a Horſe upon the *Hand* and *Heels*, is to make him take the Aids of the *Hand* and the *Heels* with a tender ſenſe.

To ride a Horſe from one *Heel* to t'other, is to make him go ſide ways, ſometimes to one *Heel*, and ſometimes to another: For inſtance; having gone ten Paces in flying from the Right-heel, you make him without ſtopping go ſtill ſide ways in flying the Left-heel; and ſo on alternately.

Inner-heel and *Outter-heel*. See *In*, and *Narrow*.

HERBER, a *French* Word us'd by the Farriers, importing the following Application. For ſome Diſeaſes, ſuch as thoſe of the Head and the Anticor, they put into the middle of a Horſe's *Counter*, a piece of *Hellebore-Root*, which makes it ſwell and ſuppurate.

HERBE, (Graſs) a Word in the *French* Academies, ſignifying a Reward, or ſome good Stuff given to a Horſe that has work'd well in the Manage.

HIP. See *Haunch*.

Hipſhot; (in *French*, *Epointé*, and *Eſhanché*), A Horſe is ſaid to be ſuch when he has wrung or ſprain'd his *Haunches* or *Hips*, ſo as to relaxate the Ligaments that keep the Bone in its due place.

HOLD. A Mare holds. See *Retain*.

HOLLOW-TOOTH'D Horſe. See *Shell-tooth'd*.

HOOF, or *Horn* of a Horſe's Foot, is a ſort of Nail, of a Finger's depth, that ſurrounds the Sole and the Coffin-bone. In ſhooeing a Horſe, the Nails are driven into the *Hoof*, in ſuch a manner, that the Shooe does not bear nor reſt upon the Sole; for the Sole being tenderer than the *Hoof*, the Shooe would injure it, and lame the Horſe. When the *Hoof* is worn, we ſay the Foot is worn. Let your Horſe's *Hoof* grow. Cracks or Clifts are apt to happen in the *Hoof*. The Diſorder call'd *Hoof-bound* happens in the *Hoof* of the Fore-feet. Your Horſe has got a Quitter-bone. See *Drive*.

Hoof-bound: A Horſe is ſaid to be *Hoof-bound*, when he has a Pain in the Fore-feet, occaſioned by the Dryneſs and Contraction

ction or Narrowneſs of the *Horn* of the Quarters, which ſtraitens the Quarters of the *Heel*, and oftentimes makes the Horſe lame. A *Hoof-bound* Horſe has a Narrow-heel, the ſides of which come too near to one another, inſomuch that the Fruſh is kept too tight, and has not its natural extent: Such Horſes ſhould be ſhooed with the Panton-Shooes. See *Panton* and *Heel*.

Hoof-caſt. A Horſe is ſaid to Caſt his *Hoof*, when a New *Horn* grows in the room of the Coffin, after its being fallen off by any Diſeaſe, ſuch as *Quitter-bones*, or *Bleymes*. A Horſe that has caſt his *Hoof*, is good for nothing but the Plough, and ſuch kind of Work. Since for want of Proper Remedies, you have ſuffer'd your Horſe to caſt his *Hoof*, tell me whether it proceeds from a *Bleyme* upon the *Cronet*, or from *Foundering*, or from a *Prick*..

HORN. See *Hoof*.

Horn. To give a Stroke with the *Horn*, is to blood a Horſe in the Roof of the Mouth, with the *Horn* of a Stag, or Roe-buck, the Tip or End of which is ſo ſharp and pointed, as to produce the effect of a Lancet. We ſtrike with the *Horn* in the middle of the Fourth Notch or Ridge of the Upper-Jaw. This Horſe is over heated, you muſt ſtrike him with the *Horn*. See *Ridge*.

A HORSE is an Animal ſo generally known, that to define him, 'tis ſufficient to ſay, he is the nobleſt and moſt uſeful of all Animals, and his Senſible Nature, Obedience, Swiftneſs and Vigor are at once the Object and the Subject of the nobleſt and moſt neceſſary Exerciſe of the Body.

HOUGH, or *Ham* of a Horſe, is the Joint of the hinder-quarter, which joins the Thigh to the Leg. Your Manage Horſe's muſt be made to bend their *Houghs*. Your Horſe has Fat, Fleſhy, Small, and conſequently defective *Houghs*; but my Horſe's *Houghs* are large, broad, well drain'd, and not inflam'd. There's no Horſe here, but what is troubled in the *Hough* with the *Houghbony*, *Spavin*, *Jardon*, *Wind-galls*, *Blood-ſpavin*, *Curbs* or *Selenders*. Look upon the *Houghs* of theſe two Horſes, the one is cloſed behind, and the other grapples. See *Grapple*.

What is it, the *Cramp* that makes your Horſe's *Hough* ſo ſtiff?

Horſe-Block. See *Montoir*.

Hough-bony; a Swelling on the Tip or Elbow of the *Hough* in a Horſe's hinder-quarters, about as big as half a Tennis-Ball.

HOUSSE, (Saddle-Caſe) is a Cover made commonly of Leather, and put upon a Saddle, to ſave it.

HOUZING, is either *Boot-houzing*, or *Shooe-houzing*: The former is a piece of Stuff made faſt to the hinder-part of the Saddle, which covers the *Groupe* of the Horſe, either for Ornament, or to cover the Horſe's Leanneſs, or to preſerve the Rider's Cloths, and keep 'em from being daub'd with the Sweat of the Horſe. The *Houzing* for ſuch as ride with Shooes, is commonly a piece of Scarlet Cloath bordered with Gold-fringe, and put round the Saddle, ſo as to cover the Croupe, and deſcend to the lower part of the Belly, to ſave the Gentleman's Silk-Stockings, when he mounts in his Shooes.

I.

JARDES, or *Jardons*, are callous and hard Swellings in the Hinder-legs of a Horſe, ſeated on the out-ſide of the Hough, as the Spavin is on the in-ſide. Jardons lame a Horſe, unleſs you give the Fire dexterouſly, and betimes,

JARRETIER, an obſolete *French* Word, ſignifying a Horſe whoſe Houghs are too cloſe together; which is now expreſſed in *French* by *Crochu*, i. e. *Crooked*, or *Hook'd*.

IN, *Inſide*, *Within*: And *Out*, *Outſide*, *Without*. The inner Heel, the outer Heel; the inner Leg, the outer Leg; the In-Rein, the Out-Rein. This way of ſpeaking relates to ſeveral things, according as the Horſe works to the right or left upon the Volts; or as he works a-long by a Wall, a Hedge, or ſome ſuch thing. Thus it ſerves to diſtinguiſh on what hand, or what ſide, the Horſeman is to give the Aids to a Horſe upon Manage. For a-long by a Wall, the outer Leg is the Leg of a ſide with the Wall; and the other Leg is the In-leg. And upon Volts, if a Horſe works to the right, the right Heel is the inner Heel, the right Leg the inner Leg; ſo that by conſequence the left Heel, and left Leg, muſt be the outſide Heel and Leg. Now the downright contrary will happen, if the horſe works to the left. Now a-days, the Riding-Maſters, to be the eaſier underſtood, uſe the Terms *right* and *left*, as for inſtance: Aſſiſt the Horſe with the right Heel, with the right Leg, with the right Rein; taking the Situation of the Heels and Legs with reſpect to the Volt. See *Enlarge*, *Gallop Falſe*, and *Large*.

In; the Head *in*, the Haunches *in*. See *Head*.

In; to put a Horſe *in*; *i. e.* to breed or dreſs him. By which Expreſſion, we underſtand putting him right upon the Hand, and upon the Heels. This Sorrel is put in (*Mis dedans*,) that is, he is broke, dreſs'd, and managed.

naged. The D. of *Newcastle* put Horses perfectly well *in*, by the means of his *Cavesson*.

INSTEP, is that part of the Hinder-leg of a Horse, that corresponds to the Shank in the Fore-leg, extending from the Ham to the Pastern-Joynt.

INTERFERE; a Horse interferes, when the side of one of his Shoes strikes against, and hurts one of the Fetlocks.

JOINTEE. See *Handful*.

K.

KICKER against the Spurs. See *Ramingue*.

KNEE of a Horse, is the Joint of the Fore-quarters that joins the Fore-Thigh to the Shank. Do not you see that your Horse has got the Malenders in the Bend of the Knee? And truly I am of the mind, that he has Selenders a coming in the Bend of the Hock or Ham. These two Horses have two scurvy Faults in their Knees; the one's Arched by fatigue; and the other is so naturally, or what we call *Brassicour*. Such a Horse has a very hard Knot, or Swelling upon his Knee. I will not buy this Peach Blossom-colour'd Horse, because his Knee is crown'd: For my part, I shall be the last Man that shall desire to be his Master.

L.

LADRE. See *Dull*.

LAME; A Horse is said to be *(Boiteux de l' Oreille) Lame of an Ear*, when he halts upon a Walk or a Trot, and keeps time to his Halting with the Motions of his Head: For all lame Horses do not keep time after that rate. *Lame of the Bridle*, is likewise us'd, by way of Raillery, to signify the same thing.

LAMPAS, or Bean, is a Swelling in the Palate of a Horse, *i. e.* an Inflammation in the Roof of his Mouth, behind the Nippers of the upper Jaw. Pray order this *Lampas* in your Horse's Mouth to be burnt.

LARGE; A Horse is said to go large, or wide, when he gains or takes in more Ground in going wider of the Center of the Volt, and describing a greater Circumference. This Horse goes too large; he launches out upon too much Ground; he does not keep subject. You must conduct that Horse large, by bringing to your inner Heel; for he goes too narrow of himself. See *Enlarge*.

LEAD; A Horse going upon a straight Line always leads, or cuts the way *(entame le chemin)* with his right foot. The Duke of *Newcastle* was the

the first that ever made use of the Term; and indeed it is very expressive. See *Gallop united*, and *Gallop false*.

LEAP, An Air of a Step and a Leap. See *Step*.

Leaping Horse, *i e.* One that works in the high Manage; A Horse that makes his Leaps in Order, and with Obedience, between two Pillars upon a strait Line, in Volts, Caprioles, Balotades, or Croupades. Use, which in most things has a sovereign Sway, excludes a Gallop, a *Terra a Terra*, and Corvets, from the number of *Leaps*; because the Horse does not rise so very high in these. Each Leap of a Leaping Horse ought to gain, or make not above a Foot and a half of Ground forward.

LEEK-HEAD, (*Poireau*). See *Wart*.

LEG of a Horse, is the Member that supports his Body, and performs the Motion when he goes. Of the Four Legs, the two before have several Parts, each of which has a peculiar Name: So that by the Name of Fore-leg, we commonly uuderstand that part of the Fore-quarters that extends from the Hough to the Pastern-Joynt; and call it the *Shank*: The part that corresponds to that in the Hinder-quarters, we call the *Instep*. But in common Discourse we confound the Fore and the Hind-quarters; and, without any distinction, say, the four Legs of a Horse. This Horse has not a Leg to go upon, that is, they are spoil'd: Which is commonly understood of the Fore-legs. Such a Horse's Leg slacks; that is, he stumbles. One of these Horses has arch'd Legs, and the others are gorged and swollen. This Sorrel has got Arrests, or Mangy Ulcers, in his Leg. Such a Horse wants the Fifth Leg; that is, he's tyr'd, and bearing upon the Bridle, lyes heavy upon the Horseman's hand.

The *French* call a Horse *Droit sur les jambes*, *i.e.* straight member'd, or straight upon his Legs, when the Fore-part of the Pastern falls perpendicularly upon the Cronet; and the Shank and the Pastern are in a straight Line. See *Straight*, and *Long-joynted*.

Legs of the Horseman: The Action of the Horseman's Legs, given seasonably and with judgment, is an Aid that consists in approaching more or less with the Calf of the Leg to the Flank of the Horse; and in bearing it more or less off, as there is occasion. This Aid, a Horseman ought to give very nicely, in order to animate a Horse: And it's so much the finer, that 'tis hidden and private: For in stretching the Ham, he makes the Horse dread the Spur; and this fear has as much effect as the Spur it self. Such a Horse knows,

knows, or is ſenſible of, the Horſeman's Legs. He takes the Aids of the Legs; he anſwers the Legs; he obeys the Legs.

The *Inſide Leg*; the *outſide Leg*' See *Inſide, Enlarge,* and *Gallop falſe.*

LENGTH; to paſſage a Horſe upon his own Length, is to make him go round in two Treads, at a Walk or Trot, upon a Spot of Ground ſo narrow, that the Horſe's Haunches being in the Center of the Volt, his own Length is much about the Semidiameter of the Volt; the Horſe ſtill working between the two Heels, without putting out his Croupe, or going at laſt faſter or ſlower than at firſt.

LESSON, is a Word us'd for the Inſtruction both of the Horſe, and the Scholar. This young Gentleman takes his Leſſon upon all ſorts of Horſe. Content your ſelf with this Leſſon of Walk and Trot; and do not attempt the Leſſon of a Gallop. The Horſe obeys this Leſſon.

LIBERTY *of the Tongue*; is a void Space left in the Middle of a Bit, to give place to the Tongue of a Horſe, made by the Bit's Arching in the middle, and riſing towards the Roof of the Mouth. The various Form of the Liberty gives Name to the Bit. Hence we ſay, a Scatch Mouth *a Pignatelle, i. e.* with the Liberty after *Pignatelli's* Faſhion: A Canon-Mouth with the Liberty like a Pigeon's Neck. In forging that Bit, don't make the Liberty too high, leſt it hurt, or at leaſt tickle the Palate, and make the Horſe carry low.

LIEU, (Place) a *French* Word imployed in the Academies, to ſignify the Situation and Poſture of a Horſe's Head. *Ce Cheval porte en beau Lieu*; *i. e.* this Horſe carries well; he holds his Neck high.

LIGHT *Horſe, (Leger)* is a ſwift nimble Runner. We likewiſe call a Horſe light that's well made, tho' he's neither ſwift nor very active: For in this laſt Expreſſion, we conſider only the Shape and Make of a Horſe, without regard to his Qualities.

Light upon the hand; A Horſe is ſaid to be ſuch, that has a good Tractable Mouth, and does not reſt too heavy upon the Bit. Your Horſes that have a thin Forehand, that is, but ſmall Shoulders, are commonly light upon the hand. This Horſe is light before, and ſubject in the Hips. We call a Coach-Horſe light, when he ſtirs nimbly, and dreads the Whip; or when he has a light Trot. All your light Coach-Horſes are good: and a hard heavy Coach-Horſe, that takes the Laſhing eaſily, is good for nothing.

Light Hand. See *Hand.*

LIGHTEN;

LIGHTEN; To lighten a Horſe, to make a Horſe light in the Fore-hand, (in *French*, *Allegerir*) is to make him freer and lighter in the Fore-hand, than behind. If you would make your Horſe light, you ought to find him always diſpos'd to a Gallop, when you put him to a Trot; and after Galloping ſome time, you ſhould put him back to the Trot again. This Horſe is ſo heavy in the Shoulders, and cleaves ſo to the Ground, that you'll find difficulty in making him light in the Fore-hand, even tho' you make uſe of the Duke of *Newcaſtle*'s Caveſſon. Your Horſe throws himſelf too much upon his Shoulders; you muſt make him light in the Fore-hand, and put him upon his Haunches. See *Haunches*, *Motion*, *Break*, and *Terraignol*.

LIGHT-BELLY'D *Horſe*, is one that commonly has flat, narrow, and contracted Sides, which makes the Flank turn up like that of a Grey-hound. Such a Horſe has but little Flank, he's light-belly'd; he travels and feeds but little, becauſe he has too much Mettle.

LINE of a Volt. See *Square*, and *Volt*.

Line of the Banquet. See *Banquet*.

LIP of a Horſe, is the Skin that covers the Sides of his Mouth, and ſurrounds his Jaws. A Horſe is ſaid to arm, or guard himſelf with his Lips, when his Lips are ſo thick that they cover their Barrs, and keep off the preſſure of the Curb.

LISTENING; A Horſe goes a Liſtening Pace. See *Ecouté*.

LOCKS, (in *French*, *Entravons*) are pieces of Leather two fingers broad, turn'd round, and ſtuff'd on the inſide, to prevent their hurting the Paſtern of a Horſe, round which they are clapp'd. An *Entrave* is compos'd of two *Entravons* joyn'd by an Iron Chain that's ſeven or eight Inches long. See *Amble*.

LONG-JOINTED *Horſe*, is one whoſe Paſtern is ſlender and pliant. This Horſe is too long-jointed; his Paſtern is ſo eaſy and weak, that the Joynt ſcarce preſſes the Earth, and ſo 'tis not proper to fatigue him. I have a Horſe, whoſe Forelegs go in a ſtraight Line, from the Knees to the Cronets; but yours is long-jointed. There are ſome long-jointed Horſes that are ſtrong and ſinewy, and do not bend the Paſtern-Joynt more than they ſhould do: And this ſort of Horſes work better in the Manage than the ſhort-jointed. Long-jointed Horſes are apt to have Wind-Galls. See *Legs*.

LOW; to carry low. See *Carry*.

LOYAL; A Horſe is

 ſaid

ſaid to be Loyal, that freely bends all his force in obeying and performing any Manage he is put to; and does not defend himſelf, or reſiſt, notwithſtanding his being ill treated.

A *Loyal* Mouth, is an Excellent Mouth, of the nature of ſuch Mouths as we call Mouths with a full reſt upon the hand.

LUMP of Fleſh. See *Bouillon*.

LUNETTE; A Half Horſe-ſhoe, ſo called; being a Shoe without the Spunges, (the Part of the Branches that runs towards the Quarters of the Foot being ſo call'd.) Your Horſe has falſe Quarters; if he is to ride in the Manage, you muſt ſhoe him with a Lunette Shoe: But if you deſign him for the Country, Panton Shoes are the fitteſt for him.

Lunettes of a Horſe, are two ſmall pieces of Felt made round and hollow, to clap upon the Eyes of a vicious Horſe that's apt to bite, and ſtrike with his Fore-feet, or that will not ſuffer his Rider to Mount him.

M.

MALENDERS are *Chops* or *Chinks* on the Bending or Joint of a Horſe's Knee, which ſometimes ſuppurate. When theſe Chops appear in the Bending of the *Hough*, they are call'd *Selenders*.

MALTWORM. See *Crepance*.

MANAGE, is a Word that ſignifies not only the Ground ſet apart for the Exerciſe of Riding the *Great Horſe*, but likewiſe the Exerciſe it ſelf.

The *Manage*, or Ground proper for managing Horſes, is ſometimes a covered place, as in your great Academies, for continuing the Exerciſe in bad Weather; ſometimes 'tis open, in order to give more liberty and pleaſure both to the Horſe and the Horſe-man. One way or other we always ſuppoſe a Center in the middle of the Manage, for regulating the *Rounds* or *Volts*. Sometimes this Center is diſtinguiſh'd by a Pillar fix'd in it, to which they tye the Horſes that are beginning to learn. Upon the ſides of the Manage other Pillars are placed, two by two, in order to teach Horſes to raiſe the Fore-quarters, by tying them with *Ropes*. See *Pillar*.

Manage or Exerciſe of a Horſe, is a particular way of working or riding him. Make your Horſes work upon the Air and the Manage that you us'd to put 'em moſt to. This Horſe is not yet capable of the Manage. There's a regular exactneſs and method in the Manage of that *Barb*, he'll work at what Manage you will.

A Horſe is ſaid to *Manage* when he Works upon *Volts* and

Airs,

Airs, which ſuppoſes him broke and bred. Such a Horſe manages well upon *Corvets*; he manages well upon *Caprioles*; he manages well, and at equal diſtances from the Center, or the Pillar that repreſents it, in the middle of the *Manage-Ground*.

To Manage a Horſe upon a *Terra à Terra*. Since this Horſe manages ſo well, he'll paſſage well from a Walk or a Trot, and will gallop well upon two Treads, and will have leſs trouble if he gallops one *Haunch* in. Make your Horſe manage upon *Corvets*.

A Horſe is ſaid to be *thoroughly managed*, or a Finiſh'd Horſe, (in *French, Achevè*) that is well broke, bred, and confirm'd in a particular Air, or Manage. Here are two Horſes but juſt initiated; but this third Horſe is *Achevè*, for he behaves well upon the hand and the heels, he's well put upon the *Haunches*, and works perfectly *Terra à Terra*.

High-manage, is the high or rais'd Airs, which are proper for leaping Horſes. See *Airs*.

Manage for a Soldier's Horſe, is a Gallop of unequal ſwiftneſs, but ſo, that the Horſe changes hands readily.

MANE (in *French, Crinieire*) is the Root of the Hair that grows on the upper part of the Neck. I like your *Rouſſin*, or your thick-bodied ſtrong *Dutch* Horſe, for he has a Narrow Mane; and mine, which I do not like, has a Broad Mane like a Coach-horſe: And you know very well, that theſe Broad Manes are generally very mangy, unleſs great care be taken to prevent it.

Mane-ſheet, is a ſort of Covering over the upper part of a Horſe's Head, and all round his Neck, which at one end has two holes for the Ears to paſs through, and then joins to the Halter upon the fore-part of the Head, and likewiſe to the Surcingle, or Long-girth upon the Horſe's Back. This Querry follows the *Engliſh* Cuſtom, in giving his Horſes Mane-Cloths all Winter, for in *France* they are ſcarce made uſe of.

MANGER is a little rais'd Bench under the Rack in the Stables, made hollow for receiving the Grain or Corn that a Horſe eats. This Horſe has the Tick, he ticks upon the Manger.

MARE. A Stud-Mare, or Mare for Breed, is one that either is with Foal, or is deſign'd to be cover'd, in order to raiſe a Breed or Race of Horſes.

MARK. A Horſe Marks, that is, he ſhews his Age by a Black Spot call'd the Bud or Eye of a Bean, which appears about five and a half in the Cavity of the Corner-Teeth, and is gone when the Horſe is eight years old; then he ceaſes to mark, and we ſay, he has raz'd. See *Teeth* and *Raſe*.

False-Mark'd; *i. e.* Counter-mark'd.

MARTINGAL, is a Broad Strap made fast to the Girths under the Horse's Belly, and runs between the two Legs, to fasten its other end under the Nose-band of the Bridle. Considering that your Horse has no *Appui* or Stay upon the hand, but tosses up to the Wind, give him a *Martingal*, that will keep him from beating upon the hand. A great many confound a *Martingal* and a *Plate-Longe*.

MASTIGADOUR, or *Slabbering-bit*, is a Snaffle of Iron, all smooth, and of a piece, guarded with Pater-nosters, and compos'd of three halfs of great Rings, made into Demi-ovals of unequal bigness, the lesser being inclosed within the greatest, which ought to be about half a Foot high. A *Mastigadour* is mounted with a *Head-stall* and two *Reins*. Now the Horse in champing upon the *Mastigadour* keeps his Mouth fresh and moist, by virtue of the Froath and Foam that he draws from his Brain. To put a Horse to the *Mastigadour*, is to set his *Croupe* to the Manger, and his *Head* between two Pillars in the Stable. Horses that use to hang out their Tongue, can't do it when the *Mastigadour* is on, for that keeps their Tongue so much in subjection, that they can't put it out.

MES-AIR, is a Manage half *Terra à Terra* and half *Corvet*.

MESMARCHURES. See *Pastern*.

MIDDLING-TEETH of a Horse, are the Four Teeth that come out at Three Years and a half, in the room of other four Foal-teeth, seated between the Nippers and the Corner-teeth; from which Situation they derive the Title of *Middling*. There's one above, and one below on each side of the Jaws. See *Teeth*.

MOLTEN-GREASE is a Fermentation or Ebullition of Pituitous and Impure Humours which precipitate and disembogue into the Guts, and oftentimes kill a Horse. This Disease does not commonly seize upon any but very Fat Horses over-rid in hot Weather. Mr. *Solleysel* has an excellent Remedy against it in his *Compleat Farrier*.

MONTER *à dos*, or *à Poil*; a *French* Expression, signifying to mount a Horse bare-back'd, or without a Saddle.

MONTOIR, or *Horse-Block*, is a Word deriv'd from *Italy*, where the Riding-Masters mount their Horses from a Stone as high as the Stirrups, without putting their Foot into the Stirrup. Now in *France* no such thing is us'd; but yet the Word *Montoir* is there retain'd, and signifies the Poise or Rest of the Horse-man's left foot

foot upon his left Stirrup. Hence *Pied de Montoir* signifies the Left or the Near-foot.

MOON-EYES: A Horse is said to have *Moon-eyes*, when the Weakness of his Eyes increases or decreases according to the Course of the Moon, so that in the Wane of the Moon his Eyes are muddy and troubled, and at New Moon they clear up; but still he's in danger of losing his Eye-sight quite.

MORE's *Head* implies the Colour of a *Roan-Horse*, who besides the mixture or blending of a *Gray* and a *Bay*, has a *Black Head*, and *Black Extremities*, as the Mane and Tail. See *Roan*.

MOTION. This Horse has a pretty *Motion*. This Expression implies the freedom of the *Motion* of the Fore-legs, when a Horse bends 'em much upon the Manage. But if a Horse trots right out, and keeps his Body straight, and his Head high, and bends his Fore-legs handsomely, then to say he has a pretty *Motion* with him, implies the liberty of the Action of the Forehand.

MOURAILLE, or *Barnacles*, is an Instrument, commonly of Iron, composed of two Branches, join'd at one end with a Hinge, for the use of the Farriers, who take hold of a Horse's Nose with it, and keep it tight, by bringing to, or almost closing the other end of the Branches, and so tying 'em with a Strap. This they do to hinder a Horse to struggle and toss, when they make any Incision upon him, or give the Fire. Some *Mourailles* are made of Wood, with a Screw, and this sort is indeed very good.

MOUTH of a Horse: The Compliance and Obedience of a Horse is owing partly to the tender or quick sense of his *Mouth*, which makes him afraid of being hurt by the Bitt; and partly to the Natural Disposition of his Members, and his own inclination to obey. Put your Horse back, and by that means you may judge in some measure of the submission and tenderness of his *Mouth*. The *Mouth* is call'd sensible, fine, tender, light and loyal. Your Horse has so fine a *Mouth*, that he stops if the Horse-man does but bend his Body backwards, and raise his hand, without staying for the Pull or Check of the Bridle. A *Mouth* is said to be fix'd and certain, when a Horse does not chack or beat upon the hand. A Fresh Foaming *Mouth*. A strong desperate spoil'd *Mouth*. A False *Mouth* is a *Mouth* that is not at all sensible, though the Parts look well, and are all well form'd. This Horse has no *Mouth*; he's without a *Mouth*. This *Mouth* is ticklish, *i.e.* the Horse is too fearful of the Bitt. You must

fix the Ticklish Mouth of this Horse with a *Canon-mouth a Trompè*, *i. e.* all of one piece, only knee'd in the middle; or rather with good Lessons, without which the *Canon-mouth* will not have its effect. You do not know how to preserve your Horse's *Mouth*; you check him too much. See *Bitt*, or *Bitt-mouth*.

A *Mouth* of a Full *Appui*, or rest upon the hand, is one that has not the tender nice sense of some Fine *Mouths*, but nevertheless has a fix'd and certain rest, and suffers a hand that's a little hard, without chacking or beating upon the hand, without bearing down, or resisting the Bitt, insomuch that he'll bear a Jerk of the Bridle, without being much mov'd. If you go to the Army, provide your self a Horse with a *Mouth* that bears a full rest upon the hand; for if you take one of a fine nice tender *Mouth*, and another Horse comes to shock or run against him in a Fight, he'll be apt to rise up upon his two hind-feet, which a Horse of a harder *Mouth* would not do. See *Appui*.

A *Mouth* that bears more than a full rest upon the hand, implies a Horse that does not obey but with great difficulty. You will not readily stop this Horse, for his Mouth is above a full *Appui* upon the hand. See *Appui*.

MUSEROLE. See *Nose-band*.

N.

NAG; little Nag, or Tit, (in *French*, *Bidet*) is a Horse of a small low Size. *France* produces a great many admirable *Bidets*, which travel and endure fatigue better than all your large Horses.

NAILS of the Bridle-hand. The different Position, or Situation of the Nails of the left Hand of the Horseman, give the Horse a Facility of changing hands, and form his Departure and Stop; by reason that the Motion of the Bridle follows such a Position of the Nails. To give a Horse head, you must turn the Nails downwards; to turn the Horse to the right, you must turn them upwards, moving your hand to the right. To change to the left, you must turn the Nails down, and bear to the left: To stop the Horse, you must turn them upwards, and lift up or raise your Hand.

Nails in the street, is a common Expression, pointing to the Wound receiv'd by a Horse in the streets, by setting his foot accidentally upon a Nail, which being planted with the Point up, sometimes runs thro' the Sole, and reaches to the Coffin-Bone, and makes the Horse lame. Your Horse has got a Street-nail that will keep him lame long enough.

NARROW; a Horſe that narrows, is one that does not take Ground enough, that do's not bear far enough out to the one hand or to the other. Your Horſe narrows too much; to enlarge him, you muſt aſſiſt him with the inſide Rein; that is, you muſt carry your hand to the outſide, and preſs him forward upon ſtrait Lines with the Calves of your Legs. Since your Horſe widens too much, you muſt narrow him, not only in turning him, but likewiſe in keeping him under, or ſubject. If he narrows too much, aſſiſt him with the Calves of your Legs; nay, prick him, and then bring to with the Out-Spur; that is, the Heel contrary and oppoſite to the Ground he has quitted, and ought to regain.

NECK of a Horſe. Your Horſe's Neck is charged with Fleſh, he has a Cock-thropled Neck, a falſe Neck, a thick Neck. Such a Horſe has a fine well-ſhap'd riſing Neck. I ſee in your Horſe's Neck the Feather which we call the *Roman* Sword. A fleſhy Neck; a Neck with the Fleſh hanging down on one ſide. A Mare's Neck; *i. e.* too ſlender and fine, and but little Fleſh upon it. See *Carry well, Carry low.*

NEIGHING is the Cry of a Horſe. Such a Horſe *neighs.*

NIPPERS are Four Teeth in the Fore-part of a Horſe's Mouth; two in the upper, and two in the lower Jaw. A Horſe puts 'em forth between the ſecond and third Year. See *Teeth.*

Nippers. Smiths or Farriers Nippers, (in *French, Tricoiſſes*) are the Pincers with which they cut the Nails they have drove in, before they rivet 'em; and which they uſe in taking off a Shooe.

NOSE-BAND, (*Muſerolle*) is the part of the Head-ſtall of a Bridle that comes over a Horſe's Noſe. Since your Horſe beats upon the hand, clap a Martingale to his Noſe-Band.

NOUER *L' Eguillette.* See *Yerk.*

O.

OBEY; A Horſe is ſaid to obey the Hands and the Heels, to obey the Aids or Helps; *i. e.* to know and anſwer 'em according to demand. Such a Horſe obeys the Spurs; *i. e.* he flies from them.

OSSELET, is a very hard Excreſcence, reſembling a little Bone, on the inſide of the Knee, (and never on the outſide) appearing to be of the ſame ſubſtance with the reſt of the Knee, and only diſtinguiſhable from the Knee by its deſcending a little lower.

OVER-DONE, *Over-rid*, or *Over-work'd*; (in *French, Outré*)

Outrè) A Horſe is ſo call'd, when his Wind and Strength are broke and exhauſted with Fatigue. An incurable purſive Horſe is called in *French*, *Pouſſif Outrè*, *i. e.* Over-done Purſive.

To Over-work a Horſe in the Manage, is call'd, *Eſtrapaſſer*.

OVER-REACH; A Horſe is ſaid to over-reach, when he brings his hinder Feet too far forwards, and ſtrikes their Toes againſt the Spunges of the Fore Shoes. A Horſe over-reaches thro' a Weakneſs in the Back, or by being ſuffered to bear too much upon the Shoulders.

OUT, *Outſide*; *Without*. See *In*.

P.

PANNELS of a Saddle, are two Cuſhions, or Bolſters, fill'd with Cow's, or Deer, or Horſe Hair, and placed under the Saddle, one on each ſide, touching the Horſes Body; to prevent the Bows and Bands to gall or hurt his Back.

PANTON-SHOE, is a Horſe-ſhoe contriv'd for recovering narrow and hoof-bound Heels; which has Spunges much thicker on the inſide than on the outſide; ſo that the part that reſts upon the Horn or Hoof runs ſlope-wiſe, to the end that the thickneſs of the inſide of the Shoe may bear up the Heel, and throw or puſh it to the outſide. Panton-Shoes are likewiſe for ſuch Horſes as have falſe Quarters.

PARE; to pare a Horſe's Foot, is to cut his Nails; that is, the Horn, and the Sole of his Foot, with a Butteris, in order to ſhoe him. This Foot is well par'd; 'tis par'd without touching the quick. In *England*, the Smith or Farrier holds the Foot of the Horſe between his Knees, and in that very poſture pares the Foot, ſets on the Shoe, drives the Nails, and rivets 'em; and this all alone, without any Aſſiſtance from the Groom.

PARER, a *French* Word us'd in former times in the Academies, implying, to ſtop: But at preſent 'tis exploded. And when the Riding-Maſters have a mind the Scholar ſhould ſtop the Horſe, they call out, *Holas*. See *Stop*, and *Half-ſtop*.

PART, or Depart; (in *French*, *Partir*) a Word us'd in the Academies to ſignify the Move and Action of a Horſe, when put on at ſpeed. Briſk up your Horſe when you part. You have no grace in your Parting. If this Horſe does but part, or take the Departure with promptneſs, he has a very juſt ſtop. From the Horſe's Parting to his Stop there's two hundred Paces of Ground. This Horſe parts upon

upon the hand very handsomely. To make your Horse part with a good grace, (*i.e.* to give him head) you must put your Bridle three fingers lower, and press gently with your Heels, or only with the Calves of your Legs. See *Echaper*.

To *Part* again. See *Repart*.

PASSADE, is a Tread or Way that a Horse makes oftner than once upon the same Extent of Ground, passing and repassing from one end of its length to the other: Which can't be done without changing the hand, or turning and making a *Demitour* at each of the Extremities of the Ground. Hence it comes, that there are several sorts of Passades, according to the different ways of turning, in order to part or put on again, and return upon the same *Piste*, or Tread; which we call Closing the Passade. See *Close*, and *Serrer*.

A *Passade* of five times, or a Demivolt of five times; is a Demitour made at the end of the straight Line, one Hip in, in five times of a Gallop upon the Haunches; and at the fifth time, ought to have clos'd the *Demivolt*, and to present upon the Passade-line straight and ready to return. The Demivolts of five Times or Periods, are the most common Airs of changing the hand, or turning, that are practis'd in the Academies.

Furious Passades, (*i.e.* upon a full Career) or *French Passades*, are such as are in Duels. To make these Passades, you put your Horse straight forwards, and towards the extremity of the Line make a Half-stop, keeping the Horse straight without traversing; then you make the Demivolt at three times, in such a manner, that the third time the Horse presents straight upon the Passade-line, and ready to set out again upon a short Gallop. You continue this short Gallop half the length of the Passade; then you put on furiously at full speed, and at the end of the Passade mark a Half-stop, and then a Demivolt of three times. This you continue to do, as long as the Horses Wind and Strength will hold. This Passade at full speed supposes that the Horse has an excellent Mouth, and requires Strength and Agility both in the Horse and Horseman. There are but few Horses that are capable of it. 'Tis said, that Monsieur *de Belleville*, one of the *French* King's Querry's, and a famous Master of the Art of Riding, was the first that christen'd these Passades with a full Career, by the name of *Passades a la Françoise*, i. e. Passades after the *French* way.

Passade of one time; A Passade in *Pirouette*, or half *Pirouette*, of one time; is a Demivolt, or Turn made by the Horse,

Horſe, in one time of his Shoulders and Haunches. To make this Paſſade, which is the perfecteſt of them all, the Horſe ſhould ſtand ſtraight upon the Paſſade-line; and then putting forwards, he forms a Half-ſtop, making falcades two or three times in ſuch a manner that he is ſtill ſtraight upon the Line; and at the laſt time he prepares to turn nimbly, and retain or fix his Haunches as a Center, ſo that the Demi-volt is perform'd in only one time of the Shoulders; and tho' the Haunches make likewiſe a time, they make it in the Center, or upon the ſame ſpot and *de ferme à ferme*, as the *French* call it.

The Rais'd or High *Paſſades*, are thoſe in which the Demi-volts are made in Corvets.

In all Paſſades, the Horſe ſhould, in making the Demi-volt, gather and bring in his Body, making his Haunches accompany his Shoulders, without falling back, or not going forward enough each time: And he ſhould go in a ſtraight Line, without traverſing or turning his Croupe out of the Line.

PASSAGE; to paſſage a Horſe, is to make him go upon a Walk or Trot upon two Piſtes or Treads, between the two Heels and ſide-ways; ſo that his Hips make a Tract parallel to that made by his Shoulders. 'Tis but of late that Paſſaging upon a Trot has been us'd; for formerly the word *Paſſage* ſignified walking a Horſe upon two Treads between the two Heels. A Horſe is paſſaged upon two ſtraight Lines, along a Wall or Hedge. He is likewiſe paſſaged upon his own length upon Volts, in going ſide-ways upon a Circle round a Center, the Semidiameter being about his own length: So that he looks into the Volt, and half his Shoulders go before the Croup. In all Paſſaging, the Horſe's outward Fore-leg muſt croſs or lap a great deal over the other Fore-leg, at every ſecond time that he marks. In a Paſſage of a Walk, and that of a Trot, the Motion of the Horſe is the ſame; only the one is ſwifter than the other. See *Haunch in*, and *Length*.

Paſſage upon a ſtraight Line, is a ſort of Manage practis'd but little in *France*, but very much in *Italy*, and yet more in *Germany*. For this Manage, they chuſe a Horſe that is not fiery, but has a good active Motion with him, and leading upon a ſtraight Line upon a Walk or Trot, teach him to lift two Legs together, one before, and one behind, in the form of a St. *Andrew*'s Croſs; and in ſetting theſe two to the Ground to raiſe the other two alternately, and keep 'em a long while in the Air; and that in ſuch a manner, that at

every

every time he gains a foot of Ground forwards. The Beauty of Paſſaging, conſiſts in holding the Legs long in the Air. The Motion of the Legs in this Paſſage, is the ſame with that of a Walk or a Trot; for they go in the ſame Order, and the only difference is, that in paſſaging upon a ſtraight Line the Legs are kept longer in the Air. Your proud ſtately Horſes, and thoſe which are accuſtom'd to this ſort of Paſſage, are proper for a Carouſel, or a Magnificent Shew. The difference of a proud ſtately Prancing, (in *French*, *Piaffer*) and Paſſaging, conſiſts only in this, that your ſtately Horſes do the former naturally, and do not keep their Legs ſo long in the Air, as in paſſaging right out. But for a *Paſſage*, there's ſo much Art requir'd, that a Horſe is two or three Years in Breeding to that Manage; and of ſix Horſes, 'tis very much if two of 'em ſucceed in it.

PASTERN-JOYNT or *Fetlock* of a Horſe's Leg, is a Joynt above the Paſtern, which ſerves for a ſecond Knee in each Fore-leg, and a ſecond Ham or Hough to each Hinder-leg. The Fetlock is apt to be cut by the ſide of one of the Shoes; and when that happens, we ſay, a Horſe cuts, or interferes. This Sorrel's Fetlock is inflam'd and ſwell'd. Upon that Horſe, I ſee a Wind-gall by the Fetlock. Sprains happen upon the Fetlock; and Cratches happen above the Fetlocks behind.

PASTERN of a Horſe, is the lower part of the Leg between the *Fetlock* or *Paſtern-Joint* and the *Cronet*. A Horſe is ſhort-jointed or long-jointed, according to the ſhortneſs or length of the *Paſtern*; and the ſhort-jointed is the beſt. All the Horſes in this Stable are out of order in the *Paſterns*. You ſee there *Crown-ſcabs*, *Cratches*, *Clefts*, *Watery Sores*, *Warts*, *Crepances*, *Ring-bones*; and in fine, there is not one of 'em that has a ſound clean *Paſtern*; nay ſome of 'em are gall'd with their Locks or Fetters.

Wrenches of the *Paſtern* are call'd in *French*, *Entorſes* and *Meſmarchures*. You muſt give the Fire to this *Entorſe*, for the Horſe is quite lame, and 'tis in vain to try any other Remedies.

PATIN-SHOOE, a *Horſe-Shooe* ſo call'd, under which is ſolder'd a ſort of half-ball of Iron, hollow within: 'Tis us'd for Hip-ſhot-horſes, and put upon a Sound Foot, to the end that the Horſe not being able to ſtand upon that Foot without Pain, may be conſtrain'd to ſupport himſelf upon the Lame Foot, and ſo hinder the Sinews to ſhrink, and the Haunch to dry up. We likewiſe clap Patin-ſhoes upon Horſes that are ſprain'd in the Shoulders.

PAW the Ground. A Horſe Paws the Ground, when his Leg being either tired, or painful, he does not reſt it upon the Ground, and fears to hurt himſelf as he walks.

PEACH-*Colour.* See *Bloſſom.*

PESATE, or *Peſade*, or *Poſade*, is the Motion of a Horſe that in lifting or raiſing his Fore-quarters, keeps his hind-legs upon the Ground without ſtirring, ſo that he marks no time with his *Haunches*, till his Fore-legs reach the Ground. This Motion is the true Means to fix his *Head* and his *Haunches*, to make him ply and bend his Fore-thighs, and to hinder him to ſtamp and clatter with his Feet. If you deſign to put your Horſe to *Corvets*, make *Peſates* his firſt Leſſon, for *Peſates* are the foundation of all Airs. See *Stop* and *Half-ſtop.*

PIAFFEUR is a proud ſtately Horſe, who being full of Mettle, or Fire, reſtleſs and forward, with a great deal of Motion and an exceſſive eagerneſs to go forwards, makes this Motion the more that you endeavour to keep him in, and bends his Leg up to his Belly: He ſnorts, traverſes if he can, and by his Fiery Action ſhews his reſtleſſneſs: Whence ſome, though very improperly, ſay, he dances. Such Horſes as theſe, or ſuch as are bred to paſſage upon a ſtraight Line, are much admir'd in Carouſels and Magnificent Feſtivals. See *Snort* and *Paſſade.*

PICKER; *Horſe-picker* is an Iron Inſtrument, five or ſix Inches long, bent or crooked on one ſide, and flat and pointed on the other, us'd by the Grooms to cleanſe the inſide of the Manage Horſes Feet, and to pick out the Earth and Sand that's got into 'em. Since your Horſes are but juſt come from the Manage, you ſhould order your Groom to take the *Horſe-picker* and pick out the Duſt that dries up their Feet, and that done, to clap Cows-dung into 'em, to keep 'em moiſt, and prevent their having *Falſe-quarters.*

PIERCE a Horſe's Shooe, Lean and Fat: To pierce Lean, is to pierce it too near the edge of the Iron: To pierce it Fat, is to pierce it further in. You have pricked my Horſe, becauſe the Shooe was pierced too fat. This Horſe's Shoe has made all the Hoof below the Rivet to ſplit, for 'was pierced too lean. This Farrier pricks the Horſes upon the Anvil; that is, he does not pierce his Shoes right, for in ſtriking the Nails through holes that are pierced ſometimes too fat, and ſometimes too lean, or too near the Quarter, he always pricks the Horſes.

PILLAR: Moſt great Manages have a Pillar fixed in the middle of the Manage-Ground,

Ground, to point out the Center: But all Manages in general have upon the side or circumference other Pillars, plac'd two and two at certain distances; from whence they are call'd the two Pillars, to distinguish 'em from that of the Center. When we speak of the latter, we call it working round the Pillar; and when we refer to the other two, we call it working between the two Pillars.

The Pillar of the Center serves to regulate and adjust the extent of Ground, to the end that the Manage upon *Volts* may be perform'd with method and justness, and that they may work in a Square by rule and measure upon the four Lines of the *Volt*, which ought to be imagin'd at an equal distance from the Center. It serves likewise to break unruly high-Mettled Horses, without indangering the Rider; the Horse being tied to a long Rope, one end of which is made fast to the Pillar, and managed by a Man plac'd by the Pillar, which keeps the Horse in subjection, and hinders him to fly out. To break such an unruly fiery Horse, and make him go forwards, put the *Cavesson* upon him, and make fast the Rope to the middle Ring, and to the Pillar, trot him round the Pillar without any person on his back, and fright him with the Shambrier or Rod, that he may know it, and fly from the least appearance of a Blow. This done, you may mount him round the Pillar, and put him on, so as that he shall not be able either to rear up, or to stop in order to do mischief; for the dread of the Shambrier will prevent all Disorders, and hinder him to stop. The Duke of *Newcastle* says this is the only case in which the use of the Pillar should be suffer'd; for in general he's so far from approving of the Pillar, that he affirms it only spoils Horses, because round it they only work by roat, and having their Eyes always fix'd upon the same Objects, know not how to manage elsewhere, but instead of obeying the hand and the heels, know nothing but the Rope and the Shambrier. In such Manages as have not this Pillar, you must imagine a place where it should be; that is, you must consider the middle of the Ground as a Center, in order to regulate and facilitate Manages upon Rounds. See *Rope* and *Ropes*.

The Two Pillars are placed at the distance of two or three Paces, the one from the other. We put a Horse between these, with a Cavesson of Leather or Cord, mounted with two big Ropes that answer from the one Pillar to the other. You must ply your Horse with the Cavesson Ropes, and make him rise between the two Pillars; when

once

once he has got a habit of curveting with eaſe, he'll give you a good ſeat on horſeback, and by the liberty of his Poſture, make you keep the Counterpoiſe of your Body, and teach you to ſtretch out your Hams. Put this Horſe between the two Pillars, to teach him to riſe before, and when that's done, you'll eaſily teach him to yerk out behind, and put himſelf upon rais'd Airs, either by the Aids, or by the Chaſtiſement of the *Shambrier*. Nay, if there be occaſion, you may make uſe of the Switch, the Poinſon, the Hand, and the Spurs.

PISTE, is the Tread or Tract that a Horſe makes upon the Ground he goes over. This Horſe-man obſerves the *Piſte*, he makes it his buſineſs to follow the Tread; that is, he follows his Ground regularly, without enlarging or narrowing, without traverſing or entabling. Such a Horſe works well upon two Treads; he works well with one *Piſte*.

PLANTED-COAT. See *Hair-ſtaring*.

PLAT-VEINS; (in *French*, *Ars*) are the Veins in which we bleed Horſes, one in the lower part of each Shoulder. When we blood a Horſe in the Shoulders, and in the flat part of the Thighs, the Vulgar People cry, He's bled in the four Plat-veins: But 'tis a miſtake, inſtead of ſaying, he's bled in all his Four limbs.

PLATE-LONGE, is a Woven Strap, four Fathom long, as broad as three Fingers, and as thick as one; made uſe of in the Manage for raiſing a Horſe's Legs, and ſometimes for taking him down, in order to facilitate ſeveral Operations of the Farrier. Some improperly give the name of *Plate-longe* to a *Martingal.*

PLUNGE. See *Eſtrapade.*

POIL; *Souffler au Poil*, *i. e.* to run upon the Hair or Skin. A *French* Expreſſion, us'd when a Horſe has a prick with a Nail, and for want of being ſufficiently open'd underneath, the Matter or Impoſthume runs between the Hoof and the Coffin-bone; and riſing above the Coffin, gains the Hair, inſomuch that it appears at the *Cronet*. See *Hair.*

POINSON, is a little Point or Piece of ſharp-pointed Iron, fix'd in a wooden Handle, which the Cavalier holds in his right Hand, when he means to prick a Leaping Horſe in the Croupe, or beyond the end of the Saddle, in order to make him yerk out behind. Put this Horſe between two Pillars, and give him the Aids of the *Poinſon*. This Horſe obeys the *Poinſon.*

POIRAU. See *Wart.*

POINTS, or Toes of a Bow of a Saddle. See *Bows.*

Point; A Horſe is ſaid to make a Point, when in working upon Volts, he does not obſerve the Round regularly, but putting a little out of his ordinary Ground, makes a ſort of Angle or Point by his Circular Tread. Your Horſe does not make Rounds well; he makes Points. You ſhould prevent it by haſtening your hand. See *Haſten.*

PONTLEVIS, is a diſorderly reſiſting Action of a Horſe, in diſobedience to his Rider; in which he rears up ſeveral times running, and riſes ſo upon his Hind-legs that he's in danger of coming over. Your Horſe makes very dangerous *Pontlevis*'s; and conſidering he is but a weak Horſe, a little Reſting upon the Bridle would bring him over. This Colt reſiſted, and was diſobedient a long time; and his Defence was to make great *Pontlevis's*: But, knowing he had ſtrength, I took the time when his Fore-feet were returning to the Ground, and clapt my Spurs ſmartly to him, which broke him at laſt. Such a Horſe doubles his Reins, and makes a *Pontlevis.*

PORTE-*Etrier.* See *Stirrup.*

PORTER, (to carry) us'd in the *French* Manage for directing or puſhing on a Horſe at pleaſure; whether forwards, upon turns, *&c.*

POSADE. See *Peſate.*

PRESS upon the hand. A Horſe is ſaid to reſiſt, or preſs upon the hand, when either thro' the Stiffneſs of his Neck, or from an ardour to run too much in Head, he ſtretches his Head againſt the Horſeman's Hand, refuſes the Aid of the Hand, and withſtands the Effects of the Bridle. My Horſe who has a thick fleſhy Neck, lyes very heavy upon the hand; but yours, who has too much fire, preſſes upon the Hand. If your Horſe is too fiery, and preſſes upon the Hand, endeavour to pacify him, by making him go more ſoftly, and pulling him backwards, If it proceeds from a Stiffneſs of the Shoulders and Neck, you muſt ſupple him with a Caveſſon made after the Duke of *Newcaſtle*'s way. See *Heavy.*

To *Preſs*, or puſh a Horſe forwards, is, to aſſiſt him with the Calves of your Legs, or to ſpur him to make him go on.

PRESTESSE; (a *French* Word, ſignifying Readineſs,) us'd in the Academies, to import the Diligence of a Horſe in working in the Manage.

PRICK, or Pinch, (in *French, Pincer*) is, to give a Horſe a gentle Touch of the Spur, without clapping them hard

hard to him. Prick with the right, pinch with the left, pinch with both. To prick or pinch is an Aid; but to *appuyer*, or bear hard with the Spur, is Correction.

Pricking of a Horse's Foot, is the Hurt receiv'd by a Nail drove too far into the Foot, so as to reach the quick, or press the Vein in the Horse's Foot when he's shod. See *Pierce*.

PUNCH; a well-set, well-knit Horse; (in *French*, *Goussaut*) is short-back'd, and thick-shoulder'd, with a broad Neck, and well lin'd with flesh. Of all the Baggage-Waggons and Carts of the Army, I see none better drawn than yours; for the Thill-Horse, and the Fore-Horse, are strong well-set Horses.

PURSYNESS, or Alteration of the Flank; is an Oppression that deprives a Horse of the Liberty of Respiration; and proceeds from some Obstruction in the Passages of the Lungs. Pursyness is a Capital and Essential Fault; so that the Seller of the Horse stands obliged to warrant him free from Pursyness, or to take him again within nine Days after sale. See *Warrant*, and *Wind*.

PUT, (in *French*, *Mettre*) is us'd for the Breaking, or Managing of a Horse: As, Put your Horse to Corvets, Put him upon Caprioles. This Horse puts, and presents himself upon Rais'd Airs. Such a Horse was not well put at first. There's a Barb very well put.

To *Put* a Horse upon his Haunches, (in *French*, *Asseoir*) is, to make him bend 'em in Galloping in the Manage, or upon a Stop. See *Haunches*.

To *Put* a Horse to the Walk, Trot, or Gallop; is to make him walk, trot, or gallop.

To *Put* a Horse under the Button. See *Button*.

PYEBAL'D Horse, is one that has White-Spots upon a Coat of another Colour. Thus there are *Pyebal'd Bays*, *Pyebal'd Sorrels*, and *Pyebal'd Blacks*; and so of the rest.

PYROET. Some are of one Tread, or *Piste*; some of two. Those of one Tread are otherwise call'd *Pirouettes de la tete a la queue*.

Pyroets de la tete a la queue, are entire and very narrow Turns made by the Horse upon one Tread, and almost in one time, in such a manner, that his Head is plac'd where his Tail was, without putting out his *Haunches*. To make Horses take this *Pyroet* with more facility, they use in the Manage to put 'em to five or six of 'em all running, without stirring off the Spot. In Duels they are of use to gain the Enemy's *Croupe*.

Pyroettes,

Pyroettes of two Pists or Treads, are Turns of two Treads upon a small compass of Ground, almost of the length of the Horse. This Horse makes his *Pyroette* of two *Pists* very readily, for he turns short and narrow, and keeps his *Haunches* low and well set.

Pyroette of one Time, or *Demi-pyroette* of one Time, or *Passade* of one Time. See *Passade*.

Q.

QUARTER. To work from Quarter to Quarter, is to ride a Horse three times in end upon the first of the four Lines of a Square; then change your hand, and ride him three times upon the second: At the third time change your hand, and so pass to the third and fourth, observing the same order.

QUARTERS of a Saddle, are the pieces of Leather or Stuff made fast to the lower part of the sides of a Saddle, and hanging down below the Saddle.

Quarters. Fore-quarters and Hind-quarters. The Fore-quarters are the Shoulders and the Fore-legs: The Hind-quarters are the Hips and the Legs behind. The two Quarters of this Horse are equally weak.

Quarters of a Horse's Foot, are the sides of the Coffin, comprehended between the Toe and the Heel on one side and t'other of the Foot. The Inner-quarters are those opposite to one another, facing from one Foot to the other; and these are always weaker than the Outside-quarters, which lie on the external sides of the Coffin. See if your Horse has not got False-quarters.

Quarter-cast. A Horse is said to cast his Quarter, when for any Disorder in the Coffin we are obliged to cut one of the Quarters of the *Hoof*: And when the *Hoof* thus cut grows and comes on anew, 'tis call'd in *French*, *Quartier neuf*; *i. e.* New Quarter.

False-Quarters, (in *French*, *Seime*) is a Cleft in the *Horn* of a Horse's *Quarters*, extending from the *Cronet* to the *Shoe*, which voids Blood, and occasions a great deal of Pain, and makes the Horse lame. Your Horse has a *False-quarter*, shoe him with *Pantofle-Shoes*, and keep his Foot fat and easie. See *Picker*, and *Lunettes*.

QUITTER-BONE. See *Cratches*.

R.

RAGOT, is a Horse that has Short Legs, a Broad Croupe, and a Strong Thick Body; differs from a *Goussaut* in this, that the latter has more Shoulders, and a thicker Neck.

RAISE: To raise a Horse upon *Corvets*, upon *Caprioles*, upon *Pesades*, is to make him work at *Corvets*, *Caprioles*, or *Pesades*. Sometimes we say, Raise the Fore-hand of your Horse. Be sure always to raise your Horse's *Fore-quarters*, after a stop form'd.

Raise is likewise us'd for placing a Horse's head right, and making him carry well, and hindring him to carry low, or to arm himself.

RAKE. A Horse Rakes when being shoulder-splait, or having strain'd his *Fore-quarter*, he goes so lame, that he drags one of his Fore-legs in a Semicircle; which is more apparent when he trots, than when he paces.

RAMINGUE. A Horse call'd in *French*, *Ramingue*, is a resty sort of Horse that resists the Spurs, or cleaves to the Spurs; that is, defends himself with Malice against the Spurs, sometimes doubles the Reins, and frequently yerks to favour his Disobedience. See *Ticklish*, and *to Double*.

RAMPIN. See *Toe*.

RASE: To *Rase* or glance upon the Ground (*razer le tapis*) is to gallop near the Ground, as our *English* Horses do. This Horse does not rise enough in his Gallop, he does not raise his Fore-quarters high enough, he goes but coldly, his Motions are too near the Ground; he gallops like an *English* Horse.

RAT-TAILS, or *Arrests*, signifie callous hard Swellings upon the hinder-legs under the *Hough*, running along the Sinew.

Rat-tail; a Horse is so call'd when he has no Hair upon his Tail.

RAZE. A Horse razes, or has razed; that is, his Corner-teeth cease to be hollow; so that the Cavity where the Black Mark was, is now filled up, the Tooth is even, smooth and raz'd, or shav'd, as 'twere, and the Mark disappears. Your Horse has raz'd, and does not mark no more; from whence we conclude that he has almost enter'd into his eighth Year. See *Teeth*.

REAR *up*; (in *French*, *Cabrer*) is said of a Horse that rises upon his hinder-legs as if he would come quite over This Horse has a Mouth too sensible, and rises before; if you cleave to the Bridle but never so little, he rises on his hinder-legs, and is in danger of coming quite over.

REINS are two Straps of Leather meeting in the Horseman's Bridle-hand, in order to make the Bitt bear, and keep the Horse subject. As soon as you are on Horseback hold your Reins even, and rest your Thumb upon both of 'em, keeping them separated by your Little finger. The Duke of *Newcastle* bestow'd the Name of Reins upon two Straps or

Ropes

Ropes of a *Cavesson*, which he ordered to be made fast to the Girths or the Pommel of the Saddle, with intent that the Rider should pull 'em with his hand, in order to bend and supple the Neck of the Horse.

False-Rein is a Lathe of Leather pass'd sometimes through the Arch of the *Banquet*, to bend the Horse's Neck: The Duke of *Newcastle* disapproves the use of it, and says it slacks the Curb, and makes the Bitt no more than a Trench that has no Curb.

REMOLADE, or *Charge*. See *Charge*.

REMOULIN: An old *French* Word, signifying a Star upon a Horse's Forehead.

RENETTE is an Instrument of Polish'd Steel, with which they sound a Prick in a Horse's Foot.

REPART, is to put a Horse on, or make him part a second time. After stopping your Horse, make him repart straight.

REPOLON, is a *Demivolte*, the *Croupe* in, clos'd at five times. The *Italians* are mighty fond of this sort of Manage. In making a *Demivolte*, they ride their Horses short, so as to imbrace or take in less Ground, and do not make way enough every time of the *Demivolte*. The Duke of *Newcastle* does not approve of the *Repolons*; alledging, that to make *Repolons*, is to gallop a Horse for half a Mile, and then to turn awkwardly, and make a false Manage.

REPRISE, is a Lesson repeated, or a Manage recommenced. To give Breath to a Horse upon the Four Corners of the *Volte*, with only one *Reprise*, that is, all with one Breath.

RESTY: A *Resty* Horse; *i. e.* a malicious unruly Horse that shrugs himself short, and will only go where he pleases. What the *French* call *Ramingue*, has much of the *Resty* in him.

RETAIN; (in *French*, *Retenir*) is what we call *hold*, in speaking of Mares that conceive and hold after covering.

To RIDE is us'd for learning the Manage. As, these two Gentlemen ride under a very good Master, but the other two ride under a *Creat*.

RIDGES, or Wrinkles of a Horse's Mouth, are the Risings of the Flesh in the Roof of his Mouth, which run across from one side of the Jaw to the other, like Fleshy Ridges, with interjacent Furrows, or sinking Cavities. 'Tis upon the third or fourth Ridge that we give the Stroke with the *Horn*, in order to blood a Horse whose Mouth is overheated.

RING-BONE, is a Hard Callous Swelling in a Horse's Pasterns, which of-

tentimes makes him very Lame.

RIPOSTE, is the Vindictive Motion of a Horſe that anſwers the Spur with a Kick of his Foot.

RIVET is that extremity of the Nail that reſts or leans upon the *Horn* when you ſhoe a Horſe. The Rivets of the Nails that you have drove into my Horſe are too great, and will certainly cut him; beſides, the Nails are ſo thick plated, that the Rivets by their bigneſs and weight will carry off the *Hoof*. Trie him once more with *Limoge* Nails, that the Rivets may be ſmaller. See *Pierce*.

ROAN: A *Roan* Horſe is one of a *Bay*, *Sorrel*, or *Black* Colour, with Grey or White Spots interſpers'd very thick. When this Particolour'd Coat is accompany'd with a Black *Head*, and Black *Extremities*, he is call'd a *Roan* with a *Blackamore's Head*; and if the ſame mixture is predominant upon a deep *Sorrel*, 'tis call'd *Claret-Roan*.

ROD, (in *French*, *Gaule*) is a Switch held by the Horſeman in his Right-hand, partly to repreſent a Sword, and partly to conduct the Horſe, and ſecond the effects of the *Hand* and *Heels*. This Horſe takes the Aids of the Switch well. Since this Gentleman has a mind to make his Horſe riſe before, give him the Aids of the *Rod*, touch him, ſwitch him upon the Legs and the Counter, and then he'll up with his Fore-hand.

ROPE, *Cord* or Strap, is a great Strap ty'd round a Pillar, to which a Horſe is made faſt when we begin to quicken and ſupple him, and teach him to fly from the *Shambrier*, and not to gallop falſe or uncompactly. In Manages that have no Pillar, a Man ſtands in the Center of of the Ground, holding the end of the *Rope*.

Ropes. *Ropes* of two Pillars are the *Ropes* or *Reins* of a *Caveſſon*, us'd to a Horſe that works between two Pillars. You muſt put your Horſe to the *Ropes*, that the conſtraint of the *Caveſſon* may make him ply his *Haunches*, and teach him to raiſe his Fore-hand. You'll never make this a good leaping Horſe, unleſs you put him to the *Ropes*, and make him anſwer the Aids of the *Poinſon*, in yerking with his hind-legs. See *Fore-thigh*.

ROUND, or *Volte*, is a Circular Tread. See *Volte*.

To *Cut a Round*. See *Cut*.

To *Round* a Horſe, or make him *round* (*arrondir*) is a general Expreſſion for all ſort of Manage upon *Rounds*: So that to *round* a Horſe upon trot, gallop, or otherwiſe, is to make him carry his Shoulders and his Haunches compactly or roundly, upon a greater or ſmaller Circle,

Circle, without traverſing or bearing to a ſide. To *round* your Horſe the better, make uſe of a *Cord* or *Strap* held in the Center, till he has acquired the habit of *Rounding* and not making *Points*. In working upon *Volts*, you ought never to change your hand, unleſs it be in preſſing your Horſe forward, and *rounding* him. See *Points*.

ROUSSIN is a ſtrong well knit, well ſtow'd Horſe, ſuch as are commonly carried into *France* from *Germany* and *Holland*; though 'tis true *France* it ſelf produces ſome ſuch.

ROWELS of a Spur. See *Spur*.

RUBICAN Colour of a Horſe is a *Bay*, *Sorrel*, or *Black*, with a *Light-gray* or *White* upon the Flanks, but ſo that this *Gray* or *White* is not predominant there.

RUN: To run a Horſe is to put him to his utmoſt ſpeed, *i.e.* a furious, quick and reſolute Gallop, as long as he can hold it. Some take Running for a Gallop, but in the Academies it ſignifies as above.

S.

SACCADE, is a Jerk, more or leſs violent, given by the Horſeman to the Horſe, in pulling or twitching the Reins of the Bridle all on a ſudden, and with one Pull; and that, when a Horſe lyes heavy upon the hand, or obſtinately arms himſelf. This is a Correction us'd to make a Horſe Carry well; but it ought to be us'd diſcreetly, and but ſeldom.

SADDLE, is a Seat upon a Horſe's Back, contriv'd for the Conveniency of the Rider. A Hunting-Saddle is compos'd of two Bows, two Bands, Fore-bolſters, Pannels, and Saddle-ſtraps: And the great Saddle has beſides theſe Parts, Corks, Hind-bolſters, and a *Trouſſequin*. The Pommel is common to both. A Horſeman that would ſit a Horſe well, ought always to ſit on his Twiſt, and never on his Buttocks; which ought never to touch the Saddle; and whatever Diſorder the Horſe commits, he ought never to move above the Saddle. This Gentleman keeps his Seat well, and never loſes the middle of his Saddle. He's always well ſet in the Saddle.

Saddle-Back'd; A Horſe is call'd Saddle-back'd, that is hard to ſit with a Saddle. You muſt beſpeak a Saddle on purpoſe for your Horſe, becauſe his Reins are low, and his Head and Neck rais'd: For all Saddle-back'd Horſes have a rais'd Head and Neck, and cover a man well.

Saddle-Caſe. See *Houſſe*.

Saddle-Roll. See *Trousse-quin.*

SAKER. See *Dock.*

SCABBARD, is the Skin that serves for a Sheath or Case to a Horse's Yard. Your Horse's Scabbard is inflam'd for want of Exercise; you must take him out, and water him.

SCAB'D *Heels*, or Frush, is an eating Putrefaction upon a Horse's Frush; which is very hard to cure, and has a Noisome Smell.

Crown Scab, is a Mealy Scurf upon the Pasterns of a Horse, that makes the Hair bristle and stare.

SCATCH-MOUTH, is a Bit-Mouth differing from a Canon-Mouth in this, that the Canon is round; whereas a Scatch is more upon the Oval. That part of the Scatch-Mouth which joyns the Bit-Mouth to the Branch is likewise different; a *Canon* being staid upon the Branch by a *Fonceau*, and a Scatch by a *Chaperon* which surrounds the *Banquet*. The effect of the Scatch-Mouth is somewhat greater than that of the Canon-Mouth, and keeps the Mouth more in subjection. Commonly your Snaffles are *Scatch-Mouths.*

SCHOOL, or *Schooling*, (*Ecole*) is the Lesson and the Labour both of the Horse and the Horseman. One of these Gentlemen has but three Months Schooling, and the other has four. This Horse manages better upon two Months Schooling, than another would have done in six. A School Pace, Gate, or Going, is the same with *Ecouté.* Which see.

SCRATCHES. See *Cratches.*

SCREW. See *Splent*, and *Through-Splent.*

SEAT, is the Posture or Situation of a Horseman upon the Saddle. Teach this Gentleman the Seat; *i. e.* direct him to place himself in a true posture. This Gentleman never loses his *Seat.* See *Counterpoise.*

Seat a Horse upon his Haunches or Hips. See *Put.*

SEEL; A Horse seels when he begins to have white Eye-brows; which happens when he's fifteen or sixteen Years old.

SELENDERS, are Chops or Mangy Sores in the Bending of a Horse's Hough, as the Malenders are in the Knees.

SERPEGER, a *French* Word us'd in the Academies, to signify the Riding of a Horse in a Serpentine way, or in a Tread with wav'd Turnings, like the Posture of a Serpent's Body. This word is now obsolete.

SERPENTINE. A Serpentine Tongue is a frisking Tongue that's always a mo-

moving, and ſometimes paſſes over the Bit, inſtead of keeping in the void Space, call'd the Liberty of the Tongue.

SEVIL of the Branches of a Bridle, is a Nail turn'd round like a Ring, with a large head, made faſt in the lower part of the Branch, call'd *Gargouille*. See *Banquet*.

SHAMBRIER is a long Thong of Leather, made faſt to the end of a Cane or Stick, in order to animate a Horſe, and puniſh him, if he refuſes to obey the Rider. To make this Horſe obedient, take a Shambrier in your hand; ſhew it him; crack it againſt the Ground; make him feel it.

SHANK of a Horſe's Leg, is that part of the Fore-leg that lyes between the Knee and the Fetlock, or Paſtern-Joynt. Your Horſe has a Thorough Splent upon his Shank. See if there is not a *Fuzee* (*i. e.* two Splents joyning to one another) on the Shank of that Sorrel. Here's a very uncommon thing upon the Shank of this Barb: He has an Arreſt, or Mangy Tumor all along the Sinew of the Shank, down to the Paſtern-Joynt. For commonly Arreſts happen in the Hind-Quarter; and Barbs are never troubled with 'em.

SHEDDING of the Hair. See *Caſt*.

SHEET. Horſe-ſheet. See *Caparaſſon*.

SHELL-*Tooth'd Horſe*, (in *French*, *Baigu*) is one, that from five Years to old Age, naturally, and without any Artifice, bears Mark in all his Fore-teeth; and there ſtill keeps that hollow place, with a black Mark which we call *Germe de Feve*, or the Eye of a Bean: Inſomuch, that at twelve or fifteen, he appears with the Mark of a Horſe that is not yet ſix. For in the Nippers of other Horſes, the hollow place is fill'd, and the Mark diſappears, towards the ſixth Year, by reaſon of the wearing of the Tooth. About the ſame Age, 'tis half worn out in the Middling Teeth: And towards the eighth Year, it diſappears in the Corner-Teeth. But after a Shell-Tooth'd Horſe has marked, he marks ſtill equally in the Nippers, the Midling, and the Corner-Teeth: Which proceeds from this; that having harder Teeth than the other Horſes, his Teeth do not wear, and ſo he does not loſe the black Spot. Among the *Poliſh*, *Hungarian*, and *Croatian* Horſes, we find a great many hollow-tooth'd Horſes; and generally the Mares are more apt to be

ſuch than the Horſes. Do not you ſee that your Horſe-Merchant is like all other Jockeys, in denying, for his own Intereſt, that he has any hollow-tooth'd Horſes? Certainly this Runner is hollow-tooth'd; for beſides that it bears Mark ſtill in all the Fore-Teeth, it ought to have its Teeth ſhort, clean, and white; whereas they are long, yellow, foul, and unfleſh'd; which betrays his Age: And I'll lay you a Wager, that in a Year's time he will have Seel, or white Eye-brows.

SHOE of a Horſe, is a piece of flat Iron with two Branches, or Wings, which being commonly forged according to the Form of the Hoof for which 'tis deſign'd, is made round at the Toe, and open at the Heel. Such a Farrier has Shoes for all Feet. Now a Shoe for all Feet, is a Shoe cut at the Toe into two equal parts; which are joyn'd by a riveted Nail, upon which they are moveable, in ſuch a manner, that the Shoe is enlarged or contracted, leſs or more, in order to fit all ſorts of Feet. This Horſe has a brittle Foot, or a brittle Hoof; you muſt ſhoe him in the Wane of the Moon. A Fore-ſhoe, a Hinder-ſhoe: A Horſe unſhod before and behind. 'Tis troubleſome to ſhoe this Horſe, for his Foot is worn, and he has gone a long time unſhod. To ſhoe a Horſe after the form of a Lunette, a Pattin, *&c.* See thoſe Words.

Shoeing Hammer, is a Hammer that the Farrier makes uſe of, to adjuſt and fit the Shoes upon the Anvil, both hot and cold.

SHORT-JOYNTED. A Horſe is ſaid to be ſhort-joynted, that has a ſhort Paſtern. When this Joynt, or the Paſtern, is too ſhort, the Horſe is ſubject to have his Fore-legs, from the Knee to the Cronet, all in a ſtraight Line. Commonly your ſhort-joynted Horſes do not Manage ſo well as the long-joynted; but out of the Manage the ſhort-joynted are the beſt for Travel or Fatigue. Your Horſe is ſhort-joynted, and *bouté*; that is, his Legs are ſtraight, from the Knee to the Cronet. See *Bouleté*, and *Bouté*.

SHOULDER of a Horſe, is the Joynt in the Fore-quarters that joyns the end of the Shoulder-blade with the extremity of the Fore-thigh.

Shoulder of a Branch, is that part of it which begins at the lower part of the Arch of the Banquet, over againſt the middle of the *Fonceau*, or *Chaperon*, and forms another Arch under the *Banquet*. The Shoulder of a Branch

Branch casts a greater or lesser Circumference, according as 'tis design'd to fortify, or weaken, the Branch. Your Horse's Bridle raises him well enough; but as for that other Bridle, that has too large a Shoulder, do not you see how it draws the Horse's Head between his Legs? See *Banquet*.

Shoulder of a Horse, is that part of his Fore-hand that lyes between the Withers, the Fore-thigh, the Counter, and the Ribs. Your Horse throws himself too much upon his Shoulders, and he sits heavy upon the hand, for want of Porting, or Seating himself upon his Haunches, and bending his Hocks. Make your Horse's Hips sustain his Shoulders, and his Fore-quarters; then you'll have him light upon the hand, and well coupled. A good Horse should be light in the Shoulders, and subject in the Hip. This Sorrel is Shoulder-splait. That Horse has got the *Spear*-Feather in his Shoulder; which is a good Mark. See *Spear-Feather*, *Shoulder-splait*, and *Supple*.

Shoulder-pegg'd, (*Chevillees*) are so call'd when they are gourdy, stiff, and almost without Motion. A Horse charg'd with Shoulders, is a Horse that has thick, fleshy, and heavy Shoulders. Your Horse over-reaches, because he goes too much upon his Shoulders.

Shoulder-splait; a Horse is said to be such when he has given his Shoulder so violent a shock as to disjoint the *Shoulder-bone* from the *Body*. Your Horse rakes as he trots, and halts so down to the Ground, that I believe he's *Shoulder-Splait*.

SIDE; to ride a Horse side-ways, is to passage him, or make him go upon two Treads, one of which is mark'd by his Shoulders, and the other by his Haunches.

SIGUETTE, is a Cavesson of Iron, with Teeth or Notches; that is, a Semi-Circle of hollow and vaulted Iron, with Teeth like a Saw, consisting of two or three pieces joyn'd with Hinges, and mounted with a Head-stall and two Ropes, as if they were the Cavessons that in former times were wont to be put upon the Nose of a fiery stiff-headed Horse, in order to keep him in subjection. There is a sort of *Siguette* that's a round Iron all of one piece, sew'd under the Nose-band of the Bridle, that it may not be in View. This Siguette we employ with a Martingale, when a Horse beats upon the hand.

SINEW; *Unsinew* a Horse, (in *French*, *Enerver*) is

is to cut the two Tendons on the ſide of his Head, about five inches under the Eyes; which two joyn in one at the Tip or end of the Noſe, in order to perform its Motion. This Tendon at the Tip of the Noſe is likewiſe cut. We unſinew, in order to dry the Head, and make it ſmaller. Upon the whole, the Word *Enerver*, or *Enervate*, is improper to be us'd on this occaſion; for 'tis not a Nerve, but a Tendon, that's cut.

Sinew-ſhrunk; A Horſe is ſaid to be Sinew-ſhrunk, when he is over-rid, and ſo born down with Fatigue, that he becomes gaunt-belly'd thro' a Stiffneſs and Contraction of the two Sinews that are under his Belly.

Sinew-Sprung, is a violent Attaint or Over-reach, in which a Horſe ſtrikes the Toe of his hinder-feet againſt the Sinew of the Fore-legs.

SKITTISH *Horſe*, (in *French*, *Ecouteux*, and *Retenu*) is one that leaps inſtead of going forward, that does not ſet out or part from the hand freely, nor imploy himſelf as he ought to do. Put that Horſe on; he's skittiſh. 'Tis too much trouble to make ſuch a Horſe go forward; he's only fit to run upon a Squadron.

SLABBERING-BIT. See *Maſtigadour*.

SLACK a Leg, (in *French*, *mollir la jambe*) is ſaid of a Horſe when he trips or ſtumbles. This Horſe ſlackens ſtrangely, after one hours hunting, and mine has kept in with the Stag till he was kill'd, without ſlackening, though he be but a Lath-back.

Slack the hand, is to ſlack the Bridle, or give a Horſe head.

SNAFFLE, after the *Engliſh* Faſhion, is a very ſlender *Bitt-mouth* without any Branches: The *Engliſh* make much uſe of 'em, and ſcarce uſe any true Bridles, but in the Service of War. The *French* call 'em *Bridons* by way of diſtinction from *Brides*; *i.e. Bridles*. See to *Reſt heavy*.

Snaffle, or ſmall Watering-bitt, is commonly a *Scatch-mouth*, accoutred with two little very ſtraight Branches, and a Curb, and mounted with a Head-ſtall and two long Reins of *Hungary* Leather. To put a Horſe in a *Snaffle* to dreſs or curry him. To lead a Horſe to Water in a *Snaffle*. To hold a Horſe with a *Snaffle* between two Pillars in the Stables. To turn a Horſe to the *Snaffle* (*au filet*) is to ſet his *Croupe* to the Manger, and his *Head* between two Pillars, to hinder him to feed.

SNORT, *Snuffler*, *Snuff*, (in *French*, *Ebrouer*) implies a certain Sound that a Horſe full of fire breathes thro' his No-

Nostrils; and sounds as if he had a mind to expel something that were in his Nose, and hinder'd him to take breath. This Noise or Sound is perform'd by the means of a Cartilage within the Nostrils, call'd in *French*, *Souris*. Horses of much Mettle snort when you offer to keep them in. 'Tis plain, your Horse is well winded, for he snorts every Turn of the Gallop, and that's a sign his Lungs are good. The Word (*Ebrouer*) is expressive in the *French*; for they have no Word of an equivalent Signification to it. See *Souris*.

SOLE: To take out the *Sole*, is to do it without touching the *Horn* of the *Hoof*; for if you take off the *Horn*, you make a *Hoof-cast*. We take out the *Sole* for several Infirmities, as you may see in Mr. *Solleysel's Compleat Horseman*. A Horse that's unsoled may recover in less than a Month.

High-sol'd, (in *French*, *Pied Comble*) a Horse whose *Sole* is round underneath, so that 'tis higher than the *Hoof*, which oftentimes makes the Horse halt, and hinders the shoeing of him, unless the Shoe's vaulted.

Sole of a Horse is a Nail, or sort of *Horn* under the *Foot*, which is much tenderer than the other *Horn* that incompasses the *Foot*, and by reason of its hardness, is properly called the *Horn* or *Hoof*. A Horse's Shoe ought to be so set upon the *Hoof*, as not to bear upon the *Sole*; for otherwise the *Sole* would be hurt, and not only make the Horse lame, but corrupt the Flesh that separates it from the *Coffin-bone*. I take it, there's a Fig in this Horse's *Sole*, but t'other is surbated.

SORREL is a Reddish Colour, with which the Mane ought to be Red or White: 'Tis distinguish'd according to the Degrees of its deepness, into a *Burnt-Sorrel*, and a *Bright* or *Light-Sorrel*; but generally speaking, 'tis a sign of a good Horse. This *Burnt-Sorrel* does not belie his Colour. He makes good the Proverb, *A Burnt-Sorrel will die before he's tyred*: For, in effect, you can never overdoe this Horse; and when other Horses are quite gone, he is still brisk; but for that *Light-Sorrel*, with the Pale Extremities, that is, the Hair of whose Extremities, *i. e.* Mane and Tale, is less tinged and whiter than the rest, he's scarce able to bear his own Tail; and though you have never so sharp Spurs, you can't ride him an hour, but he becomes insensible and heedless both of the Spur and the Whip; for these pale out-parts are frequently a sign of Weakness, and lower the Value of a Horse.

SOUND, (in *French*, *Droit*); a Horse is such that does not halt. When a Jocky sells

ſells a Horſe, he warrants him *ſound* hot and cold; that is, that he does not halt neither when you mount him, nor when he's heated, nor yet after alighting, when he ſtands and cools.

SOURIS is a *Cartilage* in the Noſtrils of a Horſe, by the means of which he ſnorts. See *Snort*.

The cutting of this *Cartilage* is call'd in *French Eſſouriſſer*.

SOUS-SUI. See *Haunches*.

SPAVIN: *Ox-Spavin*, is a Swelling in the lower part of the inſide of the *Hough*, which for the moſt part makes the Horſe lame.

Dry-Spavin, or *String-halt*, is a Stiffneſs or Gourdineſs of the Ham, ſo that the Horſe can't bend it, but is conſtrain'd to lift the *Hip* and the *Leg* all at once, without bending the *Ham*.

Blood-Spavin is a Tumor on the inſide of the Thigh, near the place where the Curb is ſeated. The *Crural-Vein* diſembogues in this part, which makes a ſoft and painleſs Swelling.

SPEAR: The Feather of a Horſe call'd the *Stroke of the Spear*, is a Mark in the Neck, or near the Shoulder of ſome *Barbs*, and ſome *Turky* and *Spaniſh* Horſes, repreſenting the Blow or Cut of a *Spear* in thoſe Places, with ſome appearance of a Scar, as 'twere. This Feather is an infallible ſign of a good Horſe.

Spear-hand, or *Sword-hand* of a Horſe-man is his Right-hand.

Spear-foot of a Horſe is the Far-foot behind. Your Horſe has a White in the Far-hind-foot. See *Foot*.

SPLENT is a Callous or hard Swelling upon the *Shank-bone* of a Horſe, on the inſide, below the Knee; and ſometimes on the outſide.

Thorough-Splent, or *Pegg'd-Splent* is a *Double Splent*, one on the outſide, and another on the inſide of the Shank, the one over-againſt the other.

SPUNGE of a Horſe-ſhoe, is the extremity or point of the Shoe that anſwers to the Horſe's *Heel*, upon which the Calkins are made. Never make Thick Spunges to your Horſe-ſhoes, for that ruins *Heels*.

SPUR is a ſmall piece of Iron, of two Branches, bended in the form of a Semicircle, for receiving the Horſe-man's Heel in their Cavity. In the middle they have a Rowel, *i. e.* a ſmall piece of Iron, with eight or ten Points or Sharp Ends, to prick the Horſe's Sides withal upon occaſion, and ſometimes to the very Quick. This Horſe knows no Spur, that is, he is not ſenſible of it. Such a Horſe obeys the Spurs; that is, he flies 'em. This Horſe is ſenſible of the Spur, he flies the Spur;

Spur; he anſwers the Aids of the Spur. A Tickliſh Horſe cleaves to the Spur. Bring to your Spur: Put on with it. Sometimes we call this Pricking; Sometimes we ſay a Horſe obeys the Heels; knows the Heels. See *Prick* and *Heel*.

SQUARE. To work in a *Square*. The *Piſte* or *Tread* of a *Volte*, inſtead of being always circular, and trac'd upon a circumference round a Center, ought likewiſe to be imagin'd ſo as if it formed four ſtraight equal Lines laid, run in a Square, and equally remov'd from the Center or the Pillar which repreſents it in the middle of the Manage-Ground: So that to work in a *Square*, is to ride along each of theſe four Lines, turning the hand at every Corner, and ſo paſſing from one Line to another.

STABLES. See *Ecurie*.

STAG-EVIL. See *Hart*.

STALLION, (in *French*, *Etalon*, or *Etelon*) is a Stone-horſe ſhut up in a Breed with Stud-Mares, for making a Breed or Race of Horſes. To give the *Stallion* to Mares; to make a *Stallion* cover a Mare in hand, is to hold the *Stallion* by the Halter or Bridle while he covers her. In our Breed, when we ſuffer the *Stallions* to cover the Mares, we always leave 'em looſe and at liberty, and never have 'em covered in hand. The Duke of *Newcaſtle* does not approve of the covering of Mares in the hand; he affirms they ought to be left to their natural liberty, by which means the Foal will be of a better make. See *Breed*.

STARTING, *Skittiſh*, *Timorous*. A Horſe is ſaid to be ſuch, that takes every Object he ſees to be otherwiſe than it is, upon which he ſtops, flies out, and ſtarts ſuddenly to one ſide, inſomuch that the Rider can't make him come near it. This Fault is more common to *Geldings* than to *Stone-horſes*. The Horſes that have bad Eyes are ſubject to it, as well as thoſe that have been long kept in a Stable without airing; but the latter are eaſily cur'd of it. When you have a skittiſh Horſe, never beat him in his conſternation, but make him advance gently, and with ſoft means to the Scare-crow that alarms him, till he diſcovers it, and gains aſſurance.

STATELY, a Horſe that goes with a proud ſtately ſtrutting Gate, is call'd in *French*, *Piaffur*: Which ſee.

STAY; to ſtay the Hand, to ſtay or ſuſtain the Horſe, is to hold the Bridle firm and high. The leaſt ſtay or ſupport of your hand will ſtop your Horſe. We likewiſe ſuſtain or ſtay a

Horſe with the In-leg, or the In-heel, when he makes his *Groupe* go before his *Shoulders* upon *Volts*. We ſtay a Horſe again, when we hinder him to traverſe, when we ride him equally, keeping him always ſubject, ſo that his *Croupe* can't ſlip out, and he can't looſe neither his Cadence, nor his Ground, but marks all his times equal.

STEP. See *Walk*.

Step and a Leap. The Air of a Step and a Leap, is the high Manage of a Horſe, that between two Leaps, or *Caprioles*, marks a *Corvet*, which upon this occaſion is call'd a Step, in ſuch a manner, that at every Leap or *Capriole*, he raiſes his Fore-legs, and his *Haunches* follow, yerking or ſtriking out his hind-legs at the end of every Leap. When a Horſe inclines to this ſort of Manage, we put him forward with the Aids of the hand, the Calves of the Legs, the *Poinſon* and the *Heel*, which ſhould be all dexterouſly employed to make him lift before and behind, and give him a good *Appui* or ſtay. See *Time*.

Two Steps and a Leap, is a Manage or Air compounded of two *Corvets*, terminating in a *Capriole*.

STIFFLE, or Great Muſcle, is the part of the hind-leg which advances towards a Horſe's Belly, and is a moſt dangerous part to receive a Blow upon.

STIRRUP, is a Reſt for a Rider's Foot, compoſed of ſome ſmall Pieces of Iron forged into Barrs, and level below, but arch'd in the upper part, by which ſide they are hung in Stirrup-Leathers. Bear vigorouſly upon your Stirrup, when you have your Foot in it; and hold the Point of your Foot higher than the Heel. When you would ſtop your Horſe, you muſt bear upon your Stirrups.

You ſhould keep your Right-ſtirrup half a Point ſhorter than the Left, for in Combat the Horſe-man bears and reſts more upon the Right; and to facilitate the mounting of your Horſe, the Left-ſtirrup ſhould be ſomewhat longer than the other. Shorten your *Stirrup* one Point; let it down a Point. Fit your *Stirrup* to the Point that ſuits you. Give your Horſe way with your *Stirrups* unbuckled and dangling, that they may ſtrike againſt his Flank, and accuſtom him to the *Spur*.

To looſe ones *Stirrups*, is to ſuffer 'em to ſlip from the *Foot*. Take care you do not loſe you *Stirrups*. You have loſt your Right-ſtirrup upon a ſmall Yerk or two.

The *Stirrup-foot*, or the *Near-fore-foot*, is the *Left-foot* before.

Stirrup-Leather is a *Lathe* or *Thong* of *Leather*, descending from the *Saddle* down by the Horse's Ribs, upon which the *Stirrups* hang.

Stirrup-bearer; (in *French*, *Porte Etrier*) is an End of Leather made fast to the End of the *Saddle*, to truss up the *Stirrups* when the Rider is alighted, and the Horse sent to the *Stable*.

STOP, is a Pause or Discontinuation of Going. To form a Stop, is to stop upon the Haunches. Acquaint these Gentlemen, that to form the Stop of a Horse, they must, in the first place, bring to the Calves of their Legs, to animate him; bend their Body backwards, raise the Bridle-hand without moving the Elbow; then vigorously extend their Hams, and rest upon the Stirrups, to make him form the Times or Motions of his Stop in *falcading* with his Haunches three or four times. I form'd the Stop of my Horse in the three or four times; *i. e.* making him falcade upon his Haunches. Your Horse forms his Stop ungracefully; for he does not bend his Haunches; he traverses, and beats upon the hand. This Horse, from parting to stopping, has held a Career of a Hundred Paces; and, after marking the Stop, made a Pesate or two at last. Do not form the Stops of your Horse so short and precipitant, unless you have a mind to spoil his Hams, and his Mouth. After stopping your Horse, make him give two or three Corvets. The opposite Term to *Stop* is Parting. In former times, the Stop of a Horse was call'd *Parade*. See *Raise*, and *Nails*.

Half a Stop, is a *Stop* not finish'd by a *Pesate*; so that the Horse, after falcading three or four times upon the Haunches, resumes and continues his Gallop, without making Pesates or Corvets. Do not you admire this *Spanish* Horse that makes such good Passades? Mind how he Gallops in a straight Line, as he is push'd, and how he forms a Half-stop, making Falcades three or four times with his Haunches very low. You say well, that if he after that made a Pesate or two, 'twould be a compleat Stop: But you see, that instead of a Pesate, he makes his Demivolt in three times, and resumes his Gallop upon a straight Line, in order to do as much at the end of the Passade. Come and see my *Barbary* Horse; who in his Galopade will make twenty Half-stops, and resume his Listening Gallop with the

ſame Cadence, without heat or diſquiet. One may truly ſay of this Barb, that the Rider has his Will in his hand. See *Falcade*, *Paſſade*, and *Peſate*.

STRAIGHT. To part or go *ſtraight* or *right out*, is to go upon a Tread traced in a *Straight-Line*. This Horſe makes *Corvets* equally well, ſtraight, and upon *Volts*. When you would puſh your Horſe forwards, make him part ſtraight, and put him back ſtraight, without traverſing or bearing ſidewiſe.

Straight-Member'd, (in *French*, *Droit ſur les jambes*) See *Legs*.

STRAIN, *Sprain*, or *Swaying*, (in *French*, *Effort*) is a violent Extenſion of the Sinews, or Relaxation of the Muſcles, that keep a Horſe's Bones tight in their Articulations: And the Word *Effort* is likewiſe us'd for a Rupture of any Vein. This Horſe is Hip-ſhot, Shoulder-ſplait, Sway'd in the Back. See *Shoulder-ſplait*.

STRANGLES, is a Collection of foul Humors form'd in the Body of a young Colt; which are voided by the Noſtrils, or by a Suppuration of ſome Glands or Knots that lye between the Bones of the lower Jaw, and are crowded with Impurities. The falſe Strangles happen in old Horſes, that have not well caſt the Strangles.

STRAPS of a Saddle, are ſmall Leather Straps, nail'd to the Bows of the Saddle, with which we make the Girths faſt to the Saddle.

STRIKE a Nail, is to drive it thro' the Horſe's Shoe, and the Horn or Hoof of his Foot; and to rivet it for holding on the Shoe. Since your Horſe has commonly much Hoof at the Toes of his Fore-feet; the Farrier may ſtrike high there without fearing the coming upon the quick: And as for the Hinder-feet, he ought to ſtrike pretty high upon the Quarter or Heel, but low at the Toe; becauſe there the Hoof is near the quick. See *Pierce*, and *Pricking*.

To *Strike* a Vein. See *Barr*.

STRING-HALT. See *Dry Spavin*.

STUB, (in *French*, *Chicot*) is a Splinter of freſh cut under-Wood, that gets into a Horſe's Foot as he runs; and piercing the Sole thro' to the quick, becomes more or leſs dangerous, according as it ſinks more or leſs into the Foot.

SUBJECT; To keep a Horſe ſubject, is an Expreſſion relating to Volts; ſignifying, to keep the Croup of the Horſe in the Round, ſo that it may not ſlip out; that he may not traverſe, and that he may work in the Manage, Croup in, marking his equal Times, without loſing his Ground.

SUPPLE a Horſe, is to make him bend his Neck, Shoulders, and Sides; and to render all the Parts of his Body more pliable. Your Horſe has a ſtiff Neck and Shoulders; he has no Motion with his Leg: You muſt try to ſupple him with a Caveſſon made after the Duke of *Newcaſtle's* way; and Trot and Gallop him, ſo as to make him turn frequently from a Trot to a Gallop.

SURBATING, is a Corruption of the Fleſh under the Sole of a Horſe, which is bruis'd and ſpoil'd by the Sole, when a Horſe runs long unſhod, and the Sole is over-dry'd, and withered.

SWEATING-IRON, or Knife, is a piece of a Sythe about a Foot long, and of the breadth of three or four Fingers, very thin, and ſuch as cuts only with one ſide. When a Horſe is very hot, and the Grooms have a mind to leſſen the *Sweat*, or make it glide off, they take this Knife or Iron in their Two Hands, and gently run the Cutting Edge along the Horſe's Skin, commonly with the Grain, or as the Hair lies, and but ſeldom againſt it; with intent to ſcrape off the *Sweat*, and dry the Horſe.

SWORD-*Hand*, or *Spear-hand*, is the Right-hand of the Horſe-man; the Left-hand is call'd the Bridle-hand.

T.

TAIL of a Horſe. A great many affirm, that the Dock of a Horſe's Tail ſerves to point out his ſixth or ſeventh Year; pleading, that about the time that the black Speck, or Eye of the Bean begins to diſappear, and the Cavity to be fill'd, the Dock of the Tail becomes longer, by reaſon that the Vigour of the young Years begins to abate, and Nature has not ſtrength enough to nouriſh and keep up the Joints or Knots that form the Dock; ſo that when the Horſe is ſix Years old, one of theſe Joints ſlackens, and begins to fall down; and a Year after, another deſcends in like manner. But this Relaxation, or Down-falling, happens ſooner in ſome than in others, according as they have been well, or ill kept, with refe-

rence to Feeding, Houſing, and Working. Accordingly we find the Marks of a Horſe's Age taken from his Tail, are ſo erroneous, that we ſee a great many Jockeys maintain, that the firſt Joynt deſcends when he is nine, and the ſecond when he is ten Years old.

TEETH, are little Bones in a Horſe's Jaws, which ſerve not only to facilitate the Nouriſhment, but likewiſe to diſtinguiſh the Age of Horſes. A Horſe has forty Teeth, including the Tuſhes; which are diſtinguiſhed as follows.

Four and Twenty of 'em are call'd the *Grinders*, which are plac'd at the bottom of the Mouth, beyond the Barrs, Twelve on each ſide of the Channel, *viz.* ſix above, and ſix below, on each ſide. Theſe Teeth continue, and do not fall, to give place to new Teeth in their room; ſo that they are of no uſe in diſtinguiſhing a Horſe's Age. However, they are ſubject to Wolves Teeth.

With reference to the other Sixteen, Twelve of 'em are call'd in their infancy *Milk* or *Foal-Teeth*, and the remaining Four go by the name of *Tuſhes*.

The Twelve *Foal-Teeth*, are ſhort, ſmall, and white Teeth, ſeated on the Fore-part of the Mouth, Six above and Six below. Theſe change and caſt, to give place to others; which, in proceſs of time, become long, large, and yellowiſh. Theſe new Teeth are diſtinguiſh'd by the different Names given them according to their putting forth; and 'tis the manner of their coming forth that gives us to know the firſt Years of a Horſe. Now of theſe Twelve, Four are call'd Nippers, Four are named Middling Teeth, and Four go by the Name of Corner Teeth.

The Four Nippers are ſeated in the Fore-part of the Mouth, Two above, and Two below. When a Horſe has put forth theſe, we conclude, that he goes from Two and a half to Three Years.

The Middling Teeth are placed near the Nippers, or Gatherers, One above, and One below, on each ſide of the Jaws. They come out and appear between Three and a half and Four Years.

The Corner-Teeth are plac'd yet more forward in the Mouth, One above, and One below, on each ſide of the Jaws. Theſe begin to ſhoot between the Fourth and the Fifth Year, and are got above the Gum at Five Years. Now, ſurmounting the Gum at that Age, they become hollow, and *mark* commonly till Seven or Eight Years.

Years. By *Marking*, we mean, that in the Hollow, or Cavity of the Corner-Teeth, a little blackiſh Speck is form'd; which, from its Reſemblance, we call the Bud or Eye of a Bean. But when the Horſe paſſes Six, this Cavity begins to fill, and the black Mark diſappears by degrees; and this Diminution of the Cavity and the Mark, continues from Six to Seven and a half. At Eight Years, the Cavity is fill'd up, and the black Mark is gone; and in regard that the Tooth is then full, even, and ſmooth as if it had been ſhav'd, we then ſay, that the Horſe has raz'd. Which happens a little before the Eighth Year; and after that the Horſe does not Mark; ſo that the ſureſt Knowledge of his Age is then took from the Tuſhes.

The Tuſhes are plac'd beyond the Corner-Teeth upon the Barrs, Two on each ſide of the Jaws, *i. e.* One above, and One below, wtihout being preceded by any Foal-Teeth. The Two Under-Tuſhes cut ſometimes at Three Years, ſometimes at Three and a half, and ſometimes Four: But the Two Upper-Tuſhes appear ſometimes at Four, and ſometimes at Four and a half; ſometimes before, and ſometimes after the Corner-Teeth, without any certain Rule: And till the Age of Six they are chamfer'd within. About Ten Years of Age the two upper Tuſhes appear much worn; which ſerves for an indication of that Age. After that, they grow out in length, and become bare of Fleſh, becauſe the Gum ſhrinks, and retires: And at laſt, about the Fifteenth or Sixteenth Year, the Horſe ſeels. A Horſe is not capable of any great Fatigue, till his Tuſhes have cut the Skin. Moſt of the *Dutch* Horſes are very ſick when their Tuſhes come forth. Mares have 'em but ſeldom, and when they have 'em they are very ſmall. See *Shell-tooth'd*, and *Counter-mark'd*.

In ſpeaking of the *Teeth*, we ſay ſuch a Horſe has caſt his under *Milk-teeth*; he has caſt his *Corner-teeth*, or his Nippers; he has put forth his Nippers. This *Sorrel* has chang'd his *Teeth*, and caſt his Nippers. This Horſe is unlucky both with his *Feet* and his *Teeth*, you muſt get him cut or gelt.

TENDON: To cut a Horſe's *Tendon*. The *Tendon* is a ſort of a Griſtle that ſurrounds one part of the *Foot*, and is ſeated between the *Hoof* and the *Coffin-bone*, near the *Cronet*. When a Horſe has a *Quitter-bone*, the Matter that

gathers between the *Coffin-bone* and the *Hoof*, spoils the *Tendon*, and makes it black; and the Cure of such a *Quitter-bone* consists in cutting and extirpating the *Tendon*.

TERRA A TERRA, is a Series of low Leaps, which a Horse makes forwards, bearing side-ways, and working upon two Treads. In this Motion the Horse lifts both his Forelegs at once; and when these are upon the point of descending to the Ground, the Hinder-legs accompany 'em with a short and quick Cadence, always bearing and staying upon his *Haunches*; so that the Motions of the Hinder-quarters are short and quick; and the Horse being always well prest and coupled, he lifts his Fore-legs pretty high, and his Hinder-legs keeps always low and near the Ground. This Manage is call'd *Terra à Terra*, because in this Motion the Horse does not lift his Legs so high as in *Corvets*. Such a Horse works very well at the *Terra à Terra*. See *Six Volts*.

TERRAIGNOL. A Horse so call'd, is one that cleaves to the Ground, that can't be made light upon the hand, that can't be put upon his *Haunches*, that raises his Fore-quarters with difficulty, that's charg'd with Shoulders, and in general, one whose Motions are all short and too near the Ground.

TERRAIN, is the Manage-ground upon which the Horse marks his *Piste* or Tread. This Horse observes his ground well; he keeps his ground well; he imbraces his ground well, without inlarging or narrowing more to one hand than to another.

TETTAR is an Ulcer almost as broad as one's hand, that appears commonly upon a Horse's *Croupe*, sometimes on his *Head*, and sometimes upon his Neck. It proceeds from *Bilious Blood*, that consumes and eats through the Hide or Skin, and causes such a violent Itching, that 'tis a hard matter to keep the Horse from scratching, and so inlarging or spreading the Ulcer. This Disorder was not known till of late.

THIGHS of a Horseman: The effect of the Rider's Thighs is one of the Aids that serves to make a Horse work vigorously in the Manage. As soon as the Cavalier closes with his Thighs, you see the Horse is enliven'd and alarm'd, as preparing himself for doing what is demanded of him, and disposing himself for the Manage. This Horse has such fine Aids, that he manages by the Aids of the Thighs alone, without needing those of the Legs.

Fore-thigh, or Arm of a Horſe, is that part of the *Fore-leg* that runs between the *Shoulder* and the *Knee*: Tho' the *Fore-thigh* do not bend or bow, yet we commonly ſay a Horſe goes fine, that bends well the *Fore-thigh*; importing thereby, that he bends well his Leg. Your Horſe bends the *Fore-thighs* ſufficiently, and lifts his *Fore-quarter* very freely, ſo that there's no occaſion to put him any longer between the two Poſts, to make him light before.

TICK, is a Habit that ſome Horſes take of preſſing their Teeth againſt the Manger, or all along the Halter or Collar, as if they would bite it. You have got a *Ticker*, that by ticking ſo often, will fill his Body with Wind, which will gripe him, and make him ſick.

TICKLISH: A Horſe is ſaid to be ticklish, that's too tender upon the Spur, and too ſenſible, that does not freely fly the Spurs, but in ſome meaſure reſiſts 'em, throwing himſelf up when they come near and prick his Skin. A *Ticklish* Horſe has ſomewhat of the *Ramingues*; *i. e.* the Kickers againſt the Spurs, but with this difference, that the latter put back, leap and kick and yerk out behind, in diſobeying the Spurs; whereas a *Ticklish* Horſe only reſiſts for ſome time, and afterwards obeys, and goes much better through the fear of a Vigorous Ham, when he finds the Horſe-man ſtretch his Leg, than he does upon being actually pricked.

TIME, is ſometimes taken for the Motion of a Horſe that obſerves Meaſure and Juſtneſs in the Manage; and ſometimes it ſignifies the Interval between two of his Motions. This Horſe-man is very attentive to all the Horſe's Times, and backs 'em ſeaſonably. Such a Horſe-man has ſlip'd two Times, and ſuffer'd the *Cadence* of the Horſe to be interrupted for want of aiding him. In the Manage of a Step and a Leap, the Horſe makes by turns a *Corvet* between two *Caprioles*; and in that caſe the *Corvet* is one time that prepares the Horſe for the *Caprioles*. Make your Horſe mark two or three times at this Stop. Here by two or three times, you are to underſtand, two or three *Falcades*. Since your Horſe is ſo very grave, if you would oblige him to give ſome *Corvets*, you ſhould put him upon a gentle Gallop, and upon his Haunches, and aſſiſt him with your Legs, to make him mark the *Times* of the *Corvets*. In this laſt Example, the Word *Time* is taken for the Motion.

The Word *Time* does likewise signifie the effect of one of the Aids. This Horse-man prepares and disposes his Horse for the effects of the Heel, in beginning with one *Time* of the Legs; and he never runs precipitantly upon his *Times*.

TIMOROUS: Fearful Horse. See *Starting*, *Skittish*. In *French*, *Peureux*, *Ombrageux*.

TITT. See *Nag*.

TOE, is the Stay of the Hoof upon the fore-part of the Foot, comprehended between the Quarters. We say commonly, (*Pince devant*, and *talon derriere*) the *Toe* before, and the *Heel* behind; implying, that in Horses, the *Toe* of the Fore-feet is stronger than the *Toe* of the *Hind-feet*: And on the other hand, the *Heels* behind are stronger than those before: And accordingly in shoing, we drive higher into the *Toes* of the *Fore-feet*, and into the *Heels* of the *Hind-feet*. See *Drive*.

And sometimes a Horse does not rest his *Hinder-feet* all equally upon the Shoe, but raises his *Heel*, and goes upon the *Toes* of the *Hind-feet*. Such a Horse is called in *French*, *Rampin*.

TONGUE: The aid of the *Tongue* or *Voice* is a sort of agreeable clacking, or a certain Sound form'd by the *Cavalier*, in striking his *Tongue* against the Roof of his Mouth, when he means to animate the Horse, and sustain him, and make him work well in the Manage. This Horse takes the Aids of the *Tongue* very well, he takes Life and Courage upon the Aids of the *Tongue*.

Tongue. A Horse that draws in, or swallows his *Tongue*, is said to do so when he turns it down his Throat, which maks him wheeze as if he were short-winded. This Fault is cured by giving him a Bitt with a Liberty for the *Tongue*. See *Liberty*.

TORCHENES, is a Long Stick with a Hole at the end of it, through which we run a Strap of Leather, the two ends of which being tied together, serve to straiten and closely tye up a Horse's Nose as long as the Stick is stay'd upon the Halter or Snaffle. This is done to keep the Horse from being unruly when they go to dress him, or upon any other occasion. In *Germany*, when the Jockeys bring to any Fair the *Roussins*, *i. e.* thick-bodied strong Stone-Horses, they clap a *Torchenes* upon their Nose, to prevent their being disorderly.

TRAMELLED: A Horse is said to be tramelled, that has Blazes or White Marks upon the fore and hind-foot of one side, as the far-foot

foot before and behind. He is so call'd from resemblance of the White-foot to the Hoses of a Half-tramel.

Cross-tramell'd Horse, is one that has White Marks in two of his Feet that stand cross-ways, like a St. *Andrew*'s Cross; as in the Far-fore-foot, and the Near-hind-foot; or in the Near-foot before, and the Far-foot behind.

TRANCHEFILE, is the Cross Chain of a Bridle that runs along the Bit-mouth, from one Branch to the other.

TRAQUENADE. See *Entrepas*.

TRAVERSE; a Horse is said to Traverse, when he cuts his Tread cross-wise, throwing his Croupe to one side, and his Head to another. When you stop your Horse, take care he does not Traverse. When you pull'd your Horse back, he Travers'd; and you do not consider, you ought to put back as straight as you advanc'd.

TRAVICE, (in *French*, *Travail*, and in some of the remoter parts of *England*, a *Break*;) is a small Inclosure, or oblong Quadrangle, plac'd before a Farrier's Shop, and consisting of four Pillars or Posts, kept together by cross Poles; the Inclosure being design'd for holding and keeping in a Horse that's apt to be unruly or disorderly, in the time of Shoeing, or of any Operation.

TREAD. See *Piste*.

TREPIGNER; A *French* Word, importing the Action of a Horse, who beats the dust with his Fore-feet in Managing, without imbracing the Volt; and who makes his Motions and Times short, and near the ground, without being put upon his Haunches. This is generally the Fault of such Horses as have not their Shoulders supple, and at liberty, and withal have scarce any Motion with them. A Horse may *Trepigner*, in going upon a straight Line.

TRIDE; A Word us'd in the Academies, signifying Short and Swift. A *Tride* Pace, is a Going of Short and Quick Motions, tho' united and easy. This Horse has a *Tride*-Career; that is, he gallops very fast, and has his Times or Motions short and nimble. A Horse is said to work *Tride* upon Volts, when the Times he marks with his Haunches are short and ready. Some apply the Word only to the Motion of the Haunches.

TRIP, or *Stumble*; A Horſe is ſaid to Trip, when he makes a falſe Step. Sometimes when a Horſe ſtumbles, the *French* ſay, *La jambe mollit*, *His foot ſlackens*. Since your Horſe is weak in the Reins and the Ridge of the Back, and his Feet are worn, he is apt to ſtumble.

TROT, is the Pace or Going of a Horſe; in which the Motion is, Two Legs in the Air, and Two upon the Ground croſs-wiſe, or in the form of a St. *Andrew's* Croſs; continuing alternately to raiſe at once the Hind-leg of one ſide, and Fore-leg of the other; leaving the other Hind and Fore-leg upon the Ground, till the former come down. A Horſe puts himſelf to a Trot, when, upon a Walk, he makes haſte, or quickens his Pace; and if he be aſſiſted by the Switch, and the Heels, he takes it yet better. This Horſe is ſure and firm, both at Step, Trot, and Gallop. This Horſe Trots freely; and in Trotting, he toſſes up his Leg; *i. e.* he bends his Fore-Thighs, and has a good Motion with him. Such a Horſe rakes as he Trots, becauſe he is ſhoulder-ſplait.

TROUSSEQUEUE. See *Saker*, or *Dock*.

TROUSSEQUIN, is a piece of Wood, cut Arch-wiſe, rais'd above the hinder Bow of a great Saddle, which ſerves to keep the Bolſters firm. There are ſome *Dutch* Saddles, call'd *Selles Razes*, which have a low *Trouſſequin*.

TROUT-COLOUR'D Horſe; is a White, ſpeckled with Spots of Black, Bay, or Sorrel; particularly about the Head, and Neck.

TRUSS'D. A Horſe is ſaid to be well Truſs'd, (in *French*, being *Gigoté*;) when his Thighs are large, and proportion'd to the Roundneſs of the Croupe. A Horſe is ſaid to be ill Truſs'd, when his Thighs are thin, and bear no proportion to the Breadth of the Croupe.

TURN, is a Word commonly us'd by the Riding-Maſters, when they direct their Scholars to change Hands. See *Change*, and *Entier*.

Turn your Thighs, *Turn* your Legs, *Turn* your Heels. You can never have fine nice Aids, or feel exactly the Motions of your Horſe; unleſs you turn your Thighs, ſo that the inſide of your Knee may touch the Saddle.

TUSHES; are the Four Teeth of a Horſe, ſeated

ſeated beyond the Corner-Teeth, upon the Barrs; where they ſhoot forth on each ſide of the Jaws, two above, and two below, about the Age of Three, and Three and a half, and ſometimes Four: And no Milk or Foal-Tooth ever comes forth in the place where they grow. See *Teeth*.

TWIST; the Inſide, or flat Part of a Man's Thigh; upon which a true Horſeman reſts upon Horſe-back.

TWISTED, (in *French*, *Biſtourné*) is a Horſe reduc'd to the ſame State of Impotency with a Gelding, by the violent Wringing or Twiſting of his Teſticles twice about; which dries them up, and deprives them of Nouriſhment.

V.

VALET, is a Stick, arm'd at one end with a blunted Point of Iron, to prick and aid a Leaping Horſe. Formerly, a Valet was call'd *Aiguillon*, (*i. e* Goad) and ſome of 'em had Spur-Rowels upon 'em, only the Points beaten down: And when a Horſe was firſt begun round a Pillar, without a Rider, they us'd to prick his Flanks with the Valet, to make him know the Spur, and obey it without reſiſting. At preſent the Valet is not employed for that Uſe in Manage-Schools; and the Name of *Goad* is ſuppreſs'd, as being only proper for Oxen.

VAULT a Shoe, is to forge it hollow, for Horſes that have high and round Soles; to the end that the Shoe, thus hollow, may not bear upon the Sole that is then higher than the Hoof. But, after all, this ſort of Shoes ſpoils the Feet; for the Sole being tenderer than the Shoe, aſſumes the Form of the Shoe, and becomes every day rounder and rounder. In Mr. *Solleyſel's* Compleat Horſeman, you may ſee the true Method of Shoeing for high and round Soles.

VEIN; to tye and ſtrike a Horſe's Vein. See *Barr*.

VESSIGON, A Wind-Gall, or ſoft Swelling on the in and outſide of a Horſe's Hough; that is, both on the right and the left of it.

VIVES, are Glands near the Throat, that ſometimes ſwell, and by their preſſure upon the Throat, do ſo cramp Reſpiration, that the Horſe

Horſe is in danger of being ſtifled.

UNCERTAIN. We call a Horſe *uncertain*, that is naturally reſtleſs and turbulent, and is not confirm'd in the Manage he's put to; ſo that he works with trouble and uncertainty. See to confirm your Horſe in his *Terra à terra*, for he's very *uncertain*.

UNITE: A Horſe is ſaid to unite, or walk in union; when in galloping, the Hind-quarters follow and keep time with the Fore. This Horſe is united, for his Hind and Fore-quarters make but one Action, without changing the Foot, or galloping falſe.

VOLTE: This Word ſignifies a Round, or a Circular Tread; and in general, when we ſay in the Academies, To make *Voltes*, to manage upon *Voltes*, we underſtand a Gate of two Treads, made by a Horſe going ſideways round a Center, in ſuch a manner, that theſe two Treads make parallel Tracts, one larger made by the Fore-feet, and another ſmaller made by the Hinder-feet, the *Croupe* approaching towards the Center, and the Shoulders bearing outwards. Sometimes the *Volte* is of one Tread; as when a Horſe makes *Volts* in *Corvets*, and in *Caprioles*, ſo that the *Haunches* follow the Shoulders, and move forward on the ſame Tread. In general, the Way or Tract of the *Volte* is made ſometimes round, ſometimes oval, and ſometimes a Square of Four ſtraight Lines; ſo that theſe Treads, whether round or ſquare, incloſe a *Terrain*, or *Manage-ground*, the middle of which is ſometimes diſtinguiſh'd by a *Pillar*, or elſe by an *Imaginary Center*, which is there ſuppoſed in order to regulate the diſtances and the juſtneſs of the *Volte*.

Renvers'd Volt is a Tract of two Treads, made by the Horſe, with his Head to the Center, and his *Croupe* out; ſo that he goes ſideways upon a Walk, Trot or Gallop, and traces out a ſmall Circumference with his Shoulders, and a larger one with his *Croupe*. This different Situation of the Shoulders and the Croupe, with reſpect to the Center, gave this *Volte* the Name of *Renverſeé*, as being oppoſite in ſituation to the former. *Renvers'd Voltes* upon a Walk, appeaſe and quiet unruly Horſes, if they are made methodically.

The Six Volts are made *Terra à Terra*, two to the Right, two to the Left, and two to the Right again; all with one Breath, obſerving the Ground with the ſame Cadence, work-

ing

ing (*tride*) ſhort and quick, and ready, the Fore-hand in the Air, the Breech upon the Ground, the Head and Tail firm and ſteady. To do the *Six Volts*, you ſhould have an excellent Horſe that's knowing and obedient, and has ſtrength to anſwer 'em.

To make a Horſe work upon the Four Corners of the Volte, is to manage him with that juſtneſs, that from Quarter to Quarter, or at each of the Corners or Angles of the *Volte*, he makes a narrow *Volt* that does not take above the quarter of the great *Volt*, the Head and Tail firm; and thus purſues all the Quarters, with the ſame Cadence, without loſing one Time or Motion, and with one Repriſe, or with one Breath.

In ſpeaking of *Volts*, we ſay, to put a Horſe upon the *Volts*; to make him work upon the *Volts*; to make good *Volts*, to embrace the whole *Volt* well; that is, to manage ſo, that the Horſe working upon *Volts*, takes in all the Ground, and the Shoulders go before the Haunches. To paſſage upon *Volts*, or ride a Horſe Head and Haunches in, is to ride him upon two Treads, upon a Walk, or a Trot.

Demi-Volt, is a Demi-round of one Tread or two, made by the Horſe at one of the Corners or Angles of the *Volt*, or elſe at the end of the Line of the Paſſade; ſo, that being near the end of this Line, or elſe one of the Corners of the *Volt*, he changes hands, to return by a Semicircle, to regain the ſame Line. When he does not return upon this Line, we ſay, he has not clos'd (*ſerrè*) his *Demi-Volte.*

Demi-Volts of the length of a Horſe, are Semicircles of two Treads, which a Horſe traces in working ſideways, the Haunches low, and the Head high, turning very narrow; ſo that having form'd a Demi-round, he changes the Hand to make another, which is again follow'd by another change of Hand, and another Demi-round which croſſes the firſt. This *Demi-Volt* of a Horſe's length, is a very pretty Manage, but very difficult; we may compare it to a Figure of 8.

Demi-Volte of five times, or *Paſſade* of five Times. See *Paſſade.*

W.

WALK, is the ſloweſt, and leaſt rais'd, of a Horſe's Goings. The Duke of *Newcaſtle* made this Motion to be two Legs diametrically oppoſite in the Air, and two upon the Ground at the ſame time, in the Form of a St. *Andrew*'s Croſs: Which, in effect,

effect, is the Motion of a Trot. But later Authors agree, that so great a Master was mistaken in this point: For, in a Walk, (as any one may observe) a Horse lifts two Legs of a side, one after the other, beginning with a Hind-leg first: As, if he lead with the Legs of his right side, then the first Foot he lifteth is his far-hind Foot; and in the time he is setting it down, (which in a *Step* is always short of the Tread of his Fore-foot upon the same side) he lifteth his far Forefoot, and setteth it down before his near Fore-foot. Again, just as he is setting down his far Fore-foot, he lifts his near Hind-foot, and sets it down again just short of his near Fore-foot, and just as he is setting it down, he lifts his near Fore-foot, and sets it down before his far Fore-foot. And this is the true Motion of a Horse's Legs upon a Walk. This Horse walks well; he raises his Legs. Such a Horse observes the Walk, or Step: He finishes the Demivolt in a Walk. Begin this Lesson in a Walk, and end it with a Walk. When you teach your Horse to turn to the right and left, or from one hand to another, make him walk at first, then teach him upon the Trot, and then upon the Gallop.

WARRANT a Horse. A Jockey that sells a Horse, is by an inviolable Custom obliged to Warrant him, that is, to refund the Money that was given for him, and re-deliver the Horse in Nine days after the first Delivery, in case he sold him when under such Infirmities as may 'scape the View of the Buyer, and as are not obviously discovered, namely, Pursiveness, Glanders, and Unsoundness, Hot and Cold. But he does not warrant him clear of such Infirmities as may be seen and discern'd. And not only a Horse-Merchant or Jockey, but Persons of what Quality soever, stand obliged by the Law of Nature, and will be constrain'd to take back the Horse, if he's affected with the Disorders first mention'd, and to repay the Money.

WART, is an Excrescence, or Superfluity of Spungy Flesh, that rises in the hinder Pasterns of Coach-Horses, almost as big as a Walnut. A Wart, (in *French*, *Poirau*) suppurates and voids red stinking Matter, and do's not cure but for a time, for it returns again.

WATERY *Sores*, (in *French*, *Mauvaises Eaux*) are a Suppuration of stinking and

and malignant Humors, which issue from the Pasterns and Fetlock-Joynts of a Horse; and that from the hinder, rather than the Fore-legs. See the proper Remedies in Mr. *Solleysel*'s Compleat Horseman.

WAY'D *Horse*, is one that's already back'd, suppled, and broken, and shews a Disposition to the Manage. Your Horse knows the Bridle, and answers the Spurs; and generally when he's put to it, he presents in such a manner, as shews he's thoroughly Way'd, (*acheminé*.)

WEAK, Easy Branch. See *Banquet*, and *Banquet-Line*.

WHITE-*Face*, or Blaze, (in *French*, *Chanfrain blane*) is a White Mark upon a Horse, descending from the Forehead almost to the Nose. See *Chanfrin*.

WHITE-*Foot*, (in *French*, *Balzane*) is a White Mark that happens in the Feet of a great many Horses, both before and behind, from the Fetlock to the Coffin. The Horses thus mark'd, are either trammell'd, Cross-trammel'd, or white of all Four. Some Horsemen place an unlucky Fatality in the white of the far Foot behind. See *Chaussé trop haut*, and *Trammel'd*.

WIND: A Horse that carries in the Wind, is one that tosses his Nose as high as his Ears, and does not carry handsomely. The difference between carrying in the Wind, and beating upon the Hand, is, that a Horse who beats upon the Hand, shakes his Head, and resists the Bridle; but he who carries in the Wind, puts up his Head without shaking, and sometimes beats upon the Hand. The opposite to carrying in the Wind, is arming and carrying low: And even between these two there's a difference. Try if a *Martingal* will bring in your Horse that carries in the Wind.

Wind. See *Breath*.

Wind-gall is a soft Swelling, occasion'd by over-working, just by the Horse's Fetlock, about as big as half a Pidgeon's Egg, and at first full of Water. A Wind-gall upon the Sinew, that grows hard, makes a Horse halt, and in the end makes him down-right lame. Your long-jointed Horses are apt to be Wind-gall'd, though they work never so little. The Wind-galls that we call *Sinewy*, happen commonly in the Hinder-legs, and nothing but Fire can cure them; nay sometimes Fire it self will not do. See also *Vessigon*.

WITHERS is the Juncture of the Shoulder-bones,

bones, at the bottom of the Neck and Mane, towards the upper part of the Shoulders. Your Horſe is hurt or wrung in the Withers. Some call this in *French*, *Encrainè*, but improperly.

Withers of the Bow of a Saddle. See *Bows*.

Wither-band, is a Band or piece of Iron laid underneath a Saddle, about four Fingers above the Withers of the Horſe, to keep tight the two pieces of Wood that form the Bow.

Wither-wrung: A Horſe is ſaid to be wither-wrung, when he has got a Hurt in the Withers; and that ſort of Hurts are very hard to cure.

WOLF's *Tooth*, is the exceſſive height of ſome of the Grinders, which ſhoot out Points as they increaſe in length, and not only prick the Tongue, but hurt the Lips in feeding.

To WORK a Horſe, is to exerciſe him at Pace, Trot, or Gallop, and ride him at the Manage. This Gentleman that has not been in the Academy three Months, works well; he'll prove a good Horſeman. Never work a Horſe but with judgment. To work in a Square. To work in a Circle. To work upon Volts. To work at the Air of *Corvets*. We do not work this day; that is, we have no Manage this day.

To work a Horſe upon *Volts*, or *Head* and *Haunches* in, or between two Heels, is to paſſage him, or make him go ſideways upon two parallel Lines.

WRIST; (in *French*, *Poing*); the Bridle-*Wriſt* is the *Wriſt* of the *Cavalier*'s Left-hand. A Horſe-man's *Wriſt* and his *Elbow* ſhould be equally rais'd, and the *Wriſt* ſhould be two or three Fingers above the Pommel of the Saddle. This Horſe follows the Bridle-hand, he does not refuſe it. To ride a Horſe from hand to hand; *i. e.* to change hands upon one Tread, you need only to turn your *Wriſt* to what ſide you would have the Horſe to turn to, without advancing your hand. If your Horſe ſtops, you muſt make uſe of both Legs. See *Hand*.

Y.

TO YERK or ſtrike with the Hind-legs (in *French*, *Nouer l' Eguillette*) is ſaid of a leaping Horſe when he flings and kicks with his whole Hind-quarters, ſtretching out the two Hinder-legs near together and even, to their full extent. Your Horſe is only

only fit for *Croupades* and *Ballotades*; for since he can't *nouer l' Eguillette*, he'll ne'r be fit for *Caprioles*. A single Yerking is call'd in *French*, *Ruade*.

YIELD, or slack the Hand, is to slack the Bridle, and give the Horse head.

Z.

ZAIN: A Horse is call'd in *French*, *Zain*, that is of a Dark Colour, neither Gray nor White, and without any White Spot or Mark upon him.

FINIS.

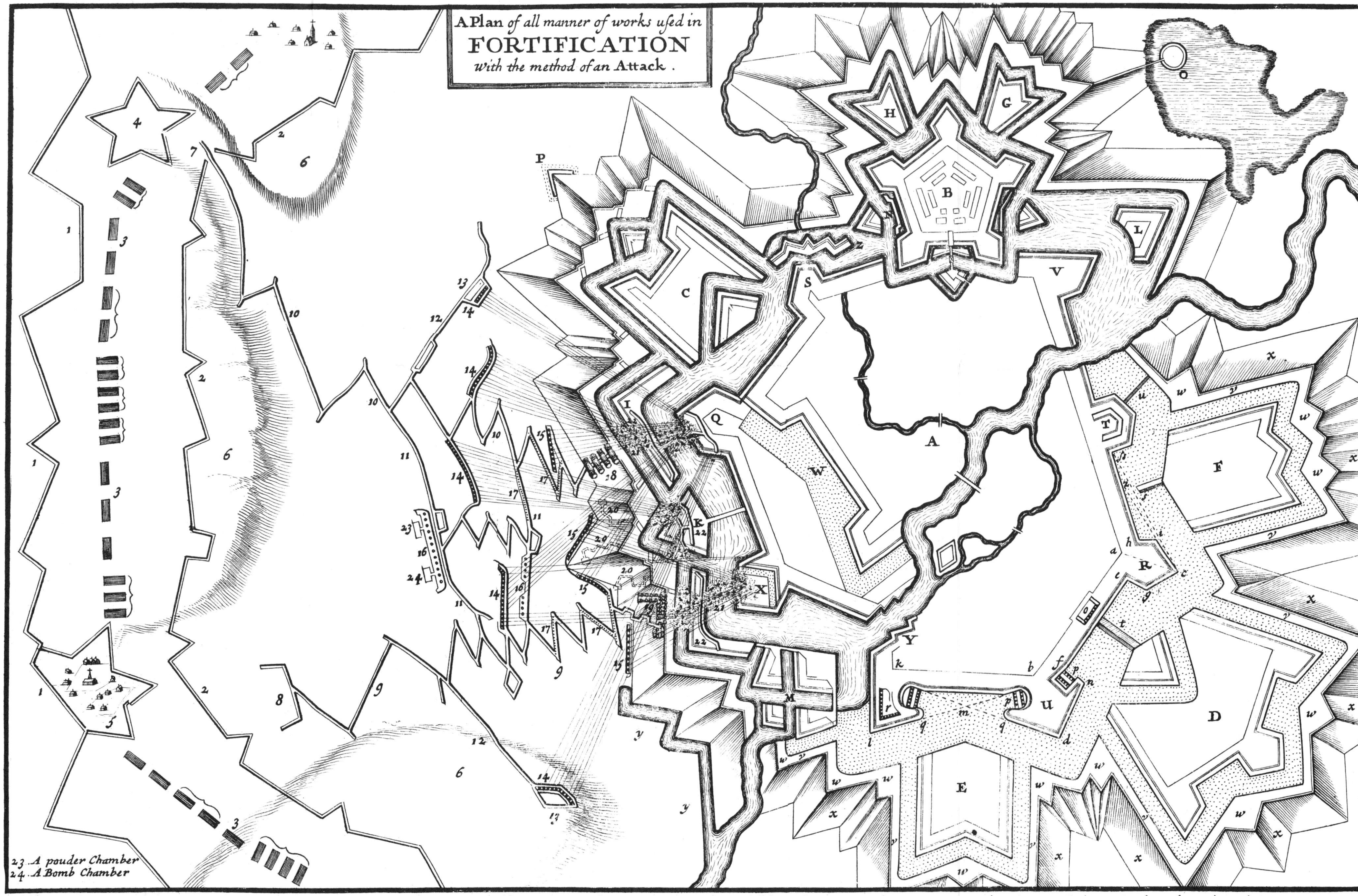

Explanation

A. An Irregular Fortification
B. A Regular Pentagon
C. A Horn Work
D. A Crown Work
E. A Single Tenaille
F. A Double Tenaille
G. A Priests Bonnet
H. A Swallows Tail
I. A Countergard
K. A Raveline
L. A Half Moon
M. Two Lunettes
N. Tenailles in yͤ Foss
O. A Horse Shoe
P. A Bonnet
Q. A Regular Bastion
R. An Irregular Bastion
S. A Deform'd Bastion
T. A Flatt Bastion
V. A Demibastion
W. A Retrenchment
Q. An Empty Bastion
T. A Double Bastion
U. A Full Bastion
X. A Retirade
Y. Redans
Z. An Envelope or Sillon
a.b. The Interior Poligone
c.d. The Exterior Poligone
e.f. A Curtine
e.g. A Flank
c.g. A Face
e.h. The Gorge
f.e.g. Angle of yͤ Curtin
e.g.c. Angle of yͤ Shoulder
g.c.i. The Flanked Angle
a.b.k. Angle of yͤ Poligon
l.m.d. Angle of yͤ Tenaille
c.z. Line of defen. Razant
c.ß. Line of defen. Fichant
ß.z. The second Flank
a.c. The Capitall
p. Caze-mates
q. Orillons
n. A Shoulder
r. A Cavalier
o. A Platform
t. A Coffre
s. A Caponiere
u. A Traverse
v. The Covert way
w. Places of Arms
x. The Glacis
y. The Advance Foss
1. Lines of Circumvallatio
2. Lines of Contrevallation
3. Part of the Camp
4. Park of Artillery
5. The Head Quarters
6. Rising Grounds
7. Opening of yͤ Trenches
8. Epaulement
9. Attack on the Right
10. Attack on the Left
11. Parralels, Lines of Communication or Boyau's
12. Likewise Boyau's
13. Places of Arms
14. Batteryes for dismounting the Enemy's Guns
15. Breach Batteryes
16. Mortar Batteryes
17. Coehorns Mortars
18. A Sapp
19. A Lodgement
20. Mines
21. Traverses to pass the Foss
22. Lodgments on yͤ works

THE
Gentleman's Dictionary.

PART II.

Of Military Affairs.

ADJUTANT, or *Aid-Major*, is an Officer appointed as an Assistant to the *Major*, to ease him in his Duty: he receives the Orders every Night from the *Brigade-Major*, which, after he has carried to his *Colonel*, he delivers out to the *Serjeants* assembled in a Ring. If there be Detachments for Convoys, Guards or Parties to be made, he gives the Number each Company must furnish, with the Hour and Place of Rendezvous: he places the Guards, and makes Detachments for what other Guards the Regiment is to furnish. He receives and distributes Ammunition to the Companies, keeps a Roll of the Officers to do them Justice in their Duty; and, by the *Major's* Order, regulates the Price of Bread, Beer, Meat, *&c.* to prevent the Mens being imposed on by the *Sutlers.* He must be vigilant and active. Each Regiment of Horse has an *Adjutant*, and a Regiment of Foot has one for each Battalion.

ADVANCE-GUARD, see *Guard*.

Advance Fosse, is a Moat or Ditch of Water round the Glacis or Esplanade of a place to prevent a Surprize; they are not much approved of late; for, being drain'd, they serve as a Trench ready made to the Besiegers. The Works raised by M. *Coehorne*, in the Year 1697, to cover *Brussels* from a second Bombardment, had a dry Moat sunk at the Foot of the Glacis, of about 7 or 8 yards over.

AID-MAJOR, see *Adjutant.*

Aid de Camp is an Officer employ'd under a *General* to carry his Orders; it is a very honourable Post for a Young Gentleman; he is obliged to a very constant Attendance, to be very well mounted, to be diligent and active. A *Lieutenant-General* is allowed two *Aids de Camp*, and a *Major-General* one.

ALARM is a sudden Calling to Arms, upon an Apprehension of Danger from the Enemy; a false *Alarm* is sometimes occasion'd by a fearful or negligent *Centinel*, and sometimes designedly by a diligent Officer, to try the Readiness of the Guards. Both in the late War, and this, the Picquet-Guard hath often been called out hastily, to see in what Readiness they were.

Alarm-Post is the Ground appointed each Regiment by the *Quarter-Master-General* for them to march to, in case of an *Alarm*; this is never done but upon an Apprehension of being attacked by the Enemy: *Alarm-Post* in a Garrison, is the Place alotted each Regiment to draw up in, upon all Occasions.

AMBLIGON, see *Triangle.*

AMBUSCADE or *Ambush*, is a Lurking Party in a Wood or other convenient Place to surprize an Enemy, either in their Convoys, Detachments, or Foragers. To discover an *Ambush*, to defeat an *Ambush*, to fall into an *Ambush*, are plain.

AMMUNITION implies all sorts of Warlike Stores.

Ammunition-Bread is carried along with an Army for the Subsistance of the Troops; each *Soldier* receiving a Loaf of six pound weight every four days.

ANGLE, as it is a Geometrical Term, is the meeting of two Lines, so that if prolong'd, they would cut one another. 'Tis either *Right-Lined*, when composed of two streight Lines; *Spherical*, when made of two crooked Lines; or *Mixt*, when composed of a streight and crooked Line: Likewise *Acute*, if under ninety Degrees; *Obtuse* if above ninety; and a *Right Angle*, if exactly ninety Degrees.

Angle, as a Term in Fortification, is explain'd in the following Definitions. (1.) *Angle of the Centre*, is that made by two Lines drawn from the Centre to the Extreams of any side of the Polygon. (2.) *Angle of the Polygon*, is the Angle made by the meeting of two sides of the Polygon, and is the same with the *Angle of the Gorge*. (3.) *Angle of the Curtin* or of the *Flank*, is the Angle form'd by the meeting of a Flank and a Curtin. (4.) *Angle of the Shoulder*, is form'd by one Face and one Flank. (5.) *Flanked Angle*, is the meeting of two Faces. (6.) *Angle of the Tenail* or *Flanking Angle*, is composed of the Line of Defence and the Curtin. (7.) *Angle forming the Flank*, is an Angle composed of one Flank and

and one Demigorge. (8.) *Angle forming the Face*, is the inward Angle composed of one Flank and one Face. (9.) *Angle of the Moat*, is that which is form'd before the Centre of the Curtin, by the Exterior Line of the Fossè or Moat. (10.) *Angle Salliant* or *Sallying Angle*, is what advances with its Point toward the Country; such is the Angle of the Counterscarp before the Point of a Bastion. (11.) *Angle Rentrant* or *Reentring Angle* is what points inwards to the Body of the Place; such is the Angle of the Counterscarp before the Centre of the Curtin.

ANSPESADE, see *Lanspesade*.

ANTESTATURE is a Traverse or Retrenchment hastily made of Gabions or Palisades, to stop an Enemy that is gaining Ground. This is call'd to Dispute Ground, or lose it Inch by Inch.

APRON is the sheet of Lead which covers the vent of a Gun.

APPOINTE was a *French* Term for a *Soldier*, who, for long Service or Bravery, was distinguish'd from the other *Centinels*, by receiving greater Pay, and being nearer Preferment: these are now suppressed: We never had any such in the *English* Army.

APPROACHES are the Trenches, places of Arms, Lodgments, Sap, Gallery; and generally all Works whereby the Besiegers advance towards a place besieged; see *Attack* and *Trench*.

Approach, Lines of Approach, see *Trench*.

ARRAIGNE is a *French* Term in Mining, see *Gallery* and *Mine*.

AREA is a Geometrical Term, signifying the superficial Content of any Figure.

ARMS are all sorts of Weapons used for Offence or Defence, such as Sword, Bayonet, Pike, Dagger, &c. Or Fire-Arms, as Firelock, Carabin, and Pistol.

Arms; *place of Arms* in a Garrison, is the place alotted for the Garrison to draw up in, upon any sudden Alarm, or other Occasion. *Places of Arms* at a Siege are small Redouts border'd with a Parapet, containing a small Body of Men, to make good the Trenches against the Sallies of the Besieged.

ARMY is a Body of Troops consisting of Horse, Foot and Dragoons, with Artillery and Provisions. It is divided into Brigades. An Army ought to be under the Command of one *General*, having *Lieutenant Generals*, *Major-Generals*, *Brigadeers*, and other Officers under him.

Flying-Army is a small Body under a *Lieutenant* or *Major-General*, sent out to harrass the Country, intercept Convoys, prevent the Enemies Incursions, cover its own Garrisons, and keep the Enemy in continual Alarm.

Wings of an *Army*, are the Troops encamped on the Flanks; they are generally Horse and Dragoons,

goons, and are called the Right-Wing and the Left-Wing.

ARQUEBUSS, called generally *Arquebuss a Croc*, is a Fire-arm used now only in Old Castles, and by the *French* in some of their Garrisons; it carries a Ball of about three Ounces and a half.

ARSENAL, see *Magazine*.

ARTILLERY is a Magazine of all sorts of Arms and Provisions for an Army, such as Cannon, Mortars, Bombs, Balls, Petards, Grenades, Small Balls, Powder, Match, all sorts of Hand-Tools, Planks, Boards, Rope, Coal, Tallow, Pitch, Rosin, Sulphur, Salt-petre, Quickmatch, all sorts of Fire works, Pontons, and many other things, too many to be named. The Attendants are *Conductors*, *Bombardeers*, *Gunners*, *Matrosses*, *Pioneers*, *Ponton-Men*, *Carpenters*, *Wheelwrights*, *Smiths*, *Coopers*, *Tin-Men*, and *Collar-makers*.

Artillery-Park is a place appointed by the *Quarter-Master-General* in the Rear of both Lines of the Army for encamping the Artillery, which is drawn up in Lines; the Guns are in one Line; the Ammunition Waggons make two or three Lines, 60 Paces behind the Guns, and 30 distant from each other; the Pontons and Tumbrils make the last Line: And all is surrounded with a Rope which forms the *Park*. The *Gunners* and *Matrosses* encamp on the Flanks, and the *Conductors*, *Bombardeers*, *Ponton-Men* and *Artificers* in the Rear.

ASSAULT or *Storm*, is the Effort made to carry a Post, and is generally made by the Regiments that have the Guard of the Trenches sustain'd by Detachments from the Army; while the Assault continues, the Batteries cease for fear of killing their own Men.

ASSEMBLY is the second beating of the Drum before a March; the first beating is called the *General*, at which the *Soldiers* get themselves ready to march; the second is called the *Assembly*, at which they strike their Tents, and roll them up, and stand to their Arms; and the third is called the *March*, and then the Army begins to move.

ATTACK; to Attack a Fort, a Post, or a Body of Troops, is the same as to assault or endeavour to carry a thing by Force; it is a much properer word than *Assault*, and more in use.

Attack; the Attacks in a Siege are the Works whereby the Besiegers come under Cover to the place besieged. There are commonly two Attacks, each mounted by a General Officer, and these have Communication one with another, by Lines or Trenches, running parallel to the Polygon of the place, that they may not be enfiladed, and are call'd the Parallels, the Boyau, or the Lines of Communication.

False Attacks are never carried on with that briskness as the other; the Design of them being to favour the real Attacks by amusing

ſing the Enemy, and obliging the Garriſon to a greater Duty. They have ſometimes been ſucceſsful, as at the Siege of *Maſtricht* by the *French*, where the D. of *Orleans* who commanded a falſe Attack againſt *Wick*, did by his Preſence and Behaviour, ſo animate his Men, that after having forced the Paliſades, he made himſelf Maſter of the Counterſcarp, and of a Half Moon.

AXES are very uſeful in an Army for cutting ways thro' Woods, for cutting Trees to make Bridges, for mending the Ways, and many other uſes. Moſt of the Foreign Regiments of Foot, eſpecially the *Danes*, have ſix Carpenters with *Felling-Axes* to march at the Head of each Battalion: and in the Artillery the *Carpenters* march with their *Axes* to mend any thing that may break down, and to help the Ways. For *ſmall Axes*, ſee *Hatchets*.

Pike-Ax is a Tool carried by the *Pioneers*, to dig up Ground that is too hard for the Spade; they are of great uſe for mending the Ways, and in Fortification.

B

BACULE is a Gate made like a Pit-fall, with a Counterpoize before the *Corps de Gardes* advanc'd near the Gates, which is ſupported by two great Stakes.

BAGGAGE-WAGGONS are thoſe in which the Officers and Regiments Baggage is carried; before a March they are appointed a Rendezvous, and are there marſhall'd by the *Waggon Maſter-General*, according to the Rank the Regiments have in the Army. On a March they are ſometimes ordered to follow their reſpective Columns of the Army, ſometimes to follow the March of the Artillery, and ſometimes to make a Column by themſelves. The *General's* Baggage is generally firſt. If the Army march from the Right, the Baggage of that Wing has the Van; if from the Left, the Baggage of the Left has the Van. Each Waggon has a diſtinguiſhing Flag to ſhow to what Regiment it belongs.

BAGONET is a ſhort broad Dagger, formerly with a round Handle, fitted for the bore of a Firelock, to be fixed there after the *Soldier* had fired; but they are now made with Iron Handles and Rings that go over the Muzzle of the Firelock, and are ſcrewed faſt; ſo that the *Soldier* fires with his Bagonet on the Muzzle of his Piece, and is ready to act againſt Horſe. Since this new ſort of Bagonets, Pikes are ſo much out of uſe, that all the *Engliſh* Regiments abroad have laid them aſide.

BAGS, ſee *Sand-bags*.

BALL, *Bullet*, or *Shot*, is of Iron or Lead, to be fired out of Cannon, Muſquet, or Piſtol Cannon-balls are of different Diameters and Weight, according to the nature of the Piece. An *Engliſh* Muſquet carries a Ball of ſix-

 teen

teen in a Pound, a Carabine of twenty four, and a Piſtol of thirty two in a Pound; I mean Ammunition-Carbines and Piſtols.

Ball, Red-hot ball, are ſuch as are made hot in a Forge ſtanding near the Gun. The Gun being loaded with Powder and wadded with a green Turf, is ſpung'd with a wet Spunge, and laid at a ſmall Elevation; that the Ball which is taken out of the Forge with a long Iron Ladle may ſlide down, the *Gunner* at the ſame time being ready to Fire: it not only fires combuſtible Matter, but Floors and Planks.

Ball, Fire balls are made of a Compoſition of Meal-Powder, Sulphur, Saltpetre, Pitch, *&c.* for Firing Houſes that incommode Trenches or advance Poſts; and are thrown by *Soldiers.*

BANDELEERS are ſmall Caſes of Wood covered with Leather, holding the Charge of Powder for a Firelock; each Muſqueteer carries twelve hanging at a Shoulder Belt, call'd a *Collar of Bandeleers.*

BAND of *Pentioners,* are a Company of 120 Gentlemen, who receive a yearly Allowance of a 100 *l.* for attending her Majeſty on ſolemn Occaſions: They are now commanded by his Grace the Duke of St. *Albans.*

Bands are likewiſe Hoops of Iron uſed about the Carriage of a Gun, ſuch as the Nave-bands, which are Hoops of Iron binding the Nave at both ends. Bands are the Bands that bind the ends of the Axiltree.

BANQUETT is a Term in Fortification, being a Foot Bank of Earth about a foot and a half high, and three foot broad, raiſed on the Rampart at the Foot of the Parapet, for the *Soldiers* to mount on to fire over.

BARACK or *Hutt,* ſee *Cazern.*

BARBE, *to fire in Barbe,* ſignifies firing over the Parapet, inſtead of uſing the Embraſures; the Parapet muſt not be above three foot and a half high.

BARM, *Berm,* or *Foreland,* ſee *Lizier.*

BARRELS are for ſeveral uſes in the Artillery, as for Powder, Small Shot, Flint, Sulphur, Salt-petre, Roſin, Pitch, Quickmatch, and many other things, ſee *Casks.*

Barrels, Thundering Barrels are filled with Bombs, Grenades, and other Fireworks, to be rolled down a Breach. At the Siege of *Namure* in the Year 1695, ſeveral *Thundering Barrels* were ſent from *Coehorne*'s Work down hill upon our Trenches, filled with Hand Grenades; but they had no great Effect, for the *Soldiers* made way for them.

BARRICADE, a Term ſometimes uſed for a Fence of Paliſades.

BARRIERE is a Gate made of wooden Bars about five foot long perpendicular to the Horizon *A,* which are kept together by 2 long Bars going acroſs, and another croſſing Di-

agonally

agonally. They are uſed to ſtop the Cut that is made through

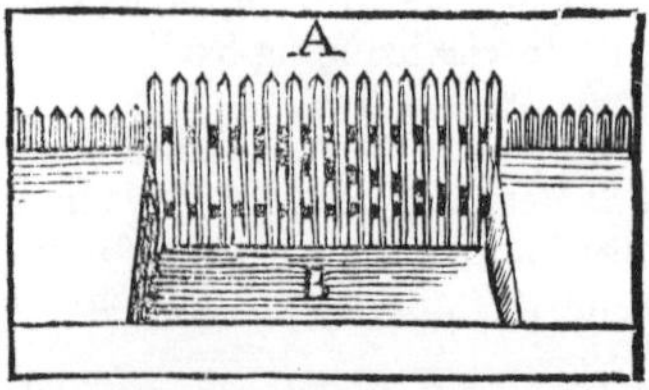

the Eſplanade before the Gate of a Town *B*.

BASE or *Baſis*, is the Foundation of a Work; *Baſis* of a Rampart, is where it joins the Ground on which it ſtands; *Baſis* of a Parapet, is where it joins the top of the Rampart.

Baſe, ſee *Cannon*.

Baſe of a Gun, is the ſame with the *Breech* of a Gun, and is that ſolid piece of Metal behind the Chace towards the Caſcabel: the great Ring behind the Touch-hole or Vent, is called the *Baſe Ring*, and the Mouldings behind that, are call'd the *Baſe* or *Breech Mouldings*.

BASKETS, ſmall *Baskets*

A, are uſed in Sieges on the Parapet of the Trench *B*, being filled with Earth. They are about a foot and half high, about a foot and a half diameter at top, and 8 or 10 inches at bottom; ſo that being ſet together, there is a ſort of Embraſures, *C*, left at their bottoms, through which the *Soldiers* fire without expoſing themſelves.

BASTION is a Maſs of Earth raiſed on the Angles of the Polygon, compoſed of two Flanks and two Faces, ſometimes fac'd with Brick or Stone. Their diſtance from one another is about 150 *Engliſh* yards: They are of ſeveral ſorts, as *Regular*, when their Faces are of an equal length, their Flanks the ſame, and the two Angles of the Shoulder equal; or *Irregular*, where that equality is not; or *Deform'd*, where the Irregularity of the Lines and Angles makes the Baſtion out of ſhape. They are *hollow*, when only ſurrounded with a Rampart and Parapet, leaving the ſpace within void and empty; or *ſolid*, when the ſpace within them is raiſed of an equal height with the Rampart of the Place.

Baſtion-Flat is when the ſide of the Polygon being double the ordinary length, a Baſtion is raiſed before the middle of the Curtin, which, becauſe its Capital is ſhort, has the flanked Angle very obtuſe, which makes the Gorge large, and the Baſtion very flat.

Baſtion-detach'd, is that which is ſeparate or cut off from the Body of the Place, and differs from a Half Moon, whoſe Rampart and Parapet are lower and not ſo

thick as thoſe of the Place, becauſe it has the ſame Proportion with the Works of the Place.

Baſtion, Double-Baſtion is a Baſtion raiſed on the Plain of another Baſtion, and is ſometimes in the nature of a *Cavalier*, ſee

Demi-Baſtion is compoſed of only one Face, one Flank, and one Demigorge.

BATTALION is a Body of Foot, generally 700, not including Officers nor Serjeants, armed with Firelocks (Pikes being quite laid aſide) Swords and Bagonets, divided into thirteen Companies, one of which is Grenadiers. The firſt Regiment of *Engliſh* Guards has four Battalions; the ſecond Regiment, the Regiment of *Scots* Guards, and the Royal *Scots*, have each two Battalions; the reſt of her Majeſty's Regiments are but one Battalion each.

BATTERY is a Parapet thrown up to cover the *Gunners* and Men employ'd about the Guns, from the Enemies Shot. This Parapet is cut into *Embraſures* for the Cannon to fire through; the height of the *Embraſures* on the inſide is about three foot, but they go ſloping lower to the outſide; their wideneſs is two or three foot, but open to ſix or ſeven on the outſide. The Maſs of Earth that is betwixt two Embraſures, is call'd the *Merlon*. The *Platform* of a Battery is a Floor of Planks and Sleepers to keep the Wheels of the Guns from ſinking into the Earth, and is always made ſloping towards the Embraſure, both to hinder the Reverſe, and to facilitate the bringing back of the Gun; ſee *Platform*.

Battery of *Mortars* differs from a Battery of Guns, for it is ſunk into the Ground and has no Embraſures; the *Dutch* call this a *Kettle*.

Battery, Croſs Batteries are ſuch whoſe Shot meet at the ſame place and form an Angle. The Advantage of ſuch Batteries is, that the one beating down what the other ſhakes, they do good Execution. *Battery d' Enfilade*, is what batters in Flank. *Battery en Echarpe*, is what batters obliquely. *Battery de Reverſe*, is what plays upon the Enemies back. *Comrade Batteries*, are thoſe which play upon the ſame place. *To raiſe a Battery*, is the Buſineſs of an Engineer. *To bring Guns upon Battery*, if it be at a Siege, muſt be in the night time by Men, having Harneſs fitted for that uſe. *To ruin a Battery*, is to blow it up, or to nail the Guns.

Battery-Maſter; his Buſineſs is to raiſe the Batteries; we had ſuch an Officer in the Reduction of *Ireland*, who was paid by the Officer of Ordnance; but that Office is now ſuppreſs'd in *England*, though the *Hollanders* have it ſtill.

BATTEL is the Engagement betwixt two Armies. To ſhun a Battel ſometimes ſhows the Conduct and Prudence of a General, as much as getting a Victory.

Prince

Prince *Vaudemont*'s Retreat from *Aarſele* to *Ghent* with a ſmall Army, in the ſight of Mareſchal *Villeroy* at the Head of upwards of 80000 Men, that had almoſt ſurrounded him in the year 1695, when the King was lying before *Namure*, was one of the greateſt Actions done during the late War.

Battel-Array is the Order in which an Army is drawn up at a Review, and is more properly call'd the *Line of Battel.*

BAYONET, ſee *Bagonet.*

BED or *Stool of a Mortar*, is a ſolid piece of Oak in form of a Parallelipiped, bigger or leſs, according to the nature of the Mortar, hollowed a little in the middle to receive the Breech, and half the Trunions: on the ſides of the Bed are fixed the Cheeks or Brackets by four Bolts of Iron, ſee *Cheeks.*

Bed of a Gun is a piece of Plank laid within the Cheeks of the Carriage upon the middle Tranſum, for the Breech of the Gun to reſt on.

BEETLES are thick round pieces of Wood of a foot and a half long, and eight or ten inches diameter, having a Handle of about four foot long; the uſe of them is for beating or rather ſetting the Earth of a Parapet, or about Paliſades, by lifting it up a foot or two, and letting it fall with its own weight; they are likewiſe call'd *Stampers*, and by Paviers *Rammers.*

BERM, ſee *Liziere.*

BILLS are a ſort of Hand-tools carried along with the Artillery for the uſe of the Army, for cutting ways through Hedges, &c. and are the ſame with our Hedge-Bills in *England.*

BIOVAC is a Night Guard perform'd by the whole Army, when there is any Apprehenſion of Danger from the Enemy.

BLINDS are properly every thing that covers the Beſiegers from the Enemy, ſuch as Wool-Packs, Faſcines, Chandeleers, Mantelets, Gabions, Sand-Bags, Earth-Baskets, &c.

BLOCKADE is the blocking up a Place by poſting Troops at all the Avenues leading to it, to keep Supplies of Men or Proviſions from getting into it, thereby propoſing to ſtarve it out, without making regular Attacks upon it; this is call'd *forming a Blockade. To raiſe a Blockade*, is to force the Troops that keep the Place block'd up from their Poſts. *To turn a Siege into a Blockade*, is plain.

BLUNDERBUSS is a ſhort Fire Arm with a large Bore, very wide at the Mouth, carrying ſeveral Piſtol Balls, proper for the Defence of a Stair-caſe or Door; the ſhorteſt ſort of them are call'd *Muſquetoons.*

BODY or *Main-Body* of an Army, are the Troops encamped betwixt the two Wings, and are generally Infantry.

BOLTS, as it is a Term in Gunnery, are of ſeveral ſorts; thoſe that

that go betwixt the Cheeks of a

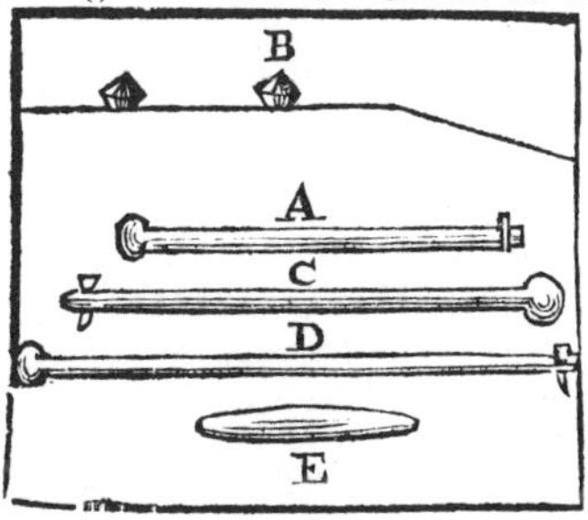

Gun Carriage to ſtrengthen the Tranſums, are call'd the *Tranſum-Bolts* A. The large knobs of Iron on the Cheek of a Carriage which keep the Hand-ſpike from ſliding when it is poiſing up the Breech of the Peece, are call'd *Priſe-Bolts*, B. The two ſhort Bolts that being put one in each end of an *Engliſh* Mortar Carriage ſerve to traverſe her, are call'd *Traverſe Bolts*, E. The Bolts that go thro' the Cheeks of a Mortar, and by the help of Coins keep her fix'd at the Elevation given her, are call'd *Bracket-Bolts*, C; and the four Bolts that faſten the Brackets or Cheeks of a Mortar to the Bed, are call'd *Bed-Bolts*, D.

BOMB is a great Shell of caſt Iron, *A*, with a large vent to receive a Fuſee. This Fuſee, *B* is made of Wood, and drove full of a Compoſition of Meal, Powder, Sulphur and Saltpetre. When the Bomb is fill'd with Powder, the Fuſee is drove into the Vent within an inch of the Head, and pitch'd over to preſerve it. When the Bomb is put in the Mortar, the Fuſee is uncaped, *E*, and ſalted with Meal powder; it takes Fire from

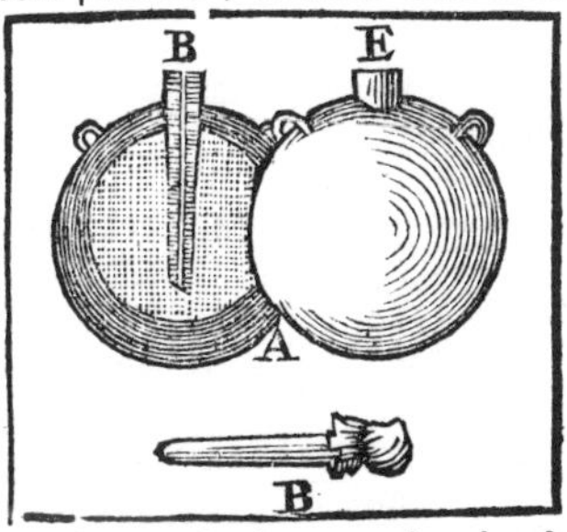

the flaſh of the Powder in the Chamber, and burns all the while the Bomb is in the Air; when the Compoſition is ſpent, it fires the Powder in the Bomb, which being confin'd, breaks the Bomb with a great Violence, blowing up whatever is about it; for the great height it goes in the Air and the force with which it falls, makes it go deep in the Earth. At the Siege of *Namure* in the year 1695, a Magazine of Hand Grenades, which were brought into a Demi-Baſtion between St. *Nicolas* Gate and the River, to be ready againſt the next Attack for our Lodgment that was upon the Paliſades, were blown up by a Bomb from Colonel *Brown*'s Battery. The *French*, *Germans*, and all other Nations, except the *Engliſh*, from whom the *Hollanders* have taken it lately, fire the Fuſee in the Mortar firſt, and then fire the Mortar at the Vent, which is a tedious, troubleſome, and uncertain way. Bombs are from 50 to 500 pound weight.

Bombardeers are thoſe employ'd about a Mortar; they drive the Fuſee,

Fuſee, fix the Shell, and load and fire the Mortar; they work with the Fireworkers on all ſorts of Fireworks, whether for War or Recreation. There is a Chief Bombardeer, and four and twenty Bombardeers eſtabliſh'd in the Office of Ordnance at a yearly Sallery.

Bomb Ketch is a ſmall Veſſel built and ſtrengthen'd with large Beams, for the uſe of Mortars at Sea.

BONNET is a ſmall Work conſiſting of two Faces, having only a Parapet with two rows of Paliſades of about 10 or 12 foot diſtance; it is generally raiſed before the Salliant Angle of the Counterſcarp, and has a Communication with the Covert way, by a Trench cut through the Glacis and Paliſades on each ſide.

Bonnet a *Pretre*, or *Prieſts Bonnet*, is a work in Fortification differing from a double Tenail in this, that as the ſides of a Tenail are parallel, thoſe of a *Prieſts Bonnet* would meet if they were prolong'd.

BOYAU or *Branch of a Trench*, is a Line or a particular Trench made parallel to the Deſence of the Place, to avoid being flanked or enfiladed. A *Boyau*, when there are two Attacks made upon a Place, ſerves as a Line of Communication betwixt them; the Parapet of a *Boyau* being ſtill turn'd towards the Place beſieg'd, it ſerves for a Line of Contravallation to hinder Sallies, and defend the Work men.

BRACKETS, ſee *Cheeks*.

BRANCH *of a Trench*, is the ſame with *Boyau*. *Branch of a Mine*, ſee *Gallery*.

BREACH is the Ruin of any part of a Fortification by the Cannon or Mines of the Beſiegers, in order to make an Attack upon the Place. To make the Attack the more difficult, the Beſieged ſow the Breach with Crow-Feet, or ſtop it with *Chevaux de Friſe*.

BREAD, ſee *Ammunition-Bread*.

BREAK *Ground*, is the firſt opening of Trenches againſt a Place, which is done in the night time, by the Advantage of ſome riſing Ground, hollow Way, or any thing that can cover the Men from the Enemies Fire.

BREAST-WORK, ſee *Parapet*.

BREECH of *Gun*, ſee *Baſe*.

BRIDGE is a conveniency made for paſſing Rivers, and is of ſeveral ſorts.

Bridge of *Communication* is a Bridge made over a River, by which two Armies or two Forts which are ſeparate by this River, have a free Communication one with another.

Draw-Bridges are made after ſeveral Faſhions, but the moſt common are made with Plyers twice the height of the Gate, and a foot diameter; the inner Square is traverſed with a St. *Andrew's* Croſs, which ſerves for a Counterpoize, and the Chains which hang from the other Extremities

of

of the Plyers to lift up or let down the Bridge, are of Iron or Braſs. *Floating or Flying Bridges* are made of two ſmall Bridges laid one upon the other, ſo that the uppermoſt by the help of Ropes and Pullies is forced forwards, till the end of it join to the place deſign'd.

Bridge of *Boats*, ſee *Pontons*.

Bridge in Gunnery is a Term given to the two pieces of Timber which go betwixt the two middle Tranſums of a Gun Carriage, on which reſts the Bed.

BRIGADE. An Army is divided into Brigades of Horſe, and Brigades of Foot; a Brigade of Horſe is a Body of eight or ten Squadrons; a Brigade of Foot conſiſts of four, five or ſix Battalions, under the Command of a Brigadeer; the eldeſt Brigade has the right of the firſt Line, and the ſecond the right of the ſecond Line, the two next take the left of the two Lines, and the youngeſt in the Centre. The Battalions which compoſe a Brigade obſerve the ſame Order.

Brigadeer is a General Officer who has the command of a Brigade; the eldeſt *Colonels* are generally advanc'd to this Poſt, they roll in Duty amongſt themſelves; he that is upon Duty, is call'd *Brigadeer* of the Day; he viſits all the Out-Guards and Poſts of the Army, and at Night takes the Orders from the *Major-General* of the Day, and delivers it to the *Majors* of Brigades, who muſt all attend at orderly time. They march at the Head of their Brigades, and are allowed a *Serjeant* and ten Men of their own Brigade for their Guard.

Brigadeers and *Sub brigadeers*, are Poſts in the Horſe Guards.

Brigade-Major, is an Officer appointed by the *Brigadeer* to aſſiſt him in the Buſineſs of his Brigade; and acts in his Brigade, the ſame as a *Major-General* does in the Army. For this Poſt are choſe the moſt ingenious and expert *Captains*. They are to wait at orderly time, to receive the Word and the Orders which they carry firſt to their proper *Brigadeer*, and afterwards deliver to the *Adjutants* of Regiments at the Head of the Brigade, where they regulate together the Guards, Parties, Detachments and Convoys, and appoint them the hour and place of Rendezvous at the Head of the Brigade, where the *Brigade Major* takes and marches them to the place of the general Rendezvous. He ought to know the State and Condition of the Brigade, and keep a Roll of the *Colonels*, *Lieutenant Colonels*, *Majors* and *Adjutants*. When a Detachment is to be made, the *Major-General* of the Day regulates with the *Brigade-Majors*, how many Men and Officers each Brigade muſt furniſh; and they again with the *Adjutants* of the Regiments, how many each Battalion is to ſend, which the *Adjutants* divide among the Companies The Complements each Regiment is to furniſh, are taken by the

Adjutant at the Head of the each Regiment at the hour appointed, who delivers them to the *Brigade-Major* at the Head of the Brigade, who again delivers them to the *Major-General* of the Day, and he remits them to the Officer who is to command the Detachment.

BRINGERS-UP the whole last Rank of a Battalion being the last Man of each File, are call'd the *Bringers-up*.

BUCKETS both of Wood and Leather, are always carried along with the Artillery in the Fire-workers Stores, where they are very useful.

BUDGE-BARRELS are

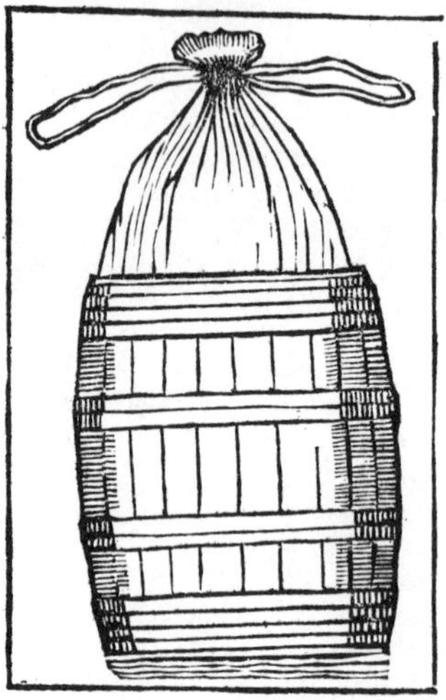

small Barrels well hoop'd, with only one head; on the other end is nailed a piece of Leather to draw together upon strings like a Purse. Their use is for carrying Powder along with a Gun or Mortar, being less dangerous and easier carried than whole Barrels; they are likewise used upon a Battery of Mortars, for holding Meal-Powder.

BULLETS, all sort of Shot for Fire Arms from a Cannon to a Pistol; those for Cannon are of Iron, the rest of Lead, see *Ball*.

BULWARK is an old Term for Rampart, see *Rampart*.

C

CADET is a young Gentleman, who, to attain to some Knowledge in the Art of War, and in expectation of Preferment, chuses at first to carry Arms as a private Man in a Company of Foot. *Cadet* differs from a Volunteer, because he takes Pay, which is no more than a private Man; but a Volunteer serves without Pay.

Cadet among the *French* signifies an Officer, who, in respect of another, is younger in Service.

CAISSON is a Chest of

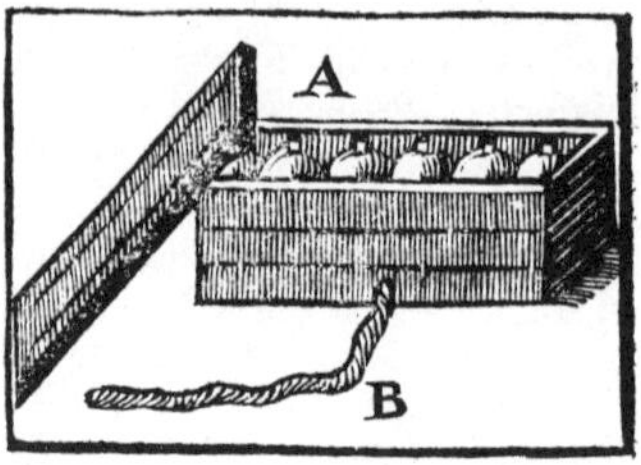

Wood holding four or six Bombs, sometimes filled only with

with Powder, and buried by the Besieged under ground, to blow up a Work which the Besiegers are like to be Masters of; as thus, after the Bonnet is blown up by the Mine, they lodge a *Caisson* under its Ruins, and the Enemy being advanc'd to make a Lodgment there, they fire the *Caisson* by the help of a Saucis or Pudding, and blow up that Post a second time.

CALIBER is a Term in Gunnery, signifying the diameter or wideness of the Bore of a piece of Ordnance.

Caliber-Compasses are Compasses

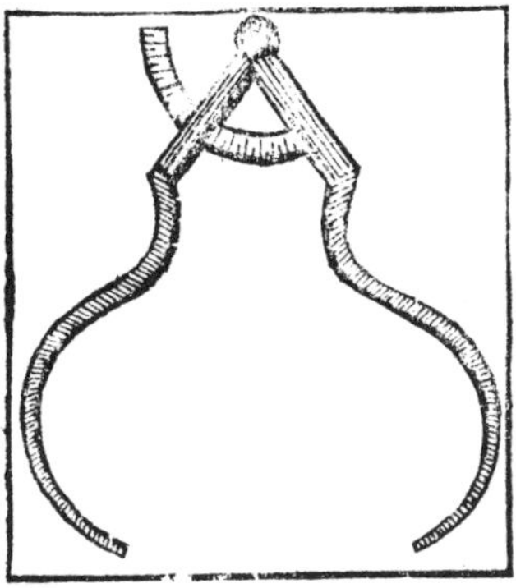

used by Gunners for taking the Diameters of the several parts of a piece of Ordnance, or of Bombs, Bullets, &c. Their Legs are therefore Circular on an Arch of Brass, whereon is mark'd the Inches and half Inches, to show how far the Points of the Compasses are opened asunder. Some are made for taking the diameter of the Bore of a Gun or Mortar.

CALTHROPS, see *Crow-Feet*

CAMP is the Ground where an Army pitch their Tents. 'Tis marked by the *Quarter-Master-General*, who appoints every Regiment their Ground. The chief Advantages to be minded in chusing a Camp for an Army, is to have it near Water, in a Country of Forage, and where the *Soldiers* may find Wood for dressing their Victuals; it ought likewise to have a free Communication with Garrisons, from whence it may have a continual supply of Provisions. The *Quarter-Master-General* in chusing the Camp, ought to consider the Advantages of the Ground; such as Hills, rising Grounds, Marshes, Woods, Rivers and Inclosures; if the Camp be near the Enemy, and no River or Marsh to cover it, the Army ought to be Entrench'd. An Army always encamps fronting the Enemy, and generally in two Lines running parallel about 500 yards distance, the Horse and Dragoons on the Wings, and the Foot in the Centre; sometimes a small Body of two, three or four Brigades is encamp'd behind the two Lines, and is call'd the *Body of Reserve*. The Artillery and Bread Waggons are generally encamp'd in the Rear of the two Lines. A Battalion of Foot is allow'd 80 or 100 Paces for its Camp, and 30 or 40 for an Interval, betwixt one Battalion and another. A Squadron of Horse is allowed 30 for its and 30 for an Interval, and more

more if the Ground will allow it. Each Battalion poſts a ſmall Guard commanded by a Subaltern Officer, about 100 yards before the Front of the Regiment call'd the *Quarter Guard*, for the ſecurity of the Regiment; and each Regiment of Horſe mounts a ſmall Guard on Foot in the Front of a Regiment under a *Corporal*, call'd the *Standard Guard*. The Grand Guard of the Army conſiſts of Horſe, and is poſted a mile and a half from the Camp towards the Enemy, on the Right and Left, by the *Lieutenant* or *Major General* of the Day, who chuſes the propereſt Places, from whence all the Avenues of the Camp may be diſcover'd. At a Siege the Army encamp with their Rear to the Place beſieged. *Flying Camp* is the Ground on which a *Flying Army* are encamp'd.

CAMPAIGN is that part of the Summer betwixt the Armies taking the Field, and their returning to Garriſon, which is commonly ſix or ſeven months; an Officer or Soldier is ſaid to have made a Campaign that has been in the Field with the Army, and may reckon as many Campaigns as he has been Summers in the Field. The Opening and Cloſe of the Campaign, is the Armies taking the Field, or returning to Garriſon.

CANNON is a piece of Ordnance of Braſs or Iron, of ſeveral ſorts and ſizes, from a *Cannon Royal* or *Cannon of* eight, which carries a Ball of ſixty three pound, to a *Baſe*, which is the loweſt Nature of Cannon. Thoſe moſt uſed in the Army or Navy, are (1.) *Demi-Cannon*; for the *Cannon Royal* and *Baſtard Cannon* are too large. It carries a Ball of 32 pound, and is uſed in the lower Tire of a Firſt Rate Man of War. (2.) *Twenty four Pounders*: (3.) *Culverins* carrying 18 pound: theſe two Natures of Guns are the beſt for Battering. (4.) *Twelve Pounders*. (5.) *Demi Culverins* carrying 9 pound Ball; theſe may be likewiſe uſed at a Siege. (6.) *Six Pounders*. (7.) *Sakers* carrying five and a quarter pound Ball. (8.) *Minions* of 4 pound, and (9.) *Three Pounders*, which are the loweſt Nature of Guns uſed in the Field or Navy: theſe ſix laſt are propereſt Field Guns for following an Army. The moſt remarkable parts about a Gun, are the Caſcabel, Mouldings, Baſe-Ring, Vent, Vent-Ring, Reinforc'd-Ring, Trunions, Dolphins, Trunion-Ring, Corniſh-Ring, Neck, Muſle, Face, and Chace, or Cylinder; ſee each in its proper place. Guns longer than ordinary, are call'd Slings, Drakes, *&c.* as thoſe ſhorter are call'd Cuts. The length of a true fortify'd Gun, is about ſeven diameters of the Metal at the Vent; the diameter of the Metal there being three diameters of the Bore; ſo that a 24 Pounder being ſix inches in diameter of her Bore, the thickneſs of the Metal at the Vent muſt be a foot and a half, and her length thirteen foot and a half.

Can-

Cannon-Basket, see *Gabions*.

Canon-Carriages, see *Carriages*.

CANVAS-BAGS, see *Sand Bags*.

CAPESQUARE is a

Term in Gunnery given to that strong Plate of Iron which comes over the Trunions of a Gun, and keeps her in her Carriage; it is fasten'd by a Hinge to the Prize-plate, that it may lift up and down; it forms a piece of an Arch in the middle to receive a third part of the Trunions, for two thirds are let into the Carriage; and the other end is fasten'd by two Iron Wedges, which are call'd the *Forelocks* and *Keys*.

CAPITAL is a Term in Fortification, signifying the Line drawn from the Angle of the Gorge to the Flanked Angle. A Bastion is said to have sixty yards of Capital, when there is sixty yards from the Angle of the Gorge to the Point of the Bastion.

CAPITULATIONS are the Articles agreed upon betwixt the Besieged and the Besiegers for surrendering a Place. The *Chamade* being beat, all Hostilities cease on both sides, and the Officer who commands in the Trenches, goes upon the Breach to hear what the Enemy proposes or would have, if it be to capitulate; the Governour sends Deputies to the General to treat, if the Capitulation be agreed to and sign'd; Hostages are delivered on both sides for the exact Performance of the Articles; one part of the Place is deliver'd to the Besiegers, and a day fixed for the Garrison to march out: The ordinary and most honourable Conditions are, to march out at the Breach, with Arms and Baggage, Drums beating, Colours flying, Match lighted at both ends, Ball in Mouth, with some pieces of Cannon and Waggons, and Convoys for their Baggage and for the Sick and Wounded.

CAPONIERE is a Work sunk on the Glacis of a Place, about four or five foot deep: the Earth that comes out of it serves to form a Parapet of two or three foot high, made with Loop-holes or small Embrasures; it is cover'd over-head with strong Planks, on which are laid Clays or Hurdles which support the Earth that covers all; it holds about 15 or 20 Men, who fire thro' these Embrasures on the Besiegers; the Clays are made sometimes in the bottom of a dry Moat. At the Siege of *Lisle* by the *French* in 1667, the Count *de Broway* Governour for the King of *Spain*, had sunk a *Caponiere* on the Salliant Angle of the Glacis, towards the Attack of

Piccardy, which was Palisaded; but was forced by the Count *de St. Paul*, who from thence threw himself into the Covert way, and made a very advantagious Lodgment, which discovered the whole breadth of the opposite Ditch. At the second Siege of *Dole* in 1674, the Besieged had a *Caponiere* in the bottom of the Fosse, which could not be ruined by the Batteries. They were much used at the Siege of *Candy*; they differ from *Coffre*; see *Cofre*.

CAPTAIN in Field Regiments is he who commands a Troop or Company; he ought to be very diligent and careful to keep his Company full of young lusty Soldiers, to know their Names and Dispositions, and what every Man is capable of; to visit their Tents and Lodgings, to see what is wanting, to pay them well, to cause them to keep themselves clean and neat in their Cloaths, Shoes, Stockings, &c. and to keep their Arms bright: He has Power in his own Company of making Serjeants, Corporals and Lanspesades; he Marches still at the Head of his Company, and Ranks according to the Seniority of his Commission.

Captain-Lieutenant, is he who commands the *Colonels* Troop or Company.

CARABINE is a Fire Arm shorter than a Firelock, and carries a Ball of 24 in the pound; they are carried by the Light Horse, hanging at a Belt over the Left Shoulder.

CARABINEERS are Regiments of Light Horse, carrying longer Carabines than the others, and are used sometimes on Foot; we have none in our Army but in Major-General *Windham*'s Regiment.

CARCASS, A, is an

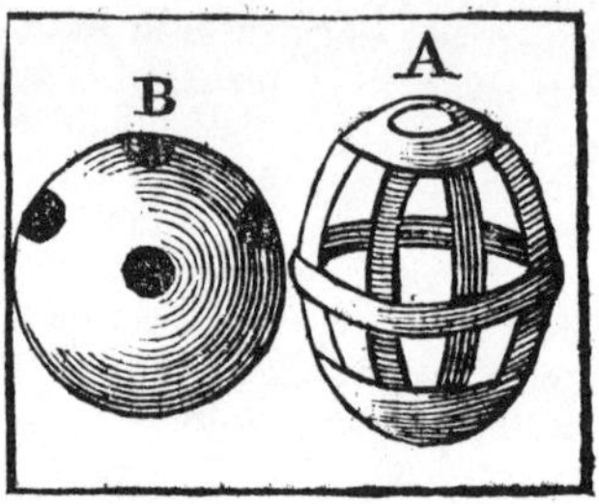

Invention of an oval Form made of ribs of Iron, afterwards filled with a Composition of Meal-Powder, Saltpetre, Sulphur, broken Glass, shavings of Horn, Pitch, Turpentine, Tallow, and Linseed Oil, and then coated over with a pitch'd Cloath; it is primed with Meal-Powder and Quickmatch, and fired out of a Mortar; the Design of it is to set Houses on fire. For lifting it up to put it in the Mortar, it has two small Cords fixed to the sides of it. Another sort there is of them, of a late Invention for the Sea Service, which differ nothing from a Bomb, save its being fill'd with a Composition as before, and having Five holes all primed with Powder and Quickmatch, which takes fire from the flash of the Mortar, and having fired the Composition, it

 burns

burns very vehemently from those Holes, *Letter B.*

CARRIAGE is a general Term for Waggons, Carts, Litters, *&c.*

Carriage of a Cannon is a long narrow Cart, invented for marching of Cannon, and for the more convenient using them in Action; it is made of two planks of Wood, commonly once and a half the length of the Gun, which are called the Cheeks, and join'd together by three wooden Transums, strengthen'd with three Bolts of Iron, and call'd the Fore, Middle, and Hind Transums; it is mounted on two Wheels, but on a march has two fore Wheels, with Shafts or Limbers added. The parts about a Carriage are the Cheeks, the Transums, the Bolts, the Plates, the Train, the Bands, the Keys and Locks, the Bridge, the Bed, the Hooks, the Trunion Holes and the Capesquare, see *each in its proper place*; see likewise the parts of Wheels and Limbers, at *Wheels* and *Limbers.*

Block-Carriage is a Cart made on purpose for carrying of Mortars and their Beds from one place to another.

Truck-Carriages are two short planks of Wood supported on two Axiltrees, having four Trucks or Wheels of solid Wood, about a foot and a half, or two foot diameter, for carrying Mortars or Guns upon Battery, when their own Carriages cannot go; and are drawn by Men.

CARTOUCH is a Case of Wood about three inches thick at bottom, girt round with Marlin, holding about 400 Musquet Balls, besides six or eight Balls of Iron of a pound weight; 'tis fired out of a Hobit, a small sort of Mortar, and is very proper for defending a Pass. A new sort was made last Spring, much better than the former, of a globulal Form, and fill'd with Ball of a pound weight; others were then made for the Guns, being of Ball of half or quarter pound weight, according to the Nature of the Gun, tied in form of a Bunch of Grapes on a Tompion of Wood and coated over; these were made in the room of the Partridge Shot formerly used, and exceed them very far, as some of the *French* Battalions experimented to their Cost, at the Battle of *Blenheim* near *Hochstet.*

CARTRIDGE is a Case of Pastboard or Parchment, holding the exact Charge of a Fire Arm; those for Musquets, Carabines or Pistols, hold both the Powder and Ball for the Charge.

Cartridge-Box is a Case of Wood or turn'd Iron cover'd with Leather, holding a dozen Musquet Cartridges; it is wore upon a Belt, and hangs a little higher than the right Pocket-hole.

CASCABEL is the knob of Metal behind the Breech of a Cannon; the diameter of it is the diameter of the Bore of the Piece; the Neck of the Cascabel,

bel, is what joins it to the Breech Mouldings.

CASERNS, see *Cazerns*.

CASKS or *Barrels* are used in the Army for carry Meal to be laid up in Magazines, or along with the Army for baking Bread for the Troops.

CASEMATE, see *Cazemate*.

CAVALIER is a great Mass or Elevation of Earth of different Shapes, some being round, and some long Squares, situate ordinarily in the Gorge of a Bastion, for those which are raised on the Curtin, are rather *Platforms*; they are bordered with a Parapet cut into Embrasures for four, six or eight Cannon, according to the Capacity of the Cavalier: They are a double Defence for the Faces of the opposite Bastion; they defend the Foss, break the Besiegers Galleries, command the Traverses in dry Moats, scowr the Salliant Angle of the Counterscarp, where the Besiegers have their Counter-Batteries, and enfilade the Enemies Trenches, or oblige them to multiply their Parallels; they are likewise very serviceable in defending the Breach and the Retrenchments of the Besieged, and can very much incommode the Retrenchments which the Enemy make, being lodg'd in the Bastion.

CAVALRY are the Regiments of Horse which serve in the Army, and may be properly call'd the *Right Arm of the Army*, for the Success of a Battle often depends on the Cavalry; their frequent Excursions to disturb the Enemy, to intercept their Convoys, and destroying the Country, are no inconsiderable Matters; the Cavalry is divided into Brigades, as the Infantry, and encamps on the Wings of the Army.

CAVIN is a natural Hollow, fit to lodge a Body of Troops: if there happen to be any near a Place besieged, they are of great use to the Besiegers; for, by the help of such a Place, they can open the Trenches, make Places of Arms, or keep Guards of Horse, without being in danger of the Enemies Shot.

CAZEMATE is a certain retired Platform in the Flank of a Bastion, for the Defence of the Moat and Face of the opposite Bastion; there are sometimes three such Platforms one behind another, the uppermost of which is on the Terre-plain of the Bastion, which makes the other two to be called *Places Basses*, or *Low Places*; they are cover'd from the Enemies Batteries by a Work of Earth added to the Angle of the Shoulder, of a circular, or sometimes of a square Form, call'd *Shoulder*, *Oreillon*, or *Epaulment*. At the famous Siege of *Candy*, the *Turks* having attack'd it regularly on the Bastions, *Bethleem* and *Panigra*, for sixteen years, after they were Masters of all the Out-Works about the Town, and these two Bastions had suffered the Ef-

fect of many Mines; the Besiegers had made terrible Breaches in both the Faces, but never could lodge themselves, for the Artillery of the Cazemates still ruin'd their Lodgments, and forc'd them to quit that Attack.

CAZERNS or *Baracks*, are lodgings built in Garrison Towns for lodging the Garrison; they are built near the Rampart, or in the wast Places of the Town; the *Baracks* or *Cazerns* lately built in *Ireland*, are of such conveniency, both for the Army and Country, that they'll perpetuate the Memory of my Lord *Galway*, who was the Projector of them.

CENTRE is the middle Point of a Circle; *Centre of a Battalion*, are the Pikes; *Centre of an Army*, is the Infantry.

CENTINEL is a private Man in a Company of Foot, armed with a Sword, Firelock, Bagonet, Bandeleers, and Cartridge-Box.

CENTRY-BOX is the same with *Guerritte*, only he one is of Wood, and the other of Stone; they are upon the flanked Angles of Bastions, and on the Angles of the Shoulder, and sometimes on the middle of the Curtin, to preserve the Centries from the Weather, see *Guerritte*.

CESSATION *of Arms*, is when a Governour of a Place besieged, finding himself reduced to the last Extremity, that he must either surrender or sacrifice himself, his Garrison and Inhabitants, to the Mercy of the Enemy, plants a *White Flag* on the Breach, or beats the *Chamade* to capitulate, at which both Parties cease firing, and all other Acts of Hostility; till the Proposals be either agreed to or rejected.

CHACE *of a Gun*, see *Cylinder*.

CHAIN is a Number of Iron or Brass Rings link'd one in another; *an Engeneers Chain* for measuring of Ground, is of a certain number of Links of an equal length; *Chains* of a Gun are of Iron, and very strong, fixed on the Draught-Hooks, and going along the Shafts of the Limbers to ease them, but they are not used for small Guns.

Chains of Draw-bridges, see *Bridge*.

Chain Shot, see *Shot*.

CHAMADE is a Signal made by beat of Drum for a Conference with the Enemy, when any thing is to be proposed, as a *Cessation of Arms* to bring off the Dead; or by the Besieged, when they have a mind to deliver up a Place upon Articles of Capitulation; and then there is a Suspension of Arms, and Hostages deliver'd on both sides.

CHAMBER *of a Mortar*, (for chamber'd Guns are out of use) is that part of the Chace where the Powder lies, and is much narrower than the rest of the Cylinder; it is of different Forms; some are like a reversed Cone or Sugar-Loaf, others globical, with

a Neck for its Communication with the Cylinder, and are called *Bottled Chambers*, but the moſt ordinary are in Form of a Cylinder *The Powder Chamber* or *Bomb-Chamber* on a Battery, is a place ſunk under Ground for holding the Powder or the Bombs, where they may be out of Danger, and preſerv'd from Rain.

Chamber of a Mine, is the place in which the Powder is confin'd, and is generally of a cubical Form, ſee *Mine*.

CHANDELEER is a Frame

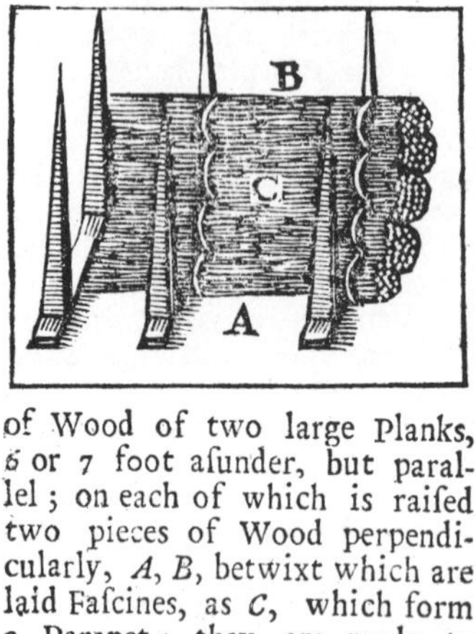

of Wood of two large Planks, 6 or 7 foot aſunder, but parallel; on each of which is raiſed two pieces of Wood perpendicularly, *A*, *B*, betwixt which are laid Faſcines, as *C*, which form a Parapet; they are made to move from Place to Place as occaſion requires, and ſerve to cover the Work-men.

CHARG'D-CYLINDER, is that part of the *Chace* of a Gun, where the Powder and Ball are contain'd.

CHAUSSE-TRAPES, ſee *Crow-Feet*.

CHEEKS *of a Gun Carriage*, ſee *Carriage*.

Cheeks of a Mortar, or *Brac-*

kets, are made of ſtrong Planks of Wood of near a ſemi-circular Form, bound with thick *Plates* of Iron *A*, and are fixed to the Bed *B*, by four *Bolts*, call'd *Bed-Bolts*, as the Figure 2 ſhows; they riſe on each ſide of the Mortar, and ſerve to keep her at what Elevation is given her, by the help of ſtrong *Bolts* of Iron as 5, which go through both *Cheeks*, both under and behind the *Mortar*, betwixt which are drove *Coins* of Wood, as 3; theſe *Bolts* are call'd the *Bracket-Bolts*, and the *Bolts* which are put one in each end of the Bed, as 4, are the *Traverſe-Bolts*, becauſe with Hand Spikes the Mortar is by theſe traverſed to the Right or Left; the *Trunion of the Mortar* is Mark'd 1.

CHESTS and *Baulks*, ſee *Pontons*.

CHEVAUX DE FRISE, or *Turn-Pikes*, are Spars of Wood about a foot diameter, and ten or twelve long, cut into ſix

Faces, and bored through; each Hole is arm'd with a ſhort ſpike

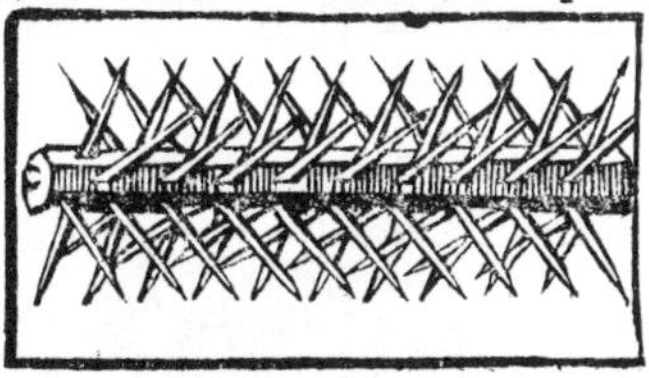

ſhod with Iron at each end, about an Inch diameter, *6* foot long, and *6* inches diſtant one from another; ſo that it points out every way, and is proper for ſtopping ſmall *Overtures*, or to be placed in Breaches; they are likewiſe a very good Defence againſt Horſe.

CHEVRETTE; among the

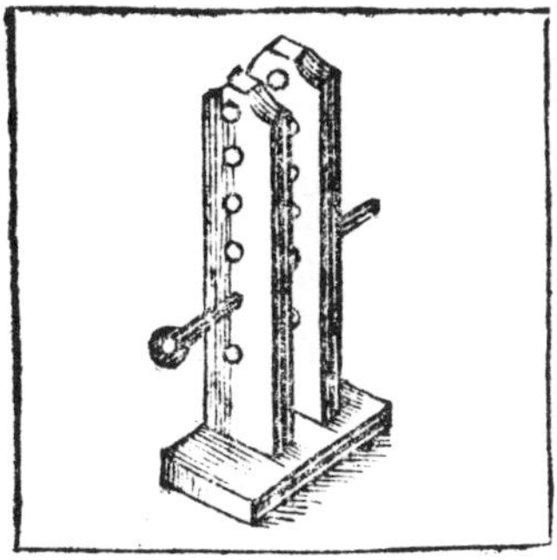

many Inventions for raiſing of *Guns* or *Mortars* into their *Carriages*, this Engine is very uſeful; it is made of two pieces of Wood of about 4 foot long, ſtanding upright upon a third which is ſquare; they are about a foot aſunder, and parallel; and are pierced with Holes exactly oppoſite to one another, having a bolt of Iron, which being put through theſe Holes higher or lower at pleaſure, ſerves with a Hand-ſpike which takes its poiſe over this Bolt, to raiſe any thing by force.

CIRCLE is a plain Figure comprehended within a crooked Line, call'd the *Circumference*, which has all its parts equally diſtant from a certain Point call'd the *Centre*.

Arch of a Circle is an undetermin'd part of the circumference of a *Circle*, being ſometimes larger and ſometimes ſmaller.

CIRCUMVALLATION is a Trench with a Parapet thrown up by the Beſiegers round their *Camp*, when they are in any apprehenſion of the Enemies attempting to relieve the Place, or raiſe the Siege. The Engeneers having made a Plan of the Country about, and marked all the Hills, Marſhes, Valleys, Rivers and Churches, and every thing which may ſerve to lodge or cover Horſe or Foot, as Vineyards, Hedges, Ruins of Houſes, &c. and agreed which way to run their Line to the beſt Advantage, ſhunning always the Foot of riſing Grounds, which the Enemy being Maſters of, may command within the Trenches; they mark it out with Picquets and Ropes, making the Foſs about 12 foot wide, and the Baſe of the Parapet eight, the height of the Parap t on the inſide being ſix foot, and on the outſide 5, with a Banquet of three foot wide, and one and a half high. The *Lines of*

Con-

Contrevallation ought to have the ſame height and breadth, and both the one and the other ought to be ſtrengthned with Forts and Redoubts, and well flanked.

CITTADEL is a Place fortified with 4, 5, or 6 *Baſtions* commanding a Town, built by Orders of the Sovereign; its being fortified with *Baſtions*, diſtinguiſhes it from a *Caſtle*, which hath only round or ſquare Towers. They are ſometimes built half within and half without the Rampart of the Town, to the end the Governour of the *Cittadel* may be Maſter of an Entry into the Town, and likewiſe to the Country, without having a Dependance on the Inhabitants. Others are built without the Town, but that muſt be upon preſſing Reaſons, ſuch as keeping Poſſeſſion of a high Ground which the Beſiegers might make uſe of, either to batter the Town, or to cover their Camp, or to preſerve Marſhes or Springs of Water, which may be uſeful to the Inhabitants, becauſe their chief Deſign is to keep the Inhabitants in ſubjection, and to hinder their correſponding with the Enemy, which cannot be done, if built at a diſtance from the Town.

CLAYES, ſee *Hurdles*.

CLOUTS, are thin Plates of Iron nailed on that part of the Axil-tree of a Gun-Carriage that comes through the Nave, through which the Linſpin goes.

CLOY *Guns*, ſee to *Nail*.

COFFER is a work ſunk in the bottom of a dry Moat, about 6 or 7 foot wide, the length of it being from one ſide of the Moat to the other, with a Parapet of about two foot high, full of Loop-holes, cover'd over-head with Joyſts, Hurdles and Earth; they ſerve to fire on the Beſiegers, when they endeavour to paſs the Moat, and differ from the Caponiere, becauſe it is longer, for the Caponiere takes not the whole breadth of the Moat; it differs likewiſe from the Traverſe and the Gallery, becauſe it is made by the Beſieged, and theſe by the Beſiegers.

COLONEL, is the *Commander in Chief* of a Regiment, whether Horſe, Foot or Dragoons, and ought to be a Perſon of Authority, to aſſume an abſolute Command, to ſee that the Companies of his Regiment be full, and kept in good Order, that they be duly exerciſed, and well payed; he may lay any Officer of his Regiment in Arreſt, but muſt acquaint the *General* with it. A *Colonel* is not allowed a Guard, but only a Centry from the Quarter Guard.

COLOURS are large Flags of Silk fixt on half Pikes, and carry'd by the Enſigns; when a Battalion is encamped, they are ſtuck before the Front of the Battalion, but if the Battalion be in Garriſon, they are lodged with the *Colonel* or commanding Officer; they are never carried on Detachments, nor with the Battalion when it mounts the Trenches. *Field Colours* are

ſmall

ſmall Flags of about a foot and a half ſquare, which are carried along with the *Quarter-Maſter-General* for marking out the Ground for the Squadrons and Battalions.

Colour-Guard, ſee *Guard*.

COLUMN *of an Army on a march*, is a long row of Troops or Baggage following one another; the firſt and ſecond Lines of the Army, as they are encamped, make generally two *Columns* on a March, filing off either from the Right or Left; ſometimes the Army marches in four, ſix, or eight *Columns*, according as the Ground will allow, and each *Column* is led by a *General* Officer.

COMMANDING *Ground* is an Eminence or Riſing Ground, overlooking a Poſt.

COMMISSARY *of the Muſters*, is an Officer appointed to Muſter the Army as often as the *General* pleaſes, to know the Strength of each Regiment, and of each Company, to receive and inſpect the Muſter Rolls, and to keep an exact State of the ſtrength of the Army.

Commiſſary of Stores, is an Officer in the Artillery, who has the charge of all the Stores, for which he is accountable to the Office of Ordnance; he is allowed an Aſſiſtant, *Clerks* and *Conductors* under him.

Commiſſary of Horſes, is likewiſe an Officer in the Artillery, appointed to have the Inſpection of the Artillery Horſes, to ſee them muſter'd, and to ſend ſuch Orders as he receives from the commanding Officer of the Artillery, by ſome of the *Conductors* of Horſes, of which he has a certain number for his Aſſiſtants.

Commiſſary of Proviſions, is he who has the Inſpection of the Bread and Proviſions of the Army.

COMMISSION is the Authority from a *Prince* or his *General*, by which an Officer officiates in his Poſt.

COMMUNICATION, *Lines of Communication*, are Trenches made to preſerve a ſafe correſpondence betwixt two Forts or Poſts, or at a Siege betwixt two Approaches, that they may relieve one another upon occaſion, it is the ſame with *Boyau*.

Bridge of Communication, ſee *Bridge*.

COMPANY is a ſmall Body of Foot commanded by a *Captain*, who has under him a *Lieutenant* and *Enſign*; it conſiſts commonly of 50 *Centinels*, 3 *Serjeants*, 3 *Corporals*, and 2 *Drums*. A *Company* in the Guards is of 80 Private Men. A Battalion of Foot conſiſts of 13 *Companies*, one of which is always Grenadiers: The Eldeſt *Company* take their Poſt next the Grenadiers, who have always the Right of the Battalion, and the Second *Company* the Left; the youngeſt has its Poſt always in the Centre.

COMPLEMEMT *of a Curtin*, is that part of it which makes the Demigorge.

Complement of the Line of Defence,

fence, is the remainder of the *Line of Defence*, after the Angle of the Flank is taken off.

COMPTROLLER *of the Artillery*, is a Post of great Trust; he inspects the Musters of the Artillery, makes the Pay-List, takes the Accompts and Remains of Stores, and is accountable to the Office of Ordnance.

CONDUCTORS are Assistants given to the Commissary of the Stores, to receive or deliver out Stores to the Army, to attend at the Magazines by turns, when in Garrison, and to look after the *Ammunition-Waggons* in the Field; they bring their Accounts every night to the Commissary, and are immediately under his Command.

CONE is a Body made by the turning of a Right angled Triangle round a *Circle*, the angular Point of the Right Angle being fixed in the *Centre*, which forms a *Pyramid*, whose *Basis* is a *Circle*.

CONTREVALLATION, is a *Trench* with a *Parapet* made by the Besiegers, betwixt them and the Place besieged, to secure them from the Sallies of the Garrison, so that the Troops which form the Siege are encamp'd betwixt the *Lines of Circumvallation* and *Contrevallation*; when the Enemy has no Army in the Field, there is no occasion for the *Lines of Circumvallation*; and when the Garrison is weak, the *Lines of Contrevallation* are seldom used; at the last Siege of *Landau* we used neither; but when the *French* besieged it in 1703, they raised *Lines of Circumvallation* and *Contrevallation* very strong, and of a large circuit.

CONTRIBUTION, is an Imposition or Tax paid by *Frontier Countries*, to redeem themselves from being plundred and destroy'd by the Enemy.

CONVERSION, is a Military Motion, which turns the Front of a Battalion where the Flank was, if the Battalion be attacked in the Flank; and as in Action this may often happen, it must be a very useful and necessary Motion.

CONVOY is a Supply of Men, Money, Ammunition or Provisions, convey'd into a Town or to an Army. The Body of Men that guard this Supply, are called likewise the *Convoy*.

CORDON is a Stone jutting out betwixt the Rampart and the Basis of the Parapet, like the Thore of a Pillar; it goes quite round the Fortification.

CORIDOR is a *French* Term for *Covert Way*.

CORNET, is the third *Commission Officer* in a Troop of Horse or Dragoons; it is a very honourable Post; he commands in the *Lieutenants* absence; his principal Duty is to carry the Standard near the middle of the first Rank of the Squadron, and should rather die than lose it, for it is a great Dishonour for a Squadron to lose their Standard to the Enemy.

COR-

CORNISH-RING, or *Astragal of a Gun*, is that ſmall Ring near the Muzzle.

CORPORAL is an *Inferior Officer* under a *Serjeant* in a *Company of Foot*; he is commonly an old *Soldier*; he receives the Word from the Rounds that paſs by the Grand Guard; he receives likewiſe the Word from ſuch as are ſtopt by his *Centries*, with his Sword drawn; he relieves the *Centries*; and when the Guard is relieved, he gives the Orders he had to the *Corporal* that is to mount, and ſhows him all the Poſts; he carries a Fuſee.

CORPS DE GARDE, ſee *Guard*.

COVERT WAY, is a ſpace of Ground level with the Country, about 3 or 4 Fathoms wide cover'd by a *Parapet* which goes quite round the Place; the greateſt Effort in Sieges, is to make a Lodgment on the *Covert Way*, which the Beſieged generally Paliſade and Undermine: this Parapet ſlopes inſenſibly towards the *Campaign*, and the *Talus* or Sloping, is called the *Glacis*, which the Beſiegers are generally oblig'd to ſap through to make a Lodgment. The Parapet of the *Covert Way* is about ſix foot high with a Banquet, and forms a *Salliant Angle*, before the *Curtin*, which ſerves for a place of Arms.

COUNCIL OF WAR, is when the *General* of an Army calls together his *Lieutenant* and *Major Generals* for their Advice upon ſome Buſineſs of Importance; he hears all their Reaſons, and having balanced and conſidered them judiciouſly, with regard to the Intereſt of his Prince and Honour of his Country, without being byaſs'd by one or another Opinion, he determines to the greateſt number of Votes.

COUNTER-APPROACHES are the Lines and Trenches made by the Beſieged, in order to attack the Works of the Beſiegers, or to hinder their Approaches. *Counter-Battery* is a Battery raiſed to play upon another to diſmount the Guns.

Counter Guard is a Work raiſed before the Point of a Baſtion, conſiſting of two long Faces, parallel to the Faces of the Baſtion, making a Salliant Angle; they are ſometimes of other Shapes, and otherwiſe ſituated.

Counter march, is an Armies turning ſuddenly their March the contrary way; which may be occaſion'd by the Enemies endeavouring to get betwixt them and their Garriſons, or may be done to diſappoint and amuſe the Enemy. A Battalion is ſaid to Counter-march, when being charged in the Rear, they change the Front by marching thoſe which are in the Front, or the *File-Leaders* to the Rear, in place of the *Bringers Up*. Ranks are ſaid to Countermarch, when the Wings of a Battalion interchange Ground.

Counter-mine is uſed when the Be-

Besiegers have, notwithstanding the Opposition of the Besieged, passed the Foss, and put the Miner to the Foot of the Rampart: They are of two sorts, being either made when the Bastion is raised, or afterwards when it is attacked. Those that are made when the Bastion is raised, are carried quite round the Faces of a Bastion; their height is from 4 to 5 foot, and broad enough for a Man to pass easily. The others, which are made in time of Necessity, when the Besiegers are undermining a Bastion, are Pits sunk deep in the Ground where the Miner is supposed to be, from whence they run out Branches in search of the Enemy's Mine, to frustrate the Effect of it, by either taking away the Powder, or cutting the Train.

Counterscarp is properly the exterior Talus or Slope of the Ditch, but it is often taken for the *Covert Way* and the *Glacis*. The Enemy in this sense are said to have attack'd the Counterscarp, or lodged themselves on the Counterscarp.

Counter-Trenches, see *Counter-Approaches*.

COURT-MARTIAL is called by a *General* to try Offenders; it is composed of a *President* named by the *General*, and of 12 *Members* and a *Judge Advocate*, who take an Oath to do Justice according to their Knowledge; the Sentence is, after being approved by the *General*, put in Execution by the *Provost-Marshal* of the Army.

COINS are wedges of Wood

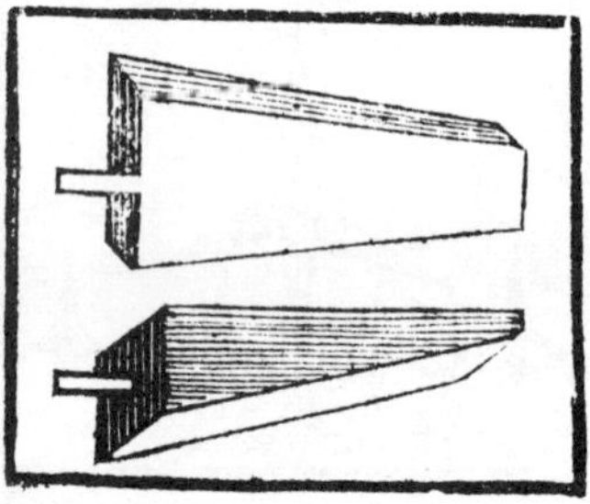

used under the Breech of a Gun, by which a *Gunner* raises or falls the Muzzle of his Piece, till he point it exactly at the Object: Each Gun has three Coins belonging to her. They are for the same use about a Mortar, see their Shape in the *Figure*.

CRAB, see *Gin*.

CRIC, see *Jack*.

CROWN-WORK is the largest of all Out-Works, and encompasses the most ground, having a very large Gorge, generally the length of the Curtin of the Place, and two long sides, terminating toward the Campaign in two Demi-bastions, each of which is join'd by a particular Curtin, to a whole Bastion, which is the Head of the Work; these Works are always raised to inclose a rising Ground, or to cover the Head of a Retrenchment, see the *Plan* at the Beginning.

CROW-FEET, *Chausse-Traps*, or *Caltrops*, are Machines of Iron, having four Points of about 3 or 4 inches long, so made that which ever

ever way they fall, there is ſtill a

Point up; they are to be thrown upon Breaches, or in Paſſes where Cavalry are to march, to whom they are very troubleſome, by running into the Horſes feet and laming them, ſee the *Figure*.

CUBE is a Geometrical Figure, being a ſolid Body bounded by ſix Squares, like a Die. *Cubical*, as a Cubical Foot or a Cubical Inch, is, when the ſides of the Squares that make the Cube are a foot or an inch long.

CUIRASSIERS are Cavalry armed with Back, Breaſt, and Head-Piece, as moſt of the *Germans* are; we have had no *Cuiraſſiers* in the *Engliſh* Army, ſince the laſt Revolution.

CULVERIN is a Cannon about 5 inches and a quarter diameter of the Bore, and from 9 to 12 foot long, carrying a Ball of 18 pound; it is a good battering Gun, but is too heavy for a Field-Piece.

CURTIN is that part of the Rampart of a Place which is betwixt the Flanks of two Baſtions, and is the beſt defended of any part of the Rampart, wherefore Beſiegers never make their Attacks on the Curtins, but on the Faces of the Baſtions, becauſe of their being defended but by one Flank.

CUNETTE or *Cuvette* is a deep Trench about three or four fathom wide, ſunk along the middle of a dry Moat, to make the Paſſage more difficult to the Enemy; it is generally ſunk ſo deep till they find Water to fill it, and is good to prevent the Beſiegers Mining.

CYLINDER or *Chace* of a Gun is the Bore or Concavity of a Piece, whereof that part which receives the Powder and Ball, is called the *Charged Cylinder*, and that which remains empty after the Gun is charged, is called the *Vacant Cylinder*.

D

DECAGON is a Figure of ten Sides or Polygons, forming ten Angles, each of which may be fortify'd with a Baſtion.

DECAMP is the breaking up from a Place where the Army has been encamped, to march to another.

DEFENCE *of a Place*, are the parts of a Wall or Rampart which flank and defend the reſt, as the Flanks, Cazemates, Parapets, and Fauſſebrays. The Face of a Baſtion, tho it has the ſimpleſt Defence of any part of the Fortification, yet it cannot be ſtormed, till the oppoſite Flank be ruined.

ined. *To be in a posture of Defence*, is to be in a Condition to resist or oppose an Enemy.

Defence, Line of Defence, see *Line*.

DEFILE is a narrow Pass which obliges an Army to file off; it is one of the greatest Obstacles that can occur in the March of an Army, especially if it happen to be betwixt Woods or Marshes, for it not only gives an Enemy an extraordinary Advantage of either attacking the Front or Rear, since they cannot come to relieve one another, because of the straightness of the Passage, but it likewise very much impedes the March of an Army. A Retreating Army puts always a Defile betwixt them and the Enemy to secure their Retreat, as the Confederate Army of the Emperour, *Spain* and *Holland*, after the Battle of *Seneff* in the year 1673, being at the Siege of *Oudenard*, and hearing that the Prince of *Conde* was advancing, they raised the Siege, and defended a Defile which happened in the March of that Prince, so that he could not without a considerable Disadvantage, hinder their saving themselves under the Cannon of *Ghent*.

To Defile, is to reduce an Army to a small Front, to march thro' such a narrow Passage.

DEGREE is properly a Term in Geometry, often used in Fortification, to measure the Angles, being the 360th part of the Circumference of a Circle; a Degree is subdivided into 60 equal Parts, called Minutes, and each Minute into 60 Seconds.

DEMIBASTION, see *Bastion*

Demi-Cannon is a Gun carrying a Ball of 32 pound weight, the diameter of its Bore is 6 inches and a half, and its length from 12 to 14 foot: they are seldom used at Sieges, because of their extraordinary Charge.

Demi-Circle is the half of a Circle, cut by a Line passing through the Centre, called the Diameter.

Demi-Culverin is a Cannon of about 9 foot long, the diameter of the Bore is 4 inches and a quarter carrying a Ball of 9 pound weight; it is a very good Field Piece, takes 13 Horses to draw it, and two *Gunners* and 4 *Matrosses* to serve it.

Demi-Gorge is that part of the Polygon which remains after the Flank is raised, and goes from the Curtin to the Angle of the Polygon, It is half of the vacant Space or Entrance into a Bastion.

DESCENT into a Moat is a deep Trench or Sap thro' the Esplanade, and under the Covert way, cover'd over-head with *Planks* and *Hurdles*, and loaded with Earth against Artificial Fires, to secure the Descent, which in Ditches that are full of Water is made to the brink of the Water, but in dry Moats the Sap is carried to the bottom of the Moat, where Traverses are made to lodge and cover the Besiegers.

DESERTER is a *Soldier*, who,

who, by running away from his Regiment or Company, abandons the Service; he is by the Articles of War punishable by Death, and after Conviction, is hang'd at the Head of the Regiment he formerly belong'd to, with his Crime writ on his Breast, and suffered to hang till the Army leave that Camp, to terrifie others.

Detachment is a certain number of Officers or Soldiers drawn out from several Regiments or Companies equally, to be employ'd as the *General* pleases, whether on an Attack at a Siege, or in Parties to scowr the Country. A Detachment of 2 or 3000 Men, is a Command for a *Brigadeer*, 800 for a *Colonel*, 4 or 500 for a *Lieutenant Colonel*; a *Captain* never marches on a Detachment with less than 50 Men, a *Lieutenant*, and an *Ensign*, and two *Serjeants*; a *Lieutenant* is allowed 30 and a *Serjeant*; an *Ensign* 20 and a *Serjeant*; and a *Serjeant* 10 or 12 Men. *Detachments* are sometimes made of entire Squadrons and Battalions.

DIAMETER of a *Circle*, is a Right Line which passes thro' the *Centre* and touches the Circumference in two Points, dividing the *Circle* into two equal parts.

DISMOUNT the Enemies *Cannon*, is to break their *Carriages*, their *Wheels*, *Axiltrees*, or any thing else so as to render them inserviceable.

Dismount, is likewise a Command to the *Dragoons* to alight or unhorse.

DISPART of a *Gun*, is the difference in the thickness of Metal between the *Base* and *Muzzle Ring*, and is found thus; take with a pair of *Caliber Compasses* the diameter of these two Rings; suppose of an *English Demiculverin*, the diameter of the Base may be about 14 inches, and that of the Muzzle about 11; the difference then is three, half of which is one and a half, which is the difference of the thickness of Metal at these two places: The *Dispart* is a piece of Stick of this length, set on the *Muzzle Ring*, which makes a Visual Ray or Line drawn from the *Base Ring* to the top of this Stick, an exact parallel to the *Axis* of the *Concave Cylinder*, or to the Soul of the Piece, and serves to direct the *Gunner's* Eye in levelling his Gun at an Object.

DITCH, see *Moat*.

DIVISIONS of a *Battalion* are the several Parcels into which a *Battalion* is divided in marching; the *Lieutenants* and *Ensigns* march before the *Divisions*.

The Divisions of an Army, are the *Brigades*.

DODECAGON, is a Figure bounded by twelve Sides, forming as many *Angles*, capable of being fortified with the same number of *Bastions*.

DONJON, is a Place of Retreat, to capitulate with more Advantage, in case of Necessity.

DOSSER, is a sort of Ba-

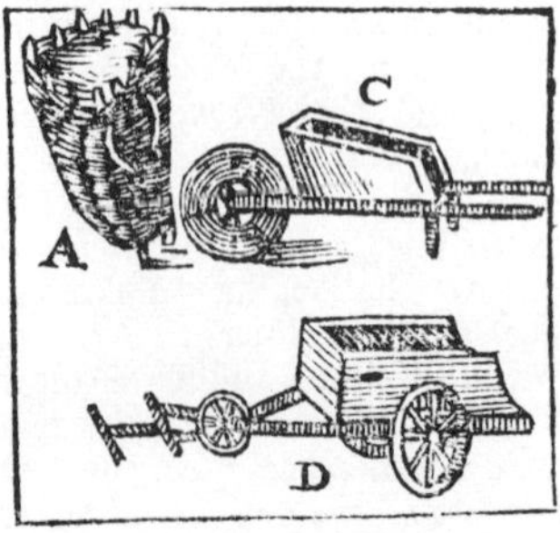

sket, shaped as the *Figure A* shows, to be carried on the Shoulders, and is used in carrying the Overplus Earth from one part of a *Fortification* to another, where it is wanted. There are likewise small *Carts* and *Wheel-barrows* for the same use, as *D. C.*

DOUBLE-TENAILLE, see *Tenaille.*

Double; to *Double*, is a Word of Command, as *Double your Ranks*, or *Double your Files*; *doubling of the Ranks*, is when the second, fourth, and sixth *Ranks* march into the first, third and fifth, so that the six *Ranks* are reduced to three, and the *Intervals* betwixt the *Ranks* become double what they were before; but, *to double by half Files*, is when the fourth, fifth and sixth *Ranks* march up to double the first, second and third, or the contrary. *To double the Files to the Right*, is when every other *File* faces to the Right, and marches into the next *File* to it; so that the six *Ranks* are turned into 12, and each *File* is 12 deep. *To double the Files to the Left*, is when every other *File* faces to the Left, and marches into the next; in doubling the *Files*, the distance betwixt the *Files* becomes double.

DRAGOON, is a Musqueteer mounted on Horse-back, fighting sometimes on Foot, and sometimes on Horse-back, according as Occasion requires: they are divided into *Brigades*, as the *Cavalry*, and each Regiment into *Troops*, each Troop having a *Captain*, *Lieutenant Cornet*, *Quarter-Master*, 2 *Serjeants* 3 *Corporals*, and 2 *Drums*, some Regiments have *Hautboyes*; they are very useful on any Expedition that requires Dispatch, for they can keep pace with the *Cavalry*, and do the Duty of *Infantry* They encamp generally on the Wings of the Army, or at the Passes leading to the Camp, and sometimes they are brought to cover the *Generals* Quarters; they do Duty on the *Generals* of Horse and Dragoons, and march in the Front and Rear of the Army.

DRAW-BRIDGE, see *Bridge.*

DRAUGHT-HOOKS, are large *Hooks* of Iron, fixt on the *Cheeks* of a *Cannon-Carriage*, two on each side, one near the *Trunion-Hole*, and the other at the *Train*, and are called the fore and hind *Draught Hooks*. Large Guns have *Draught Hooks* near the middle *Transum*, to which are fixt the *Chains* which serve to ease the *Shafts* of the *Limbers* on a march. The fore and hind *Hooks* are used for

for drawing a Gun backwards or forwards by Men, with ſtrong Ropes called *Draught-Ropes*, fixt to theſe *Hooks*.

DREIN, is a Trench made to draw the Water out of a Moat, which is afterwards filled with Hurdles and Earth, or with Faſcines or Bundles of Ruſhes and Planks, to facilitate the Paſſage over the Mud.

DRUM, is a martial Inſtrument uſed by the *Foot* and *Dragoons*, in form of a *Cylinder*, hollow within and cover'd at the two ends with Vellum, which is ſtretch'd or ſlackned at pleaſure by the means of ſmall Cords and ſliding Knots: ſome are made of Braſs, but they are commonly of Wood The *Drum-ſticks* are made of hard firm Wood, about 15 or 18 Inches long, tapering towards the point, where there is a ſmall Knob, which by beating againſt the *Drum head* makes the Sound, which is different as the ſeveral Occaſions require: As *To beat the General*, is a ſignal for the Army to make ready to march; *The Aſſembly* is the next Beat, which is an Order for the Soldiers to repair to their *Colours*; and *the March*, is to command them to move. *To beat the Retreat*, is for the Army to keep to their Tents, and not to fire till next morning, for fear of alarming the Camp; this is about Sun-ſet, at the firing of a Gun call'd the *Warning-piece*, after which the *Centries* challenge, and the *Picquet-Guard* is relieved. *The Reveille* is beat at break of day, to warn the Soldiers to riſe, and the *Centries* to forbear challenging. *The Alarm* is to call the Soldiers to Arms on notice of ſome ſudden Danger. *To beat a Parley* or *Chamade*, is to deſire a conference with the Enemy. *To beat a Call*, is to advertiſe the Soldiers to ſtand to their Arms, when a *General Officer* is paſſing by. See *Kettle Drum.*

DRUMMER, is he that beats the *Drum*; each Company of Foot has two *Drummers*; and a Battalion has a *Drum-Major*, who has the command over the other *Drums*. They are diſtinguiſh'd from the Soldiers by Cloaths of a different faſhion, generally laced with a Worſted Livery-lace; their Poſt, when a Battalion is drawn up, is on the Flanks; and on a March it is betwixt the Diviſions.

DUTY, is the exerciſe of thoſe Functions that belong to a Soldier; with this diſtinction, that *Mounting Guards* and the like, where there is not an Enemy directly to be engaged, is called *Duty*; but their marching to meet and fight an Enemy is called *Going upon Service.*

E.

EARTH-BAGS, see *Sand-Bags*.

ECHARPE, *To batter en Echarpe*, is to batter obliquely or side-ways. The *Flanks* of Count *Pagan's* construction may be batter'd *en echarpe*, because the *Angles* of the *Curtin* being so obtuse are too much discover'd.

Echaugette, see *Guerritte*.

Embrasures, are the Cuts made through the Parapet of a *Battery* for the Muzzles of the Guns, and for the Passage of the Shot. When a *Battery* is brought on the *Glacis* of a Place, there are thick planks of Wood Musquet-proof to stop the *Embrasures*, as soon as the Gun is fired, to save the *Gunners* and *Matrosses* that are employed about the Guns from the small Shot, which plays continually upon them from the Besieged. See *Battery*.

EMINENCE, is a High or Rising Ground, which over looks and commands the low places about it; such places within *Cannon-Shot* of a fortified Place are a great Disadvantage, for if the Besiegers become Masters of them, they can from thence fire into the Place.

EMPATEMENT, see *Talus*.

ENCIENTE, is the Wall or Rampart which surrounds a Place, sometimes composed of *Bastions* and *Curtins*, either faced or lined with Brick or Stone, or only made of Earth. The *Enciente* is sometimes only flanked by round or square Towers, which is called a *Roman-Wall*

ENFANS-PERDUS, see *Forlorn-hope*.

ENFILADE; to *Enfilade*, is to be Masters of a Ground from whence a Post or an Enemy is flanked, so that it may be battered all along a Right Line. At the second Siege of *Dole* in the year 1674, the King of *France* caused raise a *Battery*, which enfiladed or scowred the whole length of the *Rampart*, and dismounted three *Faulconets*, which the Enemy had planted there. In conducting the *Approaches* at a Siege, care must be had that they be not enfiladed from any Work of the Place, but that they be carried on with Windings and Turnings till they are brought to the *Glacis*, and then they are carried strait forwards, being sunk deep in the Ground, and cover'd over-head.

ENNEAGON, is a *Figure* or *Polygon* of nine Sides, and as many *Angles*, each capable of being fortified with a *Bastion*.

ENSIGN, is the Officer that carries the *Colours*, being the lowest Commission'd Officer in a *Company* of *Foot*, subordinate to the *Captain* and *Lieutenant*; it is a very honourable and proper Post for a Young Gentleman, at his first coming into the Army: He is to carry the *Colours* both in Assault or Day of Battel, and should be no ways daunted with Danger, nor should he quit his

Colours, but with his Life. In entring into a Place, mounting Guard, passing a Review, or going to Battel, he should carry his *Colours* himself on his Left Shoulder; but upon a March may have them carried by a Soldier, tho' this is not practised in the *English* Army.

ENVELOPE, is a Work of Earth made sometimes in the *Ditch* of a Place, sometimes without the *Ditch*, sometimes in fashion of a simple *Parapet*, and at other times like a small *Rampart* with a *Parapet*. *Envelopes* are often made to inclose a weak Ground, when it is to be done with simple Lines, to shun the great Charge of *Hornworks*, *Tenails*, or the like, or when they have not Ground for such large Works. The Castle of *Namure* has two *Envelopes* on the *South-West* side of the *Donjon*, one before the other, composed of two *Demi-bastions* and a *Curtin*, and call'd the first and second *Envelopes*, and without both these, a large Work extending itself on the top of the Hill with two *Demi-bastions*, call'd the *Terre Neuve*, or *New Land*. The *Cittadel* of *Besanson*, which is situated on a high steep Rock, has three *Envelopes* one before another towards the *Campaign*, which serve as so many *Covert Ways* before the Moat. The Fort *Niewerburg* in *Holland*, is famous for its *Envelope* which goes quite round the *Fort*, and is fraised and palisaded with Stakes as thick as a Man's Body, yet it was taken by the Duke of *Luxemburg*. There is a very good *Envelope* before the Port of St. *Servais* at *Maestricht*. *Douay* is environ'd with an *Envelope*, so is the Town of *Zratzein* in the Palatinate, yet were both taken by the Marquis of *Cezan*, the day that Mareschal *Turenne* gain'd the Battel over the Duke of *Lorrain* and Count *Caprara*, in the year 1674. Some give the Names of *Sillon*, *Counter-Guard* and *Conserves*, to such *Envelopes* as are made in the Moat; sometimes they are call'd *Lunettes*. See *Lunette* and *Sillon*.

EPAULE, or *Shoulder* of a *Bastion*, is the place where the *Face* and *Flank* meet, and form the *Angle* call'd the *Angle of the Shoulder*.

EPAULMENT, is a Work raised either of Earth, *Gabions* or *Fascines*, loaded with Earth to cover side Ways. The *Epaulments* of the Places of *Arms* for the *Cavalry* at the entering of the *Trenches*, are generally of *Fascines* mix'd with Earth.

Epaulment, or *Square Orillon*, is a Mass of Earth, of near a square Figure, faced with a Wall to cover the Cannon of a *Cazemate*.

EPTAGON, or *Heptagon*, is a Figure of seven *Sides* and seven *Angles*, capable of being fortified with seven *Bastions*.

EQUILATERAL Figure, whether Triangle, Square, *&c.* is a Figure whose Sides are all equal.

ESCALADE, see *Scalade*.

ESCARPE,

ESCARPE, see *Scarpe*.

ESPLANADE, is the sloping of the *Parapet* of the *Covert Way* towards the *Campaign*; it is the same with *Glacis*, but begins to be antiquated, and is more properly the empty space betwixt a *Citadel* and the Houses of a Town.

ESTOILE, see *Starr-Redoubt*.

ETAPPE, is a *French* Term for the Distribution of Provisions and Forrage to an Army in their Rout through a Kingdom, going to Winter Quarters, or returning to take the Field.

ETAPPIER, or *Undertaker*, is he that contracts with a Country or Territory, for furnishing *Troops* in their March with Provisions and Forrage. The *Etappier* is to deliver the *Etappe* to the *Majors* of Horse and Foot, or in their absence, to the *Quarter-Masters* of Horse, and *Serjeants* of Foot; they are not to give the *Soldiers* Money for their *Etappe*, because it would create Abuses. The Officers take sometimes a sum of Money from the *Etappiers*, and oblige the *Soldiers* to march two days march in one, which harasses both Men and Horse extreamly, and is a great Abuse.

EVOLUTION, is the Motion made by a Body of *Troops*, when they are obliged to change their Form and Disposition, in order to preserve a Post, or occupy another to attack an Enemy with more Advantage, or to be in a condition of defending themselves the better; and consists in *Doublings*, *Counter marches*, *Conversions*, &c. A Battalion *doubles the Ranks* when attacked in *Front* or *Rear*, to prevent its being flanked or surrounded; for then a *Battalion* fights with a larger *Front*. *The Files are doubled*, either to accommodate themselves to the necessity of a narrow Ground, or to resist an Enemy which attacks them in *Flank*; but if the Ground will allow it, *Conversion* is much preferable, because after *Conversion* the *Battalion* is in its first Form, and opposes the *File Leaders*, which are generally the best Men, to the Enemy; and likewise because doubling the *Files* in a new or not well disciplined Regiment, they may happen to fall in Disorder.

EXAGON, is a Figure bounded by six *Sides* or *Polygons*, making as many *Angles* capable of *Bastions*.

EXERCISE, is the Practise of all those Motions, Actions and Management of Arms, whereby a *Soldier* is taught the different Postures he is to be in under Arms, and the different Motions he is to make to resist an Enemy, which he must be perfect in, before he be fit for the Service.

EXTERIOR-POLYGON, see *Polygon*.

F.

FACE *of a Bastion*, is the streight Line comprehended between the *Angle* of the *Shoulder* and the *Flanked Angle*, which is composed of the meeting of the two *Faces*, and is the most advanced part of a *Bastion* towards the *Campaign*. The *Faces of a Bastion* are the weakest parts of a Fortification, being the least flanked. It is therefore there where the Breach is generally made; for the opposite *Flank* being ruined, which should defend the Passage of the Moat, the Besiegers meet with less Opposition than they could expect in attacking any other part of the Rampart which is better flanked.

Face of a place, is the *Front* comprehended between the *Flanked Angles* of two neighbouring *Bastions*, composed of a *Curtin*, two *Flanks* and two *Faces*, and is likewise called the *Tenaille of a place*. In a Siege the Attacks are carry'd against both *Bastions* when the whole *Tenaille* is attack'd.

Face Prolong'd, is that part of the *Line of Defence Razant*, which is betwixt the *Angle* of the *Shoulder* and the *Curtin*, or the *Line of Defence Razant*, diminished by the length of a Face.

Face is likewise a word of Command; as *Face to the right* is to turn upon the left Heel a quarter round to the right. *Face to the left*, is to turn upon the right Heel a quarter round to the left.

Face of a Gun, is the Superficies of the Metal at the extremity of the muzzle of the Piece.

FAGGOTS, are Men hired by Officers, whose Companies are not full, to muster, and so cheat the Queen of so much Pay. In the late War, by a Proclamation all over the Army, any that could give notice of such an Abuse was promised a Reward in Money, and his Discharge if he desired it; and the Officer, who was guilty, was to lose his Commission, and be rendred uncapable of ever serving his Majesty; and the *Faggot*, who was muster'd, was punishable with Death.

FALCON and *Falconette*, are small Guns; the first about two inches three quarters diameter of the Bore, and the other two inches; they are too small to be of use in the Field with an Army, and are now out of use.

FALSE *Alarm*, see *Alarm*.

False Attack, see *Attack*.

FANIONS, are small Flags carry'd along with the Baggage. See *Flags*.

FASCINES, are Faggots of small Wood, of about a foot di-

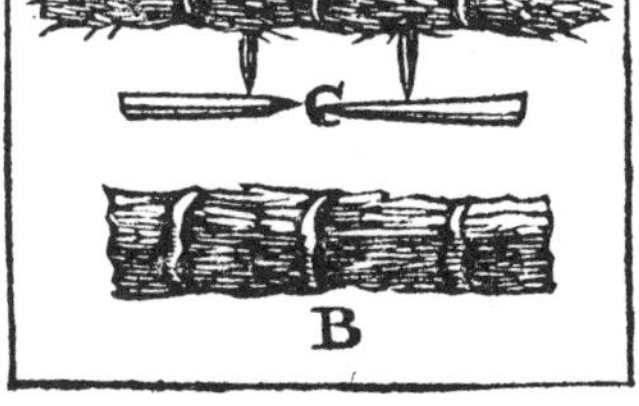

ameter

ameter and 6 foot long *A*, bound in the middle and at both ends; they are brought by the *Cavalry* to the entrance of the *Trenches*, from whence the Workmen carry them along to raiſe *Batteries* or other Works, to make *Chandeleers*, *Epaulments*, or to fill up the *Moat* to facilitate the paſſage to the foot of the Wall; they are likewiſe uſed in making the Sap or Deſcent into the Ditch, in making *Caponeers* and *Coffers*, and many other things; and being uſed ſo, are cover'd over with Earth or Raw-Hides, to prevent their being ſet on fire. They are uſed in fortifying a Place, eſpecially where the Earth is bad, to bind the *Rampart*, where they are laid athwart ways and drove down with Stakes *C*, with a Bed of Earth above them, the *Faſcines* again and Earth again, till the *Rampart* be finiſhed; or to keep up the Earth of the *Parapet*, and then they're laid length-ways, and drove faſt with Stakes of 3 or 4 foot long. There are ſhorter *Faſcines* or *Bavins*, about a foot and a half long, which being pitch'd over, are to be thrown upon a Gallery or other Work of the Enemy, to ſet it on fire. In the year 1644, when his Royal Highneſs *Gaſton* of *France*, Duke of *Orleans*, beſieged *Graveline*, having made a Paſſage or Bridge over the Moat with *Faſcines*, a *Neapolitan Soldier* of the Garriſon endeavouring to burn this *Traverſe*, his *Fireworks* rebounded on himſelf and burnt him, the *Faſcines* of the *Traverſe* being covered with Raw-Hides. *Faſcines* differ from *Sauciſſons*, the former being made of ſmall Wood, and the latter of Branches of Trees *B*. *Faſcines* are ſometimes ordered to be made by the *Cavalry*, before a march over bad Ground, and are carried by them, each *Trooper* having one to mend the Ways.

FAUSEBRAYE, is a ſmall *Rampart* about 3 or 4 fathom wide, bordered with a *Parapet* and *Banquett*, the Deſign of it is to defend the *Foſs*; they are not reckoned ſo uſeful where there is a dry Moat, becauſe the Beſiegers may make better Works for the Defence of it than *Fauſebrayes*, ſuch as *Traverſes*, *Scillons*, and *Coffers*. But in places ſurrounded with a wet Ditch, they may be more uſeful, provided they be made only before the *Curtin* and *Flanks*, for lying low, they cannot be eaſily hurt by the Enemies Cannon, and defend the *Foſs* better, becauſe of their low Situation, than the *Rampart*, which, becauſe of its height, cannot ſo well diſcover the *Foſs*. They ought never to be made before the *Faces*, eſpecially in places faced with Brick or Stone, becauſe the *Breach* being generally made in the *Face*, the Ruins and Rubbiſh of the *Rampart* are ſtop'd in the *Fauſebray*, which facilitates the Aſcent of the *Breach*; and in places lined with Brick or Stone, the pieces of Stone or Brick fly among the *Soldiers* that are in the *Fauſebray*.

FELLING-AX, see *Ax.*

FELLOWS, are 6 pieces of Wood, each whereof form a piece of an Arch of a Circle of 60 Degrees, and join'd all together by Duledges, make an entire Circle, which, with the addition of a *Nave* and 12 *Spokes*, make a *Wheel*. See the Proportion of them at *Wheel*.

FICHANT. *Line of Defence Fichant*, see *Line*.

FIELD-OFFICERS, are those that have the Power and Command over a whole *Regiment*, such are the *Colonel*, *Lieutenant-Colonel* and *Major*, but such whose Commands reach no further than a *Troop* or *Company*, are not *Field-Officers*. See *Officers*.

Field-Pieces, are small *Guns* proper to be carried along with the Army into the Field, such as 3 *Pounders*, *Minions*, *Sakers*, 6 *Pounders*, *Demi Culverins*, and 12 *Pounders*, which, because of their smalness, are easier drawn, are a less Charge, require lesser quantities of Ammunition, and are easier served.

Field Staff, is a Weapon carried by the *Gunners*, about the length of a *Halbert*, with a Spear at the end, having on each side Ears screw'd on, like the Cock of a *Match Lock*, where the *Gunners* screw in lighted Matches when they are upon Command; and then the *Field-Staffs* are said to be armed.

FILE, is the Line of *Soldiers* standing one behind another, which is the depth of the *Battalion* or *Squadron*. The *Files* of a *Battalion* of Foot, are generally six deep, and those of a *Squadron* of Horse three. The *Files* must be streight and parallel one to another; to *double the Files*, is to put two *Files* into one, which makes the depth of the *Battalion* double of what it was in number of Men. The *File Leaders* are the foremost Men in each *File*; the *Bringers up* are the last Men of each *File*, or the last Rank of the *Battalion*; the *half File-Leaders*, is the fourth Rank, because the fourth, fifth and sixth Ranks are call'd the *Rear half Files*, as the first, second and third are the *Front half Files*.

To File off, is the same as *to Defile*, or to file off from a large Front to march in length. An Army is said to file off from the Right or from the Left, when the *Squadrons* begin to move from the Right or Left, marching one after another, and so reducing the two *Lines* or *Ranks* of the Army to two long *Files*. A *Battalion* is said to file off, when it marches by Divisions or Sub-divisions.

FIRE, is a Word of Command, to the Foot *Soldiers* to discharge their *Musquets*, to the *Cavalry* to discharge their *Carabines* or *Pistols*, to the *Grenadiers* to fire their *Grenades*, and to the *Gunners* to fire the *Guns*. *Running Fire*, is when a Rank of Men drawn up, fire one after another, or when the Lines of an Army are drawn out to fire Victory,

for which each *Squadron* and *Battalion* takes it from another, from the Right of the first Line to the Left, and then from the Left to the Right of the second Line.

Fire-Arms; under this Name are comprehended all sorts of Arms that are charged with Powder and Ball, as *Cannon*, *Musquets*, *Carabines*, *Pistols*, *Blunderbusses*, &c.

Fire-Ball, is a Composition of

Meal-Powder, *Sulphur*, *Salt petre*, *Pitch*, &c. about the bigness of a *Hand Grenade A*, coated over with Flax, and primed with a slow Composition of a Fuze, to be thrown into the Enemies Works in the Night time, to discover where they are; or to fire Houses, Galleries, or other Blinds of the Besiegers; but they are then armed with Spikes or Hooks of Iron that they may not roll off, but stick or hang where they are designed to have any Effect.

Fire-Lock, or *Fusil*, is a Fire Arm carried by a Foot *Soldier*; the Barrel of it is about three foot eight inches long, the Stock is about 4 foot and eight inches, and the Bore is fit to receive a Bullet of Lead of an ounce weight.

Fire-Master, is an Officer who gives the Directions and Proportions of Ingredients for each Composition required in *Fire-works*, whether they be for Pleasure-Works, such as *Balloons*, *Rackets*, &c. or for the War, as *Fuses*, *Fire-balls*, *Carcasses*, *Quick-match*, &c. or firing of *Bombs*, *Petards*, and *Hand Grenades*; these Orders are given to the *Fire-Workers* and *Bombardeers* who execute them. The chief *Fire-Master* of *England*, is a Post belonging to the Office of Ordnance.

Fire-Pots, are small Earthen Pots, into which is put a *Grenade* filled with Powder, and then the Pot is filled with fine Powder till the *Grenade* be covered, then the Pot is covered with a piece of Parchment, and two pieces of Match across lighted; this Pot being thrown by a handle of Match where it is design'd, it breaks and fires the Powder and burns all that is near it, and likewise fires the Powder in the *Grenade*, which ought to have no Fuse, to the end its Operation may be the quicker. See the *Figure* at *Fire Ball*, B.

Fire-Workers, are Officers subordinate to the *Fire-Masters*, but command the *Bombardeers*; they receive the Orders from the *Fire-Master*, and not only see them executed, but work themselves along with the *Bombardeers*; there are twenty four *Fire Workers* establish'd in the Office of Ordnance.

Fireworks, are the Works made by the *Fire-Workers*, whether for War or Recreation.

FLAGS, in *French Fanions*, are ſmall Banners of Diſtinction ſtuck in the *Baggage-Waggons* of the Army, to diſtinguiſh the Baggage of one *Brigade* from another, and of one *Battalion* from another, that they may be marſhalled by the *Waggon-Maſter-General*, according to the Rank of their *Brigades*, where they are to keep during the March, to avoid the Confuſion that otherwiſe would be.

FLANK, is that part of the *Rampart* which joins the *Face* and the *Curtin*, comprehended betwixt the *Angle* of the *Curtin* and the Angle of the *Shoulder*, and is the principal Defence of a Place. *Engineers* have differed very much about raiſing the *Flank*, ſome make it perpendicular on the *Face*, ſome on the *Curtin*, others perpendicular on the *Line of Defence*, ſome give it an Angle of 98 Degrees with the *Curtin*, and *Vauban* makes it the Chord of a *Segment*, whoſe Centre is the Angle of the *Shoulder* of the next *Baſtion*; its uſe is to defend the *Curtin*, the *Flank* and *Face* of the oppoſite *Baſtion*; to defend the Paſſage of the Moat, batter the *Sailliant Angles* of the *Counterſcarp* and *Glacis*, from whence the Beſieged generally ruin the *Flanks* with their *Artillery*, for the *Flanks* are the parts of a *Fortification* which the Beſiegers endeavour moſt to ruin, becauſe of taking away the Defence of the Face of the oppoſite *Baſtion*.

Flank Oblique, or *Second Flank*, is that part of the *Curtin* from whence the Face of the oppoſite *Baſtion* may be diſcovered. This ſhows in a Plan upon Paper to be a good Defence, but is rejected by moſt *Engineers* as no ways uſeful for its being ruined at the beginning of a Siege, eſpecially if it be of a ſandy Earth; the ſecond *Parapet* which may be raiſed behind the former, is of no uſe, for it neither diſcovers nor defends the Face of the oppoſite *Baſtion*, beſides it ſhortens the *Flank* which is the true Defence, and the continual Fire of the Beſiegers Cannon will never ſuffer them to raiſe a ſecond *Parapet*. This ſecond *Flank* defends very obliquely the oppoſite Face, and is to be uſed only in a Place which is to be attacked by an Army without Cannon, as being only a conceited Opinion of *Cabinet Engineers*.

Flank Retired, or *Low Flank*, is one of the *Platforms* of the *Cazemate*, and is ſometimes call'd the *Cover'd Flank*. This is generally called the *Cazemate*, when there is only one *Platform* retired towards the Capital of the *Baſtion*, and covered by an *Orillon*; theſe retired *Flanks* are a great Defence to the oppoſite *Baſtion*, and to the Paſſage of the Moat, becauſe the Beſiegers cannot ſee nor eaſily diſmount their Guns.

Flanks of an Army, are the Troops encamped on the Right and Left. *Flanks of a Battalion*, are the Files on the Right and Left.

To flank, is to attack and fire upon the Flank of an Enemy; it is a very common Term, and very essential in *Fortification*, for all Works that have only their Defence right forwards are deficient, and must have each place to flank another, and be flanked reciprocaly, otherwise it is not in Perfection. The *Curtin* is the strongest part of a *Fortification*, because it is flanked at both ends by the *Flanks*, and the *Face* having but one Defence from the opposite *Flank*, is counted the weakest.

Flanked-Angle, see *Angle*.

Flask, is a Horn or such a thing made for carrying of Powder, and has a measure of the Charge of the Piece for the top.

FLYING-ARMY, see *Army*.

Flying-Bridge, see *Bridge*.

FOOT, are such as list themselves under a *Captain* to serve on Foot, and are armed with a *Sword*, *Bagonet*, *Fire lock* or *Pike*, *Collar* of *Bandeleers*, *Cartridge-Box*, *&c.* The Foot are formed into *Companies*, and according to the Articles of War, a *Soldier* is not to leave his *Company* without leave from his Officer to go about his own Business, without being reputed a Deserter, and tryed for his Life. These *Companies* are formed into *Regiments*, called *Regiments of Foot*.

Foot, as it is a Measure, contains 12 inches, and each inch 12 Lines; 6 foot make a Fathom, which is a Measure equal to the *Toise* in *France*; 5 foot make a *Geometrical Pace*, 3 foot an *English Yard*, and two foot and a half make a Common *Pace*. To be on the same *Foot*, is to be in the same Circumstances with another. A *Regiment* is said to be on the same *Foot* with another, when it has the same number of Men and the same Pay. To *gain* or *lose* Ground *Foot by Foot*, is to dispute a Post resolutely with an Enemy, losing it by degrees, and defending every bit of it to the utmost.

Foot-Bank, is a small step of Earth, on which the *Soldiers* stand to fire over the *Parapet*; there are generally two, and sometimes three, but the height of the *Parapet* from the uppermost *Foot-Bank*, ought always to be four foot and a half. See *Banquett*.

FORELAND, see *Liziere*.

FORGE, is an Engine carried along with the *Artillery* for the *Smiths*, and is a travelling *Smiths Forge*. *Forge for hot Ball*, is the place where the Ball are made hot before they be fired off; it is built of Brick, and hath a Furnace below, over which are Bars of Iron; it is cover'd over-head, and the Balls laid upon the Bars till they be red hot, and are taken out with long Ladles to be put in the Gun. The Materials for such *Forges* are carried along with the *Artillery*, when there is any design of burning Magazines or the like with hot Ball.

FORELORN in *French Enfans Perdus*; are Men detached from

from ſeveral *Regiments*, or otherwiſe appointed to make the firſt Attack in Day of Battel, or at a Siege to be the firſt in ſtorming the *Counterſcarp*, mounting the *Breach* or the like. They are called ſo from the eminent Danger they are expoſed to; but this Word is old, and begins to be obſolete.

FORMERS are of ſeveral ſorts, but the chief is for making *Cartridges* for *Cannon*; they are round pieces of Wood fitted to the diameter of the Bore of a *Gun*, on which the *Paper*, *Parchment*, or *Cotton* which is to make the *Cartridge*, is rolled before it be ſewed.

FORRAGE, is the *Hay*, *Oats*, *Barley*, *Wheat*, *Graſs*, *Fitches*, *Clover*, &c. which is cut down and brought into the Camp by the *Troopers*, for the Subſiſtance of their Horſes; it ought to be chiefly conſidered by the *Quarter-Maſter General* in encamping an Army, that it be in a Country of *Forrage*; it is he that orders the Method of the forraging, and poſts the Guards for the ſecurity of the *Forragers*. *Dry Forrage*, is the *Hay*, *Oates* and *Straw* which is delivered out from the Magazines, to the Army in Garriſon, or when they take the Field before the *Forrage* be grown up.

FORT, is a Work inviron'd on all ſides with a *Moat*, *Rampart*, and *Parapet*; the Deſign of it is to ſecure ſome high Ground, or the Paſſage of a River, to make good an Advantagious Poſt, to fortifie the Lines and Quarters of a Siege, and ſeveral other things. They are of different Figures, and are made ſmaller and greater, according as the Ground requires; ſome are in the ſhape of *Baſtions*, as the *Fort* raiſed on the ſide of the Hill of *Bouge* at *Namure*, or the *Fort* raiſed in 1701, by *Coehorne* on St. *Peter*'s Hill at *Maſtricht*, called St. *Peter's Fort*; ſome are fortify'd with entire *Baſtions*, others with *Demi-baſtions*; ſome are raiſed on a Square, and others on a Pentagon. The Fort *de la Lippe* near *Weſel*, is a Square with four large *Baſtions*, but was taken in leſs than one hour by the Prince of *Conde*, in 1672. The Fort St. *Andrew* in the Iſle of *Bommel*, is a *Pentagon*. A *Fort* differs from a *Citadel*, becauſe this laſt is always raiſed by the Orders of the Soveraign. Small *Forts* are made in form of a Star, having five, ſix or ſeven *Angles*, and are raiſed for the ſecurity of the Lines of *Circumvallation*.

FORTIFICATION, is an Art by which a Place is put in a poſture of Defence, ſo that every one of its parts defends and is defended by another, and diſcovers the Enemy in *Front* and *Flank*, oppoſing to an Enemy the breadth and depth of a *Foſs*, and the height and ſolidity of the *Rampart*, ſo that a ſmall Body of Men behind this Wall may be able to reſiſt a great Army. *Fortification* is likewiſe taken for all the

the Works that serve to cover or defend a strong Place. It is also the Art by which an *Engineer* makes Plans and Designs, raises different sorts of Works, digs the *Foss*, faces the *Ramparts*, and conducts the *Approaches*, either in the Attack or Defence of a Place: In short, it requires an *Engineer* to be a good Designer, Architect, Miner, and Mechanick, and to understand Gunnery.

Fortification *Ancient*, were Walls or Defences made of Trunks and Branches of Trees mix'd with Earth, to secure them from the Violence of an Enemy: this they altered afterwards to Walls of Stone, to secure them from surprizes; and to resist the Efforts of an Enemy, they raised little *Walls* or *Parapets* on the top of the other, behind which they made use of their Darts in Security, being cover'd from the Enemies sight by these small *Parapets*, which they cut into Loop-holes and Battlements to facilitate the Effect of their Darts, and these Walls were flanked by round or square Towers. *Modern Fortification*, is that which is flanked and defended by *Bastions* and *Outworks*, and whose Works are so solid, that they are Proof against the force of Cannon, and cannot be beat down, but by a continual Fire from several Batteries of Cannon.

Fortification *Artificial*, are the Works raised by an *Engineer*, to strengthen the natural Situation of a Place, by repairing and supplying its Defects; such as *Ravelins*, *Half Moons*, *Horn-works*, *Redoubts*, &c. *Natural* Fortification consists in a place's being strong by Nature, such as being situated on a Hill or in a Marsh, or any other way, which makes it of difficult Access, whether it be occasion'd by Rivers, Marshes, strong Defiles, or the like.

Fortification *Defensive*, regards the Precautions and the Industry by which a weak Party opposes a stronger, and concerns particularly the *Governours* of Places, who knowing the weakness and the strength of the Place intrusted to them, ought to endeavour to secure it from Surprizes and Attacks. *Fortification Offensive*, has a regard to the several Ways of annoying an Enemy, and concerns particularly the *General* of an Army in the Field, who designs to lay Siege to some of the Enemies Towns: this teaches him how to take all Advantages in marching or encamping his Troops, the Order and Dispositions of Battels, and the manner of carrying on a Siege, and of taking of Towns, *&c.*

Fortification *Regular*, consists in a Place's being regularly fortified, the sides of the *Polygon* being of a length, and the *Angles* equal; in its being defended by *Bastions* and other Works, whose relative parts are equal and uniform. *Irregular Fortification*, is when a Town has such an Irregular Situation, as renders it incapable of being regularly fortified, both because of the

the difference of its sides, some being too long, others too short; as likewise because of its being surrounded with Precipices, Valleys, Ditches, Rivers, Hills, Rocks or Mountains, and must therefore be fortified with Works suitable to the Situation.

FOSS, see *Moat*.

FOUCADE, *Fougade*, or *Fougasse*, is a small *Mine* under a *Post*, which is in danger of falling into the Enemies hands, to blow it up; it is thus made; they dig a hole or pit in the ground about eight or ten foot wide, and ten or twelve deep, which is fill'd with Sacks of Powder, upon which are laid pieces of Wood cross ways, with Stones and Earth and whatever else can make a great destruction; this is fired by the help of a *Saucisss* or *Train*, which has a communication with the *Counterscarp* or some other *Post*. We could not keep footing in the *Half-Moon* we had carry'd, because of two *Fougades* sprung by the Enemy, which ruined the Lodgment we had made in the *Gorge*.

FOURNEAU, *Powder-Chamber*, or *Chamber of a Mine*, is a Hole or Cavity made under a Work, the Top of which is sometimes cut into several Points like Chimneys to make more passages for the Powder, to the end it may have its Effects on several sides at the same time; and sometimes it is in form of a hollow Cube, about five or six foot large: the Charge of a *Chamber* or *Fourneau* is about 1000 *ll.* of Powder put into Sacks or Barrels, so that the *Saucisss* or *Pudding* may fire them all at once. It is left to the Judgment of the Engineer or Miner to augment or diminish this Quantity as he thinks convenient, and to proportion it to the Nature of the Ground or Rock on which the Work is raised, which is to be blown up; for if a great deal of Powder meet with little Resistance, it makes only a hole by raising the Earth above it with a great Violence. Sometimes they make four or five Chambers under a Work, and put but a small Quantity of Powder in each, as 100 *ll.* or the like. At the Siege of *Candy*, the *Turks* and *Venetians* made *Fourneaus* under the Rock of the *Demi bastion* of St. *Andrew*, which were charged with 70 Barrels of Powder. A *Fourneau* ought not to be charged till it be ready to spring, because the Powder lying too long in the Humidity of the Earth, loses its Force. When the Powder is put in Barrels, one of the Staves must be taken out, and a quantity of Powder scatter'd round; if it be in Sacks, they must be ript and Powder strowed about, that they may fire all at once. The Mouth of the *Fourneau* is to be stopp'd with great Planks and pieces of Wood, and the Vacancy which is left, after the *Fourneau* is charged, must be fill'd with Stones and pieces of Wood, and all the Turnings well stopp'd.

FRAISES are pieces of wood of six or seven foot long, planted under

under the *Cordon*; and in places which are not faced with Stone or

Brick, they are planted at the Base of a *Parapet*, being let about half way into the *Rampart*; they are not laid parallel to the *Base* of the *Rampart*, but a little sloping downwards with their Points, that Men cannot stand on them; their chiefest use is to hinder the Garrison from Deserting, which would be easie without them, especially in Places with dry Moats. They likewise prevent Surprizes and Escalades. See *A* in the Figure.

To Fraise a Battalion, is to line the *Musquetters* round with the *Pikes*, that if they should be charged with a Body of Horse, the *Pikes* being presented, may cover the *Musquetteers* from the shock of the Horse, and serve as a *Barricade*.

FRONT *of a Battalion* is the first Rank or the File-Leaders; it is likewise called the *Face* or *Head* of a *Battalion*. *Front* of a Squadron is the first Rank of Troopers. *Front* of an Army is the first row of Tents in the first Line, which are the *Quarter-Masters* Tents in the Horse, and the Serjeants in the Foot.

Front of a Place, is the same as the *Face of a Place*, or the *Tenaille*, being all that is contained between the *Flanked Angles* of two neighbouring *Bastions*, *viz*. the two *Faces*, two *Flanks* and the *Curtin*.

FUSE is a Pipe of Wood drove full of a Composition of *Meal Powder*, *Salt Petre* and *Sulphur*, having some Threads of *Quick-match* fixed in the top of it. When it is used it is drove into the *Bomb*, being first cut according to the Distance the *Bomb* is to be thrown, and the time it may be supposed to stay in the Air, that it may be spent, and the *Bomb* break as it falls; when the *Bomb* is put in the *Mortar*, the Cape of the *Fuse* is cut off, and the *Quickmatch* spread out of the *Fuse* upon the *Bomb*, and salted with a hand-full of Meal-powder, which takes fire from the flash of the *Mortar*, and fires the *Fuse*. See *Bomb*.

FUSIL, see *Firelock*.

FUSELEERS, are Foot *Soldiers* armed with *Fusees*, with Slings to fling them. There are four *Regiments* in our Army, which have always been called *Fuseleers*, and go by the Name of the *English*, *Scotch*, *Irish* and *Welch* *Fuseleers*, but now we have none but *Fuseleers* abroad, for the *Pikes* are quite laid aside. The first Design of *Fuseleers*, was to guard the *Artillery*, for which End the *Regiment* of *English* *Fuseleers*

leers now commanded by Sir *Charles O Hara* were firſt raiſed to ſupply the want of *Pikes*, and to ſecure themſelves againſt *Horſe*. The *Fuſeleers* uſed to carry *Turn-pikes* along with them, which in a *Camp* were placed along the front of the *Battalion*, and on a March were carried by the *Soldiers*, each carrying one of the ſhort Pikes, and two by turns the Sparr through which they are thruſt, ſo that they were quickly put together.

G.

GABIONS, are *Baskets* of 5

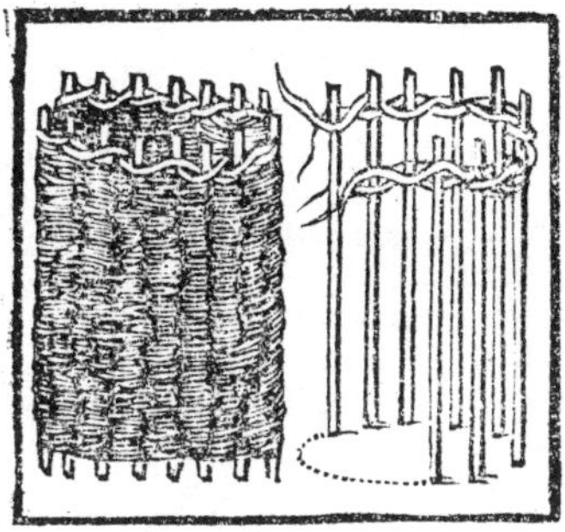

or 6 foot high, and 4 or 5 broad, equally wide at top and bottom; they are made of pieces of Willow of about ſix foot long, ſtuck in the Ground in a Circle, which they work round with ſmall Branches, Leaves and all, and afterwards fill them with Earth to make a *Cover* or *Parapet* betwixt them and the Enemy; they are ſometimes uſed in making *Batteries*, and are brought empty to the place, and placed three a breaſt, which makes the diſtance between the *Embraſures*; two are placed behind theſe, ſo as to cover the Joinings of the firſt three, and one behind the two, which make the *Embraſure* wide enough at the outſide, theſe ſix *Gabions* being the *Merlon*; the *Pioneers* or *Soldiers* employed for that uſe, fill them with Earth, but they are never ſo good as a *Battery* raiſed of Earth or Faſcines, becauſe, if there be a *Counterbattery* to play upon them, they are eaſily ruined. Sometimes they are uſed in making *Lodgments* on a *Poſt*, and ſometimes in making the *Parapet* of the *Approaches*, eſpecially when the Attack is carrying on, thro' a rocky Ground. At the Siege of *Namure* in the year 1695, the third night after opening the Trenches at the Village of *Bouge*, the Ground was ſo ſtrong and full of Rocks, that we were forced to uſe *Gabions*, and to bring the Earth a great way to fill them. When the *Approaches* are got near the *Covert Way*, the Beſiegers endeavour to ſet the *Gabions* on fire by ſmall *Faſcines* or *Bavins* pitch'd over, which they throw upon them. See the *Figure*.

GALLERY *for paſſing a Moat*, is a covered Walk made of ſtrong Beams, and cover'd over head with Planks, and loaded with Earth: 'twas formerly uſed for putting the *Miner* to the Foot of the *Rampart*: ſometimes the Gallery is

is covered over with *Raw-Hides*,

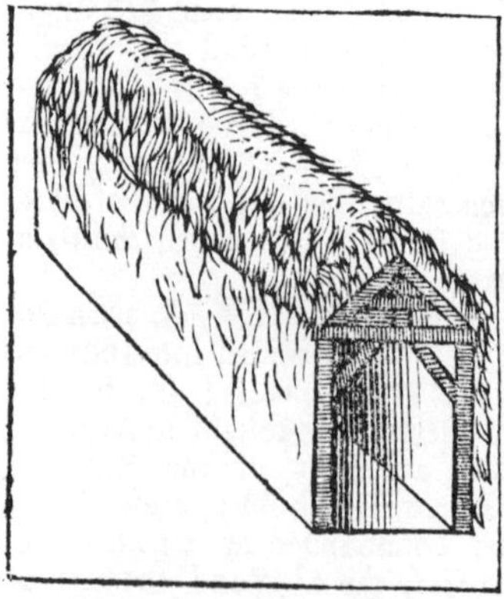

to defend it from the Artificial Fires of the Besieged. The *Gallery* ought to be very strong of double Planks on that side towards the *Flank*, to make it Musquet Proof. It is made in the *Camp*, and brought along the *Trenches* in pieces to be join'd together in the *Foss*; it ought to be eight foot high, and ten or twelve wide; the Beams ought to be half a foot thick, and two or three foot asunder; the Planks or Boards nailed on each side, and filled with Earth or Planks in the middle; the covering to rise with a ridge, that what is thrown upon it by the Besiegers with a Design to burn it, may roll off. See the *Figure*.

Gallery of a Mine, is the same as *Branch of a Mine*, and is a Passage under ground of three or four foot wide under the Works, where a *Mine* or *Countermine* is carried on. The Besieged and the Besiegers carry each of them Branches under ground in search of each others *Mines*, which often meet and destroy one another, or at least disappoint the Effect of the *Mine*. Our *Miners* having heard those of the Enemy, fixed a *Petard* in their *Gallery*, which pierced into the Enemies, and cast such a Smoak as stifled most of the Work-men. See *Mine*.

GARRISON *Town*, is a strong Place, in which *Troops* are quartered, and do Duty for the Security of the Place; keeping strong Guards at each *Port*, and a Main Guard in the *Market Place*. The *Troops* that are put into a Town, either for their Security or Subsistance in the Winter time, or are there in the Summer for the Defence of the Place, are called *the Garrison of that Town*.

GATE, is made of strong Beams and Planks, with Iron Bars, and turns upon Hinges, to secure the Entry of a Town against an Enemy. The *Gate* is generally in the middle of the *Curtin*, and has the Defence of both *Flanks*; those which are in the *Flank*, weaken the most necessary part of the *Fortification*, and when they are in the *Face*, they are still more prejudicial to the *Bastion*, which ought to be clear, to make *Retrenchments* upon Occasion. The *Gates* of a strong Place are shut about five a Clock in the Winter, six in the Spring, and seven or eight in the Summer, and are opened at seven, six and five. At the opening of the

Gates,

Gates, a Party of Horſe is ſent to *Patrouille* in the Country round the Place, to diſcover *Ambuſcades* or *Lurking Parties* of the Enemy, and to ſee if the Country be clear. In ſome Garriſons the Guard mounts at the opening of the *Gates*, ſo that in caſe of a Surprize, both the Old and the New Guards being under Arms, they are in a Condition of making a good Defence. The Word nor the Orders ought never to be given, till after the *Gates* are ſhut, for fear of Spies lurking in the Town, that may carry Intelligence to the Enemy.

GAZONS, are Sods or pieces

of freſh Earth covered with Graſs, about a foot long, and half a foot broad, cut in form of a Wedge to line the *Parapet*; if the Earth be fat and full of Herbs, it is the better; they are made ſo, that their Solidity makes a Triangle; to the end, that being mixt and beat with the reſt of the Earth of the *Rampart*, they may eaſily ſettle together and incorporate in a Maſs with the reſt of the *Rampart*. The firſt Bed of *Gazons* is fixed with pegs of Wood; the ſecond Bed ought to be laid to bind the former that is over the Joints of it, and ſo continued till the *Rampart* be finiſhed; betwixt theſe Beds they generally ſow all ſorts of binding Herbs to ſtrengthen the *Rampart*

GENS-D'ARMES, are a Body of Horſe divided into Independant Troops, called ſo, becauſe formerly they fought in Armour; they are part of the King of *France*'s Houſhold; theſe *Troops* are commanded by *Captain Lieutenants*, the King and Princes of the Blood being their *Captains*; the King's *Troop*, beſides a *Captain-Lieutenant*, has two *Sub Lieutenants*, three *Enſigns*, and three *Guidons*. The other *Troops* which are thoſe of the *Scots Gendarmes*, the *Queen's*, the *Dauphin's*, the *Gendarmes* of *Anjou*, *Burgundy*, the *Engliſh* and *Flemiſh Gendarmes*, and thoſe of the Duke of *Orleans*, are called the *Small Gendarmery*, and have each a *Captain-Lieutenant*, *Sub-Lieutenant*, *Enſign*, *Guidon*, and *Quarter-Maſter*. They carry a *Standard* longer than the Light Horſe, and divided into two Points a little rounded, generally adorn'd with ſome Device or Cypher in Embroidery and a Fringe; each *Troop* has a pair of *Kettle-Drums*, and two *Trumpets*.

GENERAL *of an Army*, is he who commands in chief, and is the ſame in an Army as the Soul is in the Body; for as all the Actions of the Body proceed from

from the Motions of the Soul, ſo that great number of *Regiments* ought to do nothing but by the Orders of the *General*; who ought to be a Man of Courage and Conduct, to have a great Experience, to be of good Quality, and Liberal; his Valour makes him a Terror to his Enemies, and upon his Conduct depends the Safety of the Army, therefore I think it the greater Qualification; for Bravery without Conduct, has often brought things to Extremity. A *General's* Conduct appears in eſtabliſhing his Magazines in convenient Places, in examining the Country, that he do not engage his *Troops* too far, without knowing which way to bring them off, and to ſubſiſt them, and in knowing to take the moſt Advantagious Poſts, either to fight or ſhun a Battel at his Pleaſure: His Experience makes his Army have ſuch Confidence in him, that they reckon themſelves ſure of Victory before they engage: His Quality begets a greater Reſpect, and augments his Authority: His Liberality gets him Intelligence of the Enemy, of their Strength and Deſigns, without which he is in the dark, and cannot know which way to take his Meaſures; he ought therefore to encourage his *Spyes*, and to have ſuch as he knows are more inclin'd to him than to the Enemy. A *General* ought likewiſe to be naturally inclin'd to great Enterprizes, to be a lover of Glory, and to have an averſion to Flattery; to make himſelf beloved by every body, by treating his Officers with Civility, hearing their Reaſons, and praiſing and rewarding good Actions, and puniſhing Crimes; he ought notwithſtanding to be rigorous and ſevere upon Occaſion, in ſeeing his Orders punctually obſerved, otherwiſe Military Diſcipline would be loſt. The Function of a *General*, is to regulate the March of the Army and their Encampment, to viſit the Poſts, to command Parties for Intelligence, to give out the Orders and the Word every night to the *Lieutenant* and *Major-Generals*; in day of Battel, he chuſes the moſt Advantagious Ground, makes the Diſpoſition of his Army, poſts the Artillery, ſends off the Baggage to a place of Security, and ſends his Orders by his *Aid de Camps*, where there is Occaſion. At a Siege, he cauſes to inveſt the Place; he views and obſerves it, orders the making of the Lines of *Circumvallation* and *Contrevallation*, and making the Attacks; he viſits often the Works, and makes Detachments to ſecure his Convoys. The Charge of a *General* is of a great Extent, and requires a particular Care, becauſe it is on him the Sovereign repoſes the Care of all his Army. His Royal Highneſs Prince *George* of *Denmark*, is now *Captain-General* of all Her Majeſty's Forces by Sea and Land, and his Grace the Duke of *Marlborough*, *Captain General* of Her Army,

 that

that acts in Conjunction with the *Emperour* and the *States General*.

General of Horse, and *General of Foot*, are Posts next under the *General* of the Army, and ought to be in the hands of Men of a singular Merit, who, by their Birth or good Qualifications, have rendered themselves deserving of so great an Employ. They have an absolute Command over all the Horse or Foot in an Army, upon all Occasions, above the *Lieutenant-Generals*.

General of the Artillery, or *Master-General of the Ordnance*, is one of the greatest Employs in the Kingdom, being a Charge of great Trust; it is generally bestowed on one of the first Peers of the Kingdom, and is at present enjoyed by his Grace the Duke of *Marlborough*; he has the Management of all the Ordnance of the Kingdom, and ought to know and consider whatever can be serviceable or useful in the *Artillery*, and to distribute the Vacancies to such as are qualified for them; and has for his Assistants in that Employ, a *Lieutenant-General*, who commands in the absence of the *General*; a *Surveyor General*, *Clerk*, *Store-keeper*, and *Clerk of Deliveries*, who are called the Principal Officers of the Ordnance; they meet three times a week at the *Tower*, about the Concerns of the Ordnance.

General, is likewise a beat of the *Drum*, see *Drum*.

GIN or *Crab*, is an Engine for mounting *Guns* on their *Carriages*, or dismounting them; it is

made of three pieces of Oak, Ash, or other strong Wood, of about 14 foot long, two of which are join'd by *Transums*, but so that they are wide asunder at bottom, as *A*, *A*, and join at the top, on a strong piece of Wood crooking forwards, called the *Head*, *B*, in which are three Pullies of Brass, over which comes a Rope, *C*, called the *Gin-Rope*, which goes likewise through other Pullies in a short strong piece of Wood, called the *Block*, *D*, and returns through the Head down the back of the *Gin*, and goes round a *Windlace*, *E*; the other piece is round, *F*, one end of it goes into the Head, and the other stands on the Ground, so that the three make a Triangle; it is called the *Pye*; when a *Gun* is to be

be mounted, there is a strong Rope tied through the *Block* and the *Trunions*, so that the *Windlace* being turned round, the *Gun* is lifted up and placed in her Carriage with ease.

To give Ground, is to retire or quit a Post, when it is attacked by an Enemy. *To get* or *gain Ground*, is to have the Advantage of the Enemy, and to force them from a Post.

GLACIS, is that Mass of Earth which serves as a *Parapet* to the *Covert Way*, which slopes easily towards the *Campaign*: the difference betwixt *Talus* and *Glacis*, is, that in the one the height is more than the Base of the Slope, and in the other, the Base of the Slope is more than the height; the breadth of the *Glacis* is generally the length of the *Flank*, but the largest are the best; it is likewise called *Esplanade*, but that Word begins to be out of use. The *Soldiers* corruptly call the top of the *Glacis* the *Counterscarp*. When the *Approaches* are brought to the foot of the *Glacis*, they are so near, that they cannot turn any way but they must be enfiladed, therefore they are carried straight forwards by Sap, unless it be resolved to carry the *Covert Way* by Assault.

GORGE *of a Bastion*, is that space which is taken equally on each side of the *Angle* of the *Figure* on the sides of the *Polygon*, which makes the Entry into the *Bastion* from the Town or Place, one half of which is called the *Demi-gorge*; there is a Proportion to be observed betwixt the *Gorge* and the other parts of the *Fortification*, as the *Curtin* and *Flanks*; if the *Gorge* be too large, the *Flanks* must be shorter, which is a great fault, they being the principal Defence of the Place, and a long *Flank* always is better than a large *Gorge*. If the *Gorge* be too small, the *Bastion* is likewise contracted so small, that there is no ground left to entrench behind the breadth. *Gorge* of a flat *Bastion*, is a right Line which terminates the distance between two *Flanks*. *Gorge* of a *Half Moon*, is the distance between the two *Flanks*, taken on the *Angle* of the *Counterscarp*. That of a *Ravelin*, is the distance between the two Sides or Faces towards the Place; the *Gorges* of all other Out-Works, are the Entry into them from the Place, or the distance between their Sides, and ought to be without a *Parapet*, only plain, for fear the Besiegers being Masters of the Work, should find there a Cover from the Fire of the Place; yet they are sometimes palisaded to prevent a Surprize. In a Siege they are generally undermined, that they may be blown up before the Enemy can make a Lodgment upon them.

GORGET, is a piece of Brass or Silver worn by Officers upon Duty upon their Breast, hanging round their Neck in a Riband; they are sometimes gilded, having some Device engraved on them,

them, as the *Colonel's Coat of Arms*, his *Crest*, or his *Cypher*.

GOVERNOUR, is a very considerable Officer, and has a great Trust reposed in him, and ought to be very vigilant and brave, knowing that it is more Honour to defend one Town, than to take two, because the abundance of Provisions and number of Men, is greater with the Besiegers than in the Garrison, the former being likewise Masters of the Country, and the others shut up. He ought to be always prepared for a Siege, to have a particular Care of the *Ramparts*, *Parapets*, and other Defences of the Place; that the *Foss* or *Moat* be kept clean and in good Order, and the Out-Works in good Repair, and well Palisaded; he ought frequently to visit the Magazines, to see that every thing be in Order, and whether there be a sufficiency for a Siege; to consider the quantity of Grain and other Provisions, and to have Registers of all the Stores, and to neglect nothing which tends to the Defence of the Place, because he must answer for it at the peril of his Head. His Charge is to order the *Guards*, the *Rounds*, and the *Patrouilles*, to give every night the Orders and the Word after the Gates are shut, to visit the *Posts*, to see that both *Officers* and *Soldiers* do their Duties, and to send frequently Parties abroad for Intelligence, and to raise Contribution.

GRENADIER, is a Foot *Soldier* armed with a *Sword*, *Firelock*, *Bayonet*, and a *Pouch* to hold his *Grenades*; they are cloathed differently from the rest of the *Battalion*, and wear high Caps; each *Regiment* has a *Company* of *Grenadiers*, which takes always the right of the *Battalion*. In *Holland* each *Company* in a *Regiment* has five or six *Grenadiers*, which being drawn out together form a *Company*. The *Grenadiers* are generally the tallest and briskest Fellows, and are always the first upon all Attacks: When there is any appearance of Action, each *Grenadier* carries three *Hand Grenades*. *Horse-Grenadiers*, called by the French *Grenadiers-Volans*, or *Flying-Grenadiers*, are such as are mounted on Horseback and fight on Foot; their Exercise is the same with the other *Grenadiers*; we had last War a *Regiment* of *Horse-Grenadiers* joined to the Guards, but since the Peace they are reduced to one *Troop* consisting of 145 Men.

GRENADE, is a Shell of Iron filled with fine Powder, which being fired, bursts the Metal in pieces amongst those that are near where it falls, who are obliged to quit their Post, or run the hazard of having their Legs and Arms broke and spoil'd. The *Grenade* has a Vent to receive a *Fuse*, which is made of the same Composition with that of a *Bomb*, that the *Grenade* may not break in the Hand of the *Grenadier* before it be delivered.

GUARD,

GUARD, is a Duty or Service which ought to be performed with a great deal of Vigilance, to secure all from the Efforts and Surprizes of an Enemy. In a *Garrison* the *Guards* are relieved every day, and it comes to every *Soldiers* turn once in three days, so that they have two nights in Bed, and a third upon *Guard*; to be upon *Guard*, to go upon *Guard*, to mount the *Guard*, to dismount the *Guard*, to relieve the *Guard*, to change the *Guard*, the *Officer* of the *Guard*, or the *Serjeant* of the *Guard*, are Words often used and well known.

Main-Guard, is the great *Guard* from whence all the other small *Guards* are detach'd; those which are to mount the *Guard*, meet at the respective *Captain's* Quarters, and are carried from thence to the *Parade*; where, after the whole *Guard* is drawn up, the small *Guards* are detach'd for the *Ports* and *Magazines*, and the Subaltern *Officers* throw Lots for their *Guards*, and are subordinate to the *Captain* of the *Main-Guard*. The *Guards* are mounted in *Garrisons* at different hours, according as the *Governour* pleases, but the most usual is at the opening of the *Gates* at ten a Clock, or at two in the Afternoon.

Advanc'd-Guard, is the Party of either Horse or Foot that march before a Body, to give them Notice if any Danger appears; when the Army is upon their March, the *Grand-Guards* that should mount that day, serve as an *Advance Guard* to the Army; if a Body of Foot be marching, their *Advanc'd Guard* are Foot. In small Parties, 6 or 8 Horse are sufficient, and are not to go above 4 or 500 yards before the Party. An *Advanc'd-Guard* is likewise the small Body of 12 or 16 Horse, under a *Corporal* or *Quarter-Master*, which are posted before the *Grand-Guard* of the *Camp*.

Rear-Guard, is that part of the Army which brings up the *Rear*, which is generally the old *Grand-Guards* of the *Camp*. The *Rear-Guard* of a Party, is 6 or 8 Horse that march about 4 or 500 paces behind the Party. The *Advanc'd-Guard* going out upon Party, make the *Rear-Guard* in their return.

Grand-Guard, are 3 or 4 *Squadrons* of Horse commanded by a Field Officer posted before the *Camp* on the *Right* and *Left Wing* towards the Enemy, for the security of the *Camp*; this *Guard* mounts every morning about 7 or 8 a Clock; see *Camp*.

Quarter-Guard, and *Standard-Guard*, see *Camp*.

Picquet-Guard, is a certain number of Horse and Foot which are to keep themselves in a readiness in case of an *Alarm*; the Horse keep their Horses saddled, and are booted all the time, in order to mount in a minute. The Foot draw up at the Head of the *Battalion*, at the beating of the

the *Tattou*; but return to their Tents, where they hold themselves in a readiness to march, upon any sudden *Alarm*; this forms a good Body able to make a Resistance, till the Army can be in a readiness.

Forrage-Guard, is a Detachment sent out to secure the *Forragers*, and are posted at all places, where either the Enemies Party may come to disturb the *Forragers*, or they may be spread too near the Enemy and be taken; this is likewise called the *Covering Party*, and marches generally the night before the forraging, that they may be posted in the morning before the *Forragers* come; they consist both of Horse and Foot, and must stay at their Posts till the *Forragers* be all come off the Ground.

Artillery-Guard, is a Detachment from the Army to secure the *Artillery*; their *Corps de Garde* is in the Front, and their *Centries* round the Park; this is a 48 hours *Guard*, and upon a March they go in the Front and Rear of the *Artillery*, and must be sure to leave nothing behind; if a *Gun* or *Waggon* break down, the *Captain* is to leave a part of his *Guard* to assist the *Gunners* and *Matrosses* in getting it up again.

Corps de Garde, are *Soldiers* entrusted with the *Guard* of a Post under the command of one or more Officers.

Guards; *the Horse-Guards*, are Gentlemen chosen for their Bravery and Fidelity, to be entrusted with the Guard of the *Queen's Body*, divided into three *Troops*, called the *Troops of Guards*; each *Troop* hath a *Colonel*, 2 *Lieutenant-Colonels*, a *Cornet*, a *Guidon*, 4 *Exons*, *Brigadeers*, and *Sub-brigadeers*, and 160 Private Men. *The Foot Guards*, are *Regiments* of Foot appointed for the *Guards* of her Majesty and her Palace; there are two *Regiments* of them, called the *First* and *Second Regiments of Guards*, the one having four *Battalions*, and the other two; the *Regiment* of *Scots Guards* is likewise two *Battalions*.

GUERITE, is a sort of

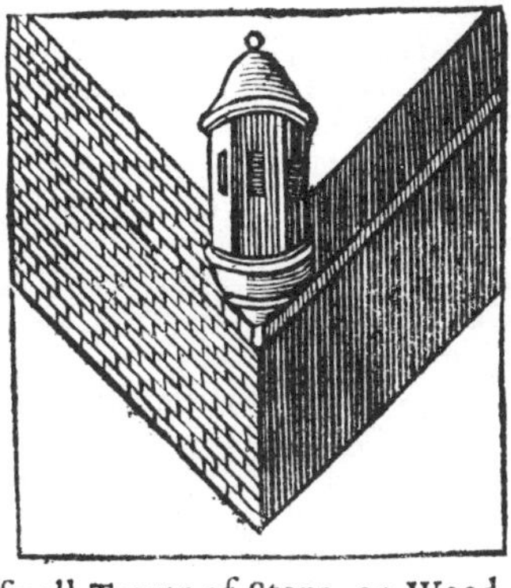

small Tower of Stone or Wood, generally on the Point of a *Bastion*, or on the Angles of the *Shoulder*, to hold a *Centinel*, who is to take care of the Foss, and to watch to hinder Surprizes; some call *Echaugette* those which are made of Wood and are of a square form, for the *Guerites* of Stone are roundish, and are built half without the Wall, and terminate

minate at a Point below, which ought to be at the *Cordon*, that the *Centinel* may discover along the *Faces*, *Flanks*, and *Curtins*, and all along the *Foss*: they ought to be about six foot high, and their breadth three and a half.

GUIDES; *Captain of the Guides*, is an Officer appointed for providing Guides for the Army, of which he ought to have always a sufficient number with him, that know the Country very well, to send out as occasion requires; such as are to guide the Army on a March, for Convoys, Parties, Baggage, Artillery, and Detachments; to provide which, he ought to have a Party of Horse to go to the Adjacent *Villages*, *Castles* or *Forts*, to demand *Boors*, whom he brings to his Quarters, and keeps under a Guard for fear they make their Escape, till the Army come to another Ground, where he can be provided with others: He ought to understand several Languages, especially that of the Country in which the Army is.

GUIDON, is a *French* Term for him that carries the *Standard* in the Guards or *Gens d'Armes*, and signifies likewise the *Standard* itself.

GUN, see *Cannon*.

GUNNER, is one appointed for the Service of the *Cannon*; he carries a *Field Staff* and a large *Powder-Horn* in a string over his Left Shoulder; he marches by the *Guns*, and when there is any Apprehension of Danger, his *Field-Staff* is armed with Match, and a Linstock stuck upon his *Gun*, with a light Match; his Business is to lay the *Gun* to pass, and to help to load and to traverse her.

H.

HAIR-CLOTHS are large pieces of Cloth made with half Hair; they are used for covering the Powder in the Wagons, or upon Battery; as likewise for covering fixed *Bombs* or *Hand-Grenades*; and are for several other uses in a *Magazine*.

HALBARD, is the Arms carry'd by the Serjeants of Foot and Dragoons; the Head of the *Halbard* ought to be a foot or 15 Inches long; one end ought to be hollow to receive the Staff, but the other broad, ribb'd in the middle, edg'd on both sides, and drawing to a Point, like the point of a two-edged Sword. On one side of the Head is likewise fixed a piece in form of a *Half-moon* or Star, and on the other a broad Point of 4 Inches long, crooked a little, which is very commodious for drawing *Fascines*, *Gabions*, or whatever Obstacle else happen in the way. The Staff of the *Halbard* is about five foot long, and an inch and half diameter, made of Ash or other hard Wood. *Halbards* are very useful in determining the Ground betwixt the Ranks, and for dressing the Ranks and Files of a Battalion, and like-

wise for chastising the Soldiers. It is carried on the Left Shoulder.

HALF-FILES, see *File.*

HALF-MOON, is properly an *Out-work* composed of two *Faces* making a *Salliant Angle*, whose *Gorge* is turned like a *Crescent*, or forming an Arch of a Circle; they were used formerly for covering the points of Bastions; but have been found found of no use, because having only the Ravelins to defend them, they are but very indifferently flank'd; the *Ravelins* that are built before the *Curtins* are now called *Half Moons*, the Name of *Ravelin* being almost laid aside by the *Soldiers.*

HALT, is to discontinue the March of Troops, either in order to rest, or upon what occasion may happen. The word of Command *Halt* is an Order to stop the March and to stand still.

HAND-BARROW, is ordinarily made of hard light Wood, and is of great use in Fortification for carrying Earth from one place to another, and in a Siege for carrying *Bombs* or *Cannon-ball* along the Trenches, and for several other Uses.

Hand-Screw, see *Jack.*

Hand-Spike, is a piece of Ash, Elm, or other strong Wood, five or six foot long, cut thin like an Edge at one End, that it may get the easier betwixt things which are to be separated, or under any thing that is to be raised; it is better than a *Crow* of Iron, because its Length allows a better poize.

HATCHET, is a small *Ax* used by the *Pioneers*, who go before to prepare the ways for an Army, in cutting down *Hedges, Bushes, Stiles* or *Gates*. The *Grenadeers* carry sometimes each a *Hatchet* by his Side, and the *French Dragoons*, who have but one Pistol, have each a *Hatchet* hanging at his Saddle-Bow. on the Right-side.

HEAD of a Work, is the *Front* of it, next to the Enemy and farthest from the Place; as the *Front* of a *Horn work* is the Distance between the *Flanked Angles* of the *Demibastions*. The *Head* of a *Double Tenaille* is the *Salliant Angle* in the middle and the two other sides which form the *Re-entring Angles*.

HEDGE; *to line a Hedge*, is to plant *Musquetteers* along it under cover, either to make fire upon an Enemy that is advancing, or to save themselves from the *Horse*, or to defend a *Pass* or *Defile.*

HEIGHT, see *Eminence.*

Height of the Bore of a Piece, is the Diameter of the Bore.

HELVE, is the Handle of a *Hatchet, Pickax, Mattock*, &c. *To helve*, is to put *Handles* to them.

HEPTAGON, is a figure capable of being fortified with seven regular *Bastions.*

HERISSON, is a *Barriere* made of one strong Beam or Plank of Wood, stuck full of Iron-spikes it is supported in the middle, and turns upon a *Pivot* or *Axis*; it

it is uſed in ſtopping a paſſage, in

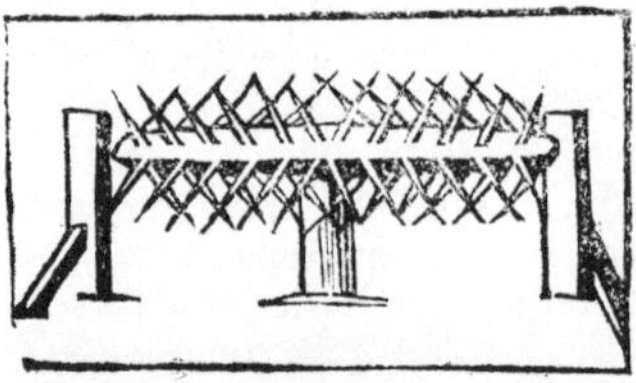

nature of a *Turn-ſtile*, for it is equally balanced upon the *Pivot*, which ſtands upright in the middle of the Paſſage, upon which it turns round, as there is occaſion to open or ſhut the Paſſage.

HERSES or *Portculliſes*, are

ſtrong pieces of Wood jointed croſſ-ways like a *Lattice* or *Harrow*, uſed formerly to hang in the middle of a *Gate-way* of fortified Towns, to be let fall to ſtop the Paſſage, in caſe the Gate had been broke down or Petarded. It is either a Stop or a Separation, if any of the Enemy have already enter'd, for before it can be broke open, the Beſieged have time to rally and repulſe them. See *Orgues*.

Herſe, is likewiſe an Engine like

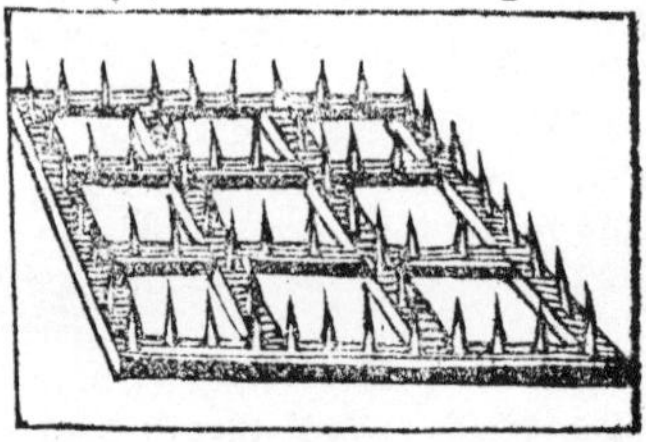

a *Harrow*, ſtuck full of Iron-ſpikes; it is uſed in place of the *Chevaux de Friſe*, to throw in the ways where *Horſe* or *Foot* are to paſs, to hinder their March, and upon Breaches to ſtop the *Foot*. *Common Harrows* are ſometimes made uſe of, and are turned with their Points upwards. See the figure.

HERSILLON, is for the

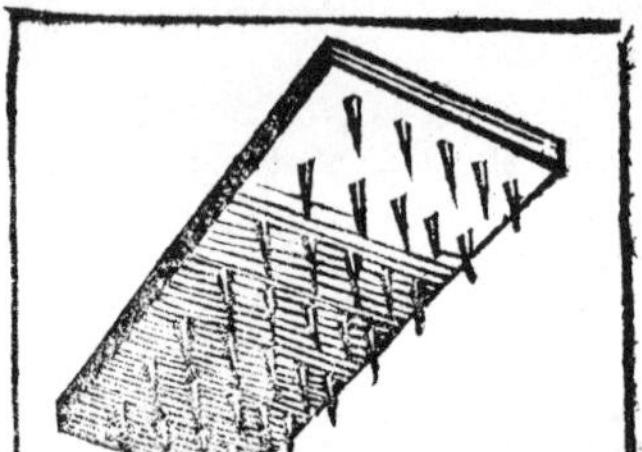

ſame uſe, as the *Herſe*, and is made of one ſtrong plank of Wood about 10 or 12 foot long, ſtuck full of Points or Spikes on both ſides, as the figure ſhews

HEXAGON, is a figure of ſix ſides, capable of being fortified with ſix *Baſtions*.

HIDES, *Tan'd Hides* are always carry'd along with an Artillery, eſpe-

especially in the *Fire-workers* stores for covering Powder or fixed Bombs from the Rain; and are very useful upon a *Battery*, or in a *Laboratory*.

HOBITS, are a sort of small *Mortars*, about 8 inches diameter, some 7, some 6; they differ nothing from a *Mortar*, but in their Carriage, which is made after the fashion of a *Gun-Carriage*, only much shorter; they march with the *Guns*, and are very good for annoying an Enemy at a distance, with small *Bombs*, which they throw two or three Miles; or in keeping a Pass, being loaded with *Cartouches*. See *Cartouch*.

HOLLOW-SQUARE, see *Square*.

HONEY-COMBS, are flaws and defects in the charged *Cylinder* of a *Cannon*; it is a fault in casting the Piece.

HOOKS of Iron placed on the *Cheeks of a Gun Carriage*, two before and two behind, are called *Draught-Hooks*, which see.

HORISONTAL *Superficies*, is a *Superficies* parallel with the *Horizon*, as a plain or level Country or Field without any Rising-Ground or Hollow.

HORN-WORK, is a Work which the *French* Engineers prefer before *Tenailles*, *Swallows Tails* or *Priests-Bonnets*, because it takes in a great deal of Ground and has a better Defence; it is composed of two Long Sides or Faces parallel, the Distance between them being the Length of one *Curtin*; their Length, measuring from the *Angle* of the *Shoulder*, is the Length of one side of the *Polygon*, or of the *Curtin* and one *Demigorge*. The *Head* or *Front* of this Work is fortified with two *Demibastions* and a *Curtin*.They have sometimes *Flanks* on their long sides, and then they are called *Horn-works* with double *Flanks* or *Shoulders*. They have generally a *Ravelin* in their *Gorge*, and a small *Ravelin* before the *Curtin*. This Work is quite rejected by *Coehorne*, being of too great a Circumference, and so small a Defence, that 'tis not worth the *Mason-work* with which 'tis faced.

HORSE, is generally taken for that Body of *Troops* that fight on *Horse-back*, as the *Horse-Guards* and the *Light-Horse*. See *Cavalry*.

Horse-Shoe, is a small round or oval Work, with a *Parapet*, made generally in a Moat or Marsh.

HOSPITAL, is a place appointed, at a Siege or with the Army, for the Sick and Wounded, and ought to be provided with *Physicians*, *Chirurgeons*, *Overseers*, *&c.* with all sorts of Medicines, Drugs, and whatever else may be wanted, in great Quantities; as likewise *Beds*, *Sheets*, *Coverlets*, *Shirts* and *Spare Linen* for *Bandages*. The Director of an *Hospital* ought to be a very careful and just Man, and to see that the Men want nothing; for there are often great Rogueries committed in an *Hospital*, because they have every thing at command, as the *Wine*, *Brandy*, *Syrrups*, *Bread*, *Butter*, *&c.* which are often misemploy'd.

The *Hoſpital* ought to be furniſhed with the beſt *Phyſicians* and *Surgeons*, that the Life of a Man which he has expoſed for the Service of his *Prince*, may not be loſt by an Evil Operation of a bad *Surgeon*, when it may be preſerved by being well dreſſed; the *Hoſpital* is generally ſettled in ſome Town near the *Army*, where the *Sick* and *Wounded* may be conveniently carried to it, only a part of it ſtays with the *Army* and moves with it, and is called the *Flying-Hoſpital.*

HURDLES, or *Clayes*, are

made of Branches or Twigs interwoven together in the figure of a long Square, about five or ſix foot long, and three or three and a half broad; the cloſer they are woven they are the better. They are for ſeveral uſes, as for covering *Traverſes* and *Lodgments*, *Caponeers*, *Coffers*, &c. and are covered over with Earth to ſecure them from the Artificial Fireworks of the Enemy, and from the Stones which might be thrown upon them, and likewiſe to lay upon marſhy Ground, or to paſs the *Foſs*, eſpecially when it is full of mud or ſlime. See their Form in the Figure.

HUTT, ſee *Cazerne.*

I.

JACK, is an Engine much

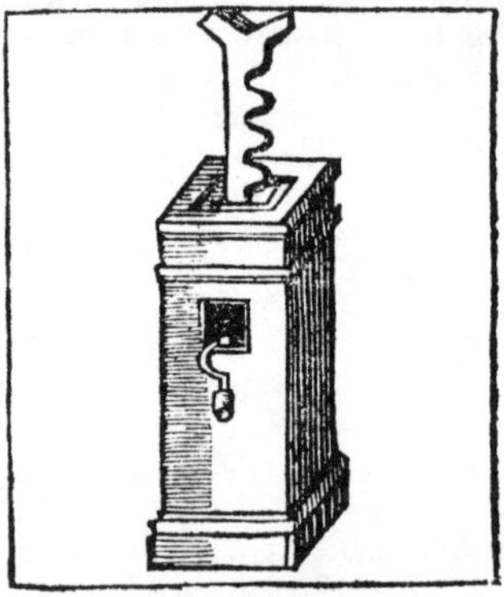

uſed about *Guns* or *Mortars*, and is always carried along with the *Artillery* for raiſing up the *Carriages*, or ſupporting the *Axletree*, if a *Wheel* chance to be broke; it is likewiſe uſed for traverſing large *Mortars*, ſuch as thoſe of 18 inches diameter, which are upon *Low Dutch Carriages*, and for elevating them; for traverſing the Sea *Mortars*, and many other uſes, too tedious to be named: with this Engine, one Man is able to raiſe more than ſix could do without it. See the *Figure*.

ICHNOGRAPHY, as Father *L'Amy* explains it in his *Perſpective*,

ſpective, is a *Greek* Word, ſignifying properly the Figure or Print which the ſole of the Foot leaves on the ground, which the *Greeks* call *Ichnos*. Amongſt *Architects*, it is the Section of a Building cut horizontally near the Ground. 'Tis likewiſe what we call a *Plan*: So the *Plan* or *Ichnography* of a *Church*, is the Mark left by this *Church*, if it were razed; or the firſt appearance in Building, when the Foundation is ready to appear above ground. The *Ichnography* of a *Cube*, or gaming *Dye*, is a Square; and that of a right *Cylinder*, is a Circle. See *Plan*.

INCAMP; *To Incamp*, is the pitching of *Tents* when the Army after a March is come to a place where it is deſigned to ſtay a night or longer; the *Serjeants Tents* in the *Foot*, and the *Quarter-Maſters* of *Horſe*, are the firſt of the *Company* or *Troop*; the *Officers* incamp in the Rear, the *Subalterns* in one Line next the *Company* fronting from it, the *Captains* in another Line at ſome diſtance, each behind his own *Company* fronting the *Subalterns*; the *Field Officers* behind them, the *Colonel* in the Centre, the *Lieutenant-Colonel* on his Right, the *Major* on his Left, and the *Sutlers* behind all. Each *Company* makes a Line in File, having an Allowance of 7 foot for a *Tent*, and 2 foot diſtance; the *Tents* of two *Companies* front one another, leaving a Street of five or ſix yards betwixt them. The *Troops* of *Horſe* incamp the ſame way, only the diſtance between the *Tents* is about 3 or 4 yards for the *Forrage*, and the ſpace between two *Troopers*, is 14 or 15 yards for the *Stables*: At two yards diſtance from the Doors of their *Tents*, is a Rope called the *Picquet Rope*, ſtretched upon pointed Stakes, to which their *Horſes* are tied. For the Ground allowed a *Battalion* or *Squadron*, ſee *Camp*

INCAMPMENT, ſee *Camp*.

INDENTED-LINE, is a *Line* running out and in, like the Teeth of a *Saw*, forming ſeveral *Angles*, ſo that one ſide defends another; they are uſed on the banks of *Rivers* where they enter a Town, likewiſe the *Parapet* of the *Covert Way* is often indented. This is by the *French Engineers* called *Redents*; ſmall Places are ſometimes fortify'd with ſuch a *Line*, but the fault of ſuch *Fortifications*, is that the Beſiegers from one *Battery* may ruin both the ſides of the *Tenaille* or *Front* of a Place, and make an Aſſault, without fear of being enfiladed, ſince the Defences are ruined.

INDEPENDANT *Troop* or *Company*, is what is not incorporated into any *Regiment*.

INFANTRY, are the *Regiments* or *Independant Companies* of *Foot* in an *Army*; they are formed into *Brigades*, as the *Cavalry* are; each *Brigade* conſiſting of 4, 5, or 6 *Battalions*; the *Regiments* of *Foot Guards* take the Poſt of all the *Infantry* in the *Army*; the other *Regiments* take Poſt by Seni-

Seniority; the oldest *Regiment* takes the Right of the first *Brigade*; the second takes the Right of the second *Brigade*; the third of the third, and so on. The next in Seniority take the Left of the *Brigades*, and leave the Centers for the youngest *Regiments*. The first *Brigade* takes the Right in the first *Line*, and the second *Brigade* the Right of the second *Line*; the third and fourth the Left of the two *Lines*, and the youngest *Brigades* in the *Centre*.

INGINEER, is an able expert Man, who, by the help of *Geometry*, delineates upon Paper or marks upon the Ground, all sorts of Forts and other Works proper for Offence or Defence; who understands the Art of *Fortification*, and cannot only discover the Defects of a Place, but find a proper Remedy for them; and who knows both how to make an Attack, and how to defend a Place *Ingineers* are extreamly necessary for both these, and ought to be not only Ingenious, but brave in proportion to their Knowledge, for the Employ requires Men expert and bold. At a Siege, when the *Ingineers* have observ'd and narrowly view'd the Place, they are to acquaint the *General* which they judge the weakest part, and where the Approaches may be made with most Ease. Their Business is to delineate the Lines of *Circumvallation* and *Contrevallation*, taking all the Advantages of the Ground; to mark out the *Trenches*, *Places of Arms*, *Batteries* and *Lodgments*, taking great care that none of their Works be flanked or discover'd from the Place. They are to make a faithful Report to the *General* of what is doing, to demand a sufficient number of Work-men and Utensils, and to foresee whatever is necessary, that there be good Provision made of *Fascines*, *Picquets*, *Gabions*, *Spades*, *Shovels*, *Pick-axes*, *Hatchets*, *Sand-bags*, *Planks*, *Boards*, *Mallets*, *Stampers*, *Dossers*, *Wheel-barrows*, &c. An *Ingineer* ought to be very perfect in *Arithmetick*, to project the Plots of Places, and calculate the Expences of a Siege; in *Geometry*, to measure his Works and raise Plans; in *Military Architecture*, to distinguish himself in his Profession; in *Civil Architecture*, to know how to construct Buildings and Works of Places; in *Mechanicks*, to know how to make Sluices, march *Cannon*, and use all sorts of *Machines*; in *Perspective*, to know how to express his Works on Paper, in their just Proportion; and without *Design*, he can neither make *Carts* nor *Plans*. These Sciences are called the *Genius*, in which consists the whole Spirit of *War* and *Fortification*.

INSULT, is the attacking a Post with open Force, without using *Trenches*, *Saps*, or any *Approaches*, but coming without Shelter to Blows with the Enemy. The Besiegers generally insult the *Counterscarp* to shun the Enemies *Mines*, that they may have pre-

prepar'd) by not giving them time to Fire them. In ſuch Attacks, the *Grenadiers* march before the reſt of the *Troops*, and the *Work-men* go prepar'd to made a *Lodgment*. In the year 1695, when our *Approaches* were advanced within 150 yards of the *Covert Way* before the *Ravelin*, which cover'd St. *Nicolas Gate* at *Namure*, the King gave Orders to inſult the *Counterſcarp* about 4 a Clock in the Afternoon, which was done with that Bravery, that we lodged upon the *Parapet* of the *Covert Way*.

INTRENCH'D; an *Army* is ſaid to be *intrench'd*, when they have raiſed Works before them to fortifie themſelves againſt the Enemy, that they may not be forced to fight at a Diſadvantage. It was a chief Maxim amongſt the *Romans*, that in their moſt haſty Marches they *intrench'd* every night.

INTRENCHMENTS, are all ſorts of Works made to fortifie a Poſt againſt an Enemy. The Word ſignifies a *Foſs* or *Trench*, with a *Parapet*, or Rows of *Faſcines* loaded with *Earth*, *Gabions*, *Sand bags*, or *Hogſheads* filled with Earth, to cover the Men from the Enemies Fire. A *Poſt* intrench'd, is when it is cover'd with a *Foſs* and *Parapet*. See *Retrenchment*.

INVALIDE, is a Man who has ſpent his time in the Wars, and is either through Age, or by reaſon of his Wounds, render'd incapable of the Service. They are diſpoſed of in *Hoſpitals*, ſuch as *Chelſey* for the Land Forces, and *Greenwich* for the Sea-men; and are there provided with Meat, Drink and Cloaths, and are allowed a little Money beſides.

INVESTING *a Place*, is when a *General* having a Deſign to beſiege it detaches a Body of *Horſe* to ſieze all the *Avenues*, blocking up the *Garriſon*, and preventing Relief getting into the Place, till the *Army* and the *Artillery* are got up to form the Siege. In the year 1695, when the King to amuſe the Enemy, march'd his *Army* towards the *French* Lines before *Ypres*, and made a Sham Siege of *Knoque*; by which means he drew the *Marſhal* of *Villeroy*, with the *Grand Army* of *France* to oppoſe him; he order'd the Count *Tſerclais de Tilly*, Velt-Marſhal of the Elector of *Cologne*, who lay with a Body of *Horſe* to cover *Liege* and *Maeſtricht*, to inveſt *Namure* on the North ſide of the *Maes*, while General Flemming did the ſame on the South ſide, till our Army got up and formed that Famous Siege, which laſted two months after the *Trenches* were opened.

ISOSCELES, ſee *Triangle*.

K.

KETTLE, is a Term the *Dutch* give to a Battery of *Mortars*, because it is sunk under ground. See *Battery*.

KETTLE-DRUMMER, is a Man on Horse-back appointed to beat the *Kettle-Drums*, from which he takes his Name. He ought to be a Man of Courage, and of a good Meen, having a neat Motion with his Arms, and a good Ear; he marches always at the Head of the *Squadron*, and his post is on the Right when the *Squadron* is drawn up.

Kettle-Drums, are two sorts of large Basins of *Copper* or *Brass*, rounded in the bottom, and covered over with Vellum or Goat-skin, which is kept fast by a Circle of Iron, and several Holes fasten'd to the body of the *Drum*, and a like number of Screws to screw up and down, with a Key for the Purpose. The two *Drums* are kept fast together by two straps of Leather which go thro' two Rings which are fasten'd the one before, and the other behind the pomel of the *Kettle-Drummer's* Saddle; they have each a Banner of Silk or Damask richly embroider'd with the *Sovereigns* Arms, or those of the *Colonel*, and are fringed with Silver or Gold, and to preserve them in bad Weather, they have each a cover of Leather. The *Drum-Sticks* are of Crab-Tree or other hard Wood, of 8 or 9 inches long, with two knobs on the ends, which beat the *Drum* head and cause the sound. The *Kettle-Drum* with *Trumpets*, is the most Martial Sound of any; each *Regiment* of *Horse* has a pair, and of late the *Dragoons* have likewise *Kettle-Drums*. The Train of *Artillery* have a very large pair on a *Charriot* drawn by two white Horses.

KLINKETS, are a sort of small Gates made through *Palisades* for Sallies.

L.

LABORATORY, or *Laboratorium*, is any sort of Work-House; but we bring it in here as a Term belonging to *Gunnery*, and it signifies the Place where the *Fire-workers* and *Bombardeers* prepare their Stores; such as driving of *Fusees*, fixing of *Shells*, making of *Quick-match*, fixing *Carcasses*, and all other Fireworks belonging to War or Recreation. There is sometimes a large Tent carried along with the *Artillery* to the Field for this Use, with all sorts of Tools and Materials, and it is call'd the *Laboratory Tent*.

LADLE *of a Gun*, is the Instrument wherewith the Powder is put into the Piece; it is made of a plate of Copper bowed in form of a half *Cylinder*, rounded at one end, the other being fixed upon a long Staff; this filled with Powder, the *Gunner* carries with his Left Hand under the end of it to keep the Powder from falling out, till he enters it

in

in the Muzzle of the Piece; when he has carried the Powder home to the charged *Cylinder*, he turns the *Ladle* that the Powder may fall out, and withdraws it. *Ladles* are fitted to the Bore of each *Gun*, and hold Powder sufficient for the Charge.

LANE; *to make a Lane*, is to draw Men up in two Ranks facing one another, which is generally done in the Streets through which a great Person is to pass, as a Mark of Honour. But oftentimes Men are drawn up so for a *Soldier* to run the *Gauntlet*, each having a small Wand or Swith in his hand to whip the Offender as he passes.

LANSPESADE, or *Anspesade*, is under a *Corporal*, and assists him in his Duty, and performs his Duty in his absence; they are generally the most vigilant and brave of the Company, they have the same Pay as a Foot *Soldier*, but in *France* they have a greater Allowance; they are excused from Common Duty, they teach the new *Soldiers* their Exercise, and post the *Centries*; their Place on a March is on the Right of the second Rank.

LANTHORNS are used at Sieges for the night time upon the *Batteries*, but these are your *Blind* or *Dark-Lanthorns*. They have always a great Provision of them in all the Store-Houses.

LEADERS, see *File-Leaders*.

LIEUTENANT of *Horse* or *Foot*, is the second Officer in the *Troop* or *Company*, and ought to be a very Intelligent Person, because in the absence of the Captain he commands and therefore he ought to be as well qualified as the *Captain*; when the *Company* is at Arms, he takes the Left of the *Captain*, but the Right, if the *Ensign* be there. He marches the *Company* in the absence of the *Captain*, but when the *Captain* is present, his Post is in the *Rear*; when the *Battalion* marches in Line of Battel, the *Lieutenants* take their Posts at the Head of the Divisions, according to their Seniority. He ought to inspect the Actions of the *Serjeants* and *Corporals*, to keep them to their Duty, and take care of every thing that is necessary to the *Company* to see them Exercise, to cause them keep their Arms clean and fit for Service, and to see that the *Soldiers* be provided of Powder and Ball.

Lieutenant Colonel, is the second Officer in a *Regiment*, and should be one who has signaliz'd himself by some brave Action; a Man of great Experience; knowing how to attack or defend a *Post*, lead the *Regiment* to Battel, and how to make a good Retreat; he is to see the *Regiment* kept to their Exercise, and is to know the Qualifications of all the Officers of the *Regiment*. In the absence of the *Colonel* he commands the *Regiment*; his Post is on the *Colonel's* Left Hand, three paces before the *Captains*, when there is but one *Battalion*; but if the *Regiment* be of

of two, the *Colonel* commands the first, and he the second. *Colonels* and *Lieutenant-Colonels*, are excused from mounting the Guard when the *Regiment* is in Garrison.

Lieutenant-Colonel of Horse, is the same; he marches at the *Head* of the second *Squadron*.

Lieutenant-General, is a Charge of great consequence, requires Persons of Courage and Conduct, and whose Ability and Fidelity has appeared on several Occasions: They ought not only to understand their own, but also the Business of a *General*, because they are often intrusted with the Command of a *Flying-Camp*, and sometimes with a part of the *Army*. The number of *Lieutenant-Generals* is not limited; they are more or less, according as the *Army* is great or small: there is every day a *Lieutenant-General* upon Command, who is called the *Lieutenant-General of the Day*. In Day of Battel, they command the *Wings* of the *Army*; and at a Siege, they command in the *Trenches* by turns. They are to execute the *Generals* Orders upon all Occasions: some are appointed over the *Cavalry*, others for the *Infantry*; sometimes on the *Advance-Guard*, others on the *Rear-Guard*, and sometimes they command a *Flying-Army*. They ought to be daily with the *General* to know his Orders; they are allowed each two *Aid de Camps*, and a *Foot Guard* mounted by a *Subaltern*, with a *Serjeant* and 30 Men.

LIFE-GUARD, see *Guards*.

LIGHT-HORSE; all the *Regiments* of *Horse*, except the *Guards*, are called the *Light-Horse*; each *Regiment* consists of six *Troops* (only *Lumley*'s has nine) and is commanded by a *Colonel*, *Lieutenant-Colonel*, *Major*, *Captains*, *Lieutenants*, *Cornets*, and *Quarter-Masters*; they rank according to Seniority. See *Regiment* and *Troop*.

LIMBERS, is a sort of advanc'd Train joined to the Carriage of a *Cannon* upon a March; it is composed of two Shafts wide enough to receive a Horse betwixt them, which is called the *Fillet-Horse*; these Shafts are joined by two bars of Wood, and a bolt of Iron at one end, and have a pair of small Wheels; upon the *Axletree* rises a strong Iron Spike, upon which the Train of the *Carriage* is put upon a March, but when a *Gun* is upon Action; the *Limbers* are run out behind her.

LINE, as it is a Term in *Geometry*, is a length without breadth, whose Extremities are Points.

Line of an Army, see *Camp*.

Line in Fortification, is sometimes applied to a *Trench*. *Line of Gabions*, *Lines of Communication*, *Circumvallation*, *Contrevallation*, and *Lines of Approaches*, are already explained. To run a *Line*, is to dig a *Trench* with a *Parapet*; *to line a Work*, is to trace it out; *to line a Work*, is likewise to face it with Brick or Stone. *Lines* are sometimes made to cover a Coun-

try, as the *French Lines* in the late War, which run from *Dunkirk* by *Ypres*, *Menin*, *Courtray*, *Lisle*, and ended at the *Scheld* near *Tournay*; they were fortified from distance to distance, with small *Forts*; these were forced by the Duke of *Wirtenberg*, in the year 1693, near *Pont Espiere*. The *French* in the beginning of this War, made a *Line* from the *Scheld* below *Antwerp* to cover that Place and all *Brabant*, which passes by *Leer*, *Arschot*, and *Leuwe*, to the *Mehaigne*, three Leagues above *Huy*.

Line-Capital, see *Capital*.

Line of Defence, is a supposed *Line* drawn from the *Angle* of the *Curtin*, or from any other Point in the *Curtin*, to the *Flanked Angle* of the opposite *Bastion*; or it is a *Line* representing the flight of a *Musquet Ball* from the place where the *Musqueteers* stand to scowr the *Face* of the *Bastion*, and ought never to exceed the reach of a *Musquet*. It is either *Fichant* or *Razant*; the *first* is when it is drawn from the *Angle* of the *Curtin* to the *Flanked Angle*, which ought never to exceed 800 foot; the *last* when it is drawn from a Point in the *Curtin*, razing the *Face* of the *Bastion*; this *Line* shows how much of the *Curtin* is taken off fort he second *Flank*: it is called likewise the *Flanking* or *Stringent-Line*.

To line Hedges, see *Hedges*.

LINSPINS, are small pins of Iron which keeps the *Wheel* of a *Cannon* or *Waggon* on the *Axletree*, for when the end of the *Axletree* is put through the *Nave*, the *Linspin* is put in to keep the *Wheel* from falling off.

LINSTOCK, is a short Staff of Wood about three foot long, upon one end of which is a piece of Iron which divides in two turning from one another, having each a place to receive Match, and a Screw to keep it fast; the other end is pointed and shod with Iron to stick in the Ground. It is used by *Gunners* to fire the *Guns*; and failing of it, a piece of cleft Stick to stick the Match in, does the Business every jot as well.

LIZIERE; *Berm*, *Foreland*, or *Relais*, is a space of Ground left at the *Foot* of the *Rampart*, on the side next the Country, design'd to receive the Ruins of the *Rampart*, to prevent its filling up the *Foss*; it is sometimes palisaded for the more security, and in *Holland* is planted generally with a *Quickset-Hedge*; when this space is covered with a *Parapet*, it is called a *Fausebray*, or *Low Wall*.

LOCKSPIT, is the small Cut or Trench made with a *Spade* of about a foot wide, to mark out the first *Lines* of a Work.

LODGMENT, is a Work raised with *Earth*, *Gabions*, *Fascines*, *Wool-packs*, or *Mantelets*, to cover the Besiegers from the the Enemies Fire. In conducting the *Approaches* at certain Distances, are made *Lodgments* or *Places of Arms* to flank the *Trenches*, capable of holding 100 Men, which

which ſerve as a guard to the *Trenches*. But *Lodgments* made on the *Glacis*, *Covert Way*, *Breach*, *&c.* are much more dangerous, being more expoſed to the Enemies Fire, and having leſs Earth. When it is reſolved upon to inſult the *Covert Way*, tho' a great many Men are loſt, and generally the braveſt; there muſt be great Proviſion made of *Faſcines*, *Sand-bags*, &c. in the *Trenches*; and during the Action, the *Pioneers* with *Mantelets*, *Faſcines*, *Wool-packs*, or *Sand-bags*, ſhould be making the *Lodgment*; covering themſelves as advantagiouſly as poſſible from the oppoſite *Baſtion*, or the Place moſt to be feared.

LOZANGE, or *Rhombe*, is a Figure of four equal ſides whoſe *Angles* are two Acute, and two Obtuſe.

LUNETTE, is a ſmall Work raiſed ſometimes in the middle of the *Foſs* before the *Curtin*, forming an *Angle*; its *Terreplein* riſing but a little above the *Surface* of the *Water*, about 12 foot broad, with a Parapet of 18 foot.

There is another ſort of *Lunettes*, which are larger, and raiſed to cover the *Faces* of a *Half Moon*; they are likewiſe compoſed of two *Faces*, a longer and a ſhorter; ſee their Figure in the general *Plan*.

M.

MACHINE, is a heap of ſeveral pieces joined together, and ſo diſpoſed, as to ſerve both for augmenting or diminiſhing the Force whereby a Body is moved, according to the different Uſes to which it is applied in *War*, *Architecture*, and other *Arts*. *Machines* are either *Simple* or *Compoſed*; the *Simple* are the *Balance*, *Crow*, *Pully*, *Wheel*, *Coin* and *Screw*; to which ſome add the *Inclining Plan*, and the plain *Surface*.

Machines Compoſed, are thoſe which are compoſed of ſeveral *Simple Machines*, which cannot be numbred, becauſe in their Conſtruction, the *Simple Machines* may be uſed a great many different ways.

MADRIERS, are long planks of Wood very broad, uſed for ſupporting the Earth in mineing, in carrying on a *Sap*, in making *Coffers*, *Caponiers*, *Galleries*, and many other Uſes at a Siege. They are likewiſe uſed to cover the mouth of *Petards* after they are loaded, and are fixed with the *Petards* to the *Gates*, or other places deſign'd to be forced open; when the Planks are not ſtrong enough, they are doubled with plates of Iron.

MAGAZINE or *Arſenal*, is the Place where all ſorts of Stores are kept, where *Guns* are founded, and where the *Carpenters*, *Wheelwrights*, *Smiths*, *Turners*, and other Handicrafts, are conſtantly employed in making all things belonging to an *Artillery*, as *Carriages*, *Waggons*, *Tumbrills*, *&c.* The place ought to be large, that

that every thing may be conveniently disposed without Confusion; the Powder by it self in a dry place, otherwise the Saltpetre grows moist and damp. The *Corselets*, *Breast-Plates*, *Helmets*, *Pikes*, *Halberts*, *Musquets*, *Carabines*, *Pistols*, *Swords*, *Bagonets*, &c. in order; the *Cords*, *Ropes*, *Match*, *Brass*, *Tin*, *Lead*, and all other things belonging either to the Attack or Defence of a Place, ought to be in particular places. Fixed *Bombs*, *Grenades*, *Petards*, &c. must be in dry Places; *Cannon Ball*, empty *Bombs* and *Grenades*, may be piled up in the Court, or in places made on purpose, with a small Wall betwixt the *Balls* of different Calibers. The *Carriages*, *Wheels*, *Limbers*, *Waggons*, and *Tumbrils* must be in Shades, to save their being spoiled or rotted by the Rain or Snow. The *Fire-workers* Stores must be allowed a Place and a *Laboratory* by themselves. A *Magazine* being the Place appointed for making and keeping all sorts of War-like Stores, ought to be well situated in a strong place, and well guarded.

MAIN-BODY *of the Army*, is the *Body of Troops* that marches between the *Advance* and the *Rear-Guard*. In a *Camp*, it is that part of the *Army* which is encamped betwixt the *Right* and *Left Wing*.

Main-Guard, or *Grand-Guard*, is a *Body of Horse* posted before the *Camp*, for the safety of the *Army*. In Garrison, it is that *Guard* to which all the rest are subordinate; see *Guard*.

MAJOR *of a Regiment of Horse or Foot*, is the next Officer to the *Lieutenant-Colonel*, and is generally made from eldest *Captain*; he ought to be a very careful vigilant Man; he is to take care that the *Regiment* be well exercised, that they be drawn up in good Order at a *Review* or upon the *Parade*, or any other Occasion; to see it march in good Order, and to rally it, in case of its being broke. He is the only Officer among the *Foot* that is allowed to be on Horseback in time of Action, that he may be the readier to execute the *Colonels* Orders, either in advancing or drawing off the *Regiment*: He is appointed an *Aid-Major*, or *Adjutant* for his Assistant.

Major of a Brigade, see *Brigade-Major*.

Major General, is a Post of great Consequence, and he who possesses it, ought to be brave, judicious, and a Man of Experience in marching and encamping of *Troops*; he is the next Officer to the *Lieutenant-General*; when there are two Attacks at a Siege, he commands that on the Left. His chief Business is to receive the Orders every night from the *General*, or in his absence, from the *Lieutenant-General of the Day*, which he is to distribute to the *Brigade-Majors*, with whom he is to regulate the *Guards*, *Convoys*, *Parties* and *Detachments*, and appoints

points the place and hour of their Rendezvous. He is to know the Strength of each *Brigade* in general, and of each *Regiment* in particular, and to have a List of all the *Field Officers*. Finally, he is in the *Army*, the same as the *Major* of a *Regiment* in the *Regiment*: He is allowed an *Aid de Camp*, and has a *Serjeant* and 15 Men for his *Guard*.

Town-Major, is the third Officer in a Garrison; his Business is to see the *Guards* mounted, the the *Rounds* and *Posts* assign'd; he regulates the *Centinels*, goes every evening to receive the Word from the *Governour*, and gives it out upon the Place of *Arms*, to the *Adjutants* and *Serjeants* of the Garrison; he goes his *Round Major*, visits the *Corps de Gardes*, and sees that all the *Soldiers Arms* are fixed and in good Order; he causes necessary Ammunition to be distributed among them, orders the Gates to be open'd and shut, and gives the *Governour* an Account of all that passes in the Place.

MALLET is so well known it needs no Description; its use in *Fortification* is for driving in the Stakes or Piquets that fasten the *Fascines* or *Gazons*; it is likewise used at Sieges, both upon the *Batteries* and in the *Trenches*, on several Occasions.

MANTELETS, are great planks of Wood of about five foot high, and three inches thick, which serve at a Siege to cover the Men from the Enemies Fire, being pushed forward on small *Trucks*; they are of two sorts, either single or double. Single *Mantelets A*, are made in joining two or three such planks together with bars of Iron, to make three foot or three and a

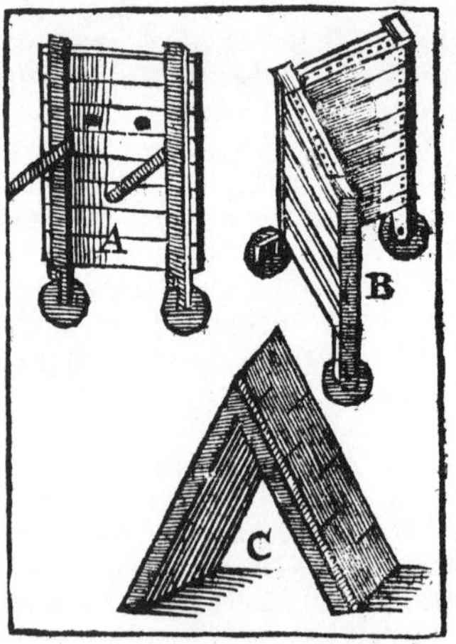

half broad, to cover those that carry them from the Enemies Fire. Double *Mantelets*, *B*, are made by putting Earth between two such rows of Planks, and are used in making *Approaches* and *Batteries* near the Place, as the others are in making *Lodgments* on the *Counterscarp*; they are covered with Letten, and are then made smaller at bottom than at the top, that they may be the more easily join'd together, to cover the *Soldiers* from the *Grenades* and *Fireworks* of the Place, as C. Some are so made, as to cover the *Soldiers* from the

Fire on *Front* or in *Flank*, as *B.*

MARCH, *in general*, is the steps made in marching, or the moving of a Body of Men from one place to another. *The Beat of the Drum*, when the *Soldiers* are upon march, or beginning to march, is likewise called *the March.* It is likewise a word of Command, when a *Battalion* is to alter its Disposition.

To March, is to move from one place to another. The Orders being given the evening before a March, that the *Drums* beat at a certain hour, the *Soldiers* are then ready drawn up at their *Standards* and *Colours*, in *Squadrons* or *Battalions*, ready to begin their March. As many Accidents may happen in the March of an *Army*, as *Defiles*, *Marshes*, *Woods* and the like, it is the Prudence of a *General* to order his March accordingly, and to take care that the *Columns* of his *Army* have a free Communication one with the other. The *March* of an *Army* is composed of an *Advance-Guard*, the *Main-Body*, and the *Rear-Guard*, and is sometimes in two, four, six or eight *Columns*, according as the *Ground* will allow.

MARSH, is a standing Water, or Water mixed with Earth, whose bottom is very dirty, which drys up and diminishes very much in the Summer.

MARSHAL *of France*, is the highest Preferment in the *Army* or in the *Fleet*: it is the same with *Captain-General*; when two or more *Marshals* are in one *Army*, the eldest Commands.

Velt-Marshal in *Germany*, *Brandenburg*, *Holland*, *&c.* is likewise the same with *Captain-General.*

Marshal de Camp, is in *France* the next Officer to the *Lieutenant-General*, and is the same as a *Major-General* with us. See *Major-General.*

Marshal de Battaille, was once a Post in the *French Armies*, but now is suppressed.

MASTER DE CAMP in *France*, is he who commands a *Regiment* of *Horse*, being the same as a *Colonel* of *Horse.*

Master de Camp General, is likewise a Post in *France*, being the second Officer over all the *Light-Horse*, and commands in the absence of the *Colonel-General.*

Master-Gunner, is an Officer who has the command over all the *Gunners* in the Kingdom, and is subordinate to the Principal Officers of her Majesty's Ordnance. Each Garrison has likewise a *Master-Gunner*, who commands the other *Gunners.*

Master-General of the Ordnance, see *General.*

MATCH, a small Rope or Twist about three quarters of an inch diameter twisted hard, which being lighted at one end, burns leisurely without going out; it was formerly used for *Matchlock Musquets*, but these are now out of use; it is now used for firing the *Cannon*, *Mortars*, *Hand Grenades*, *&c.* it is sometimes used in blowing up *Mines*, when it is to be done

done ſome hours after; the Match being cut according to Judgment to burn ſix, eight or ten hours, at the end of which time the Match being burned out, fires the Powder and ſprings the Mine; the hardeſt and dryeſt Match is generally the beſt.

MATROSSES, are *Soldiers* in the *Artillery* next to a *Gunner*; their Buſineſs is to aſſiſt the *Gunners* about the *Gun*, to Traverſe, Spunge and Fire, to aſſiſt him in Loading, *&c.* they carry *Fire-locks*, and march along with the *Store Waggons*, both as a Guard, and to help in caſe a *Waggon* ſhould break down.

MAXIMS, are certain Propoſitions, ſo clear in themſelves, that they cannot be denied, without going againſt Senſe and Natural Reaſon; they are generally ſerviceable in all manner of Demonſtrations, and deſerve to be eſtabliſh'd as Infallible.

Maxims in *Fortification*, are certain general Rules eſtabliſh'd by *Engineers*, founded on Reaſon and Experience, which being exactly obſerved, a Place fortified according as they direct, will be in a good Poſture of Defence. The chiefeſt *Maxims* are,

1. *There ought to be no part in the Fortification of a place, but what is diſcovered and flanked by the Beſieged.* For if there be any part of a Place which is not well flanked, the Enemy being there under cover, will with the more eaſe attack it in that place, and carry it.

2. *A Fortreſs ſhould command all the Country round it*; that the Beſiegers may not cover themſelves; nor find places to favour their *Approaches* and *Attacks*, nor to overlook the Works of the Place, to batter them with more Advantage.

3. *The Works furtheſt diſtant from the Centre of the Place, muſt be ſtill loweſt, and commanded by thoſe that are nearer*; to the end, they may be defended by the higher Works, and thoſe nearer the Place; that ſo the Enemy by being expoſed, may be obliged to quit them, after they have been poſſeſs'd of them, becauſe of the Fire of the Beſieged; and likewiſe, that the Enemy by being Maſters of ſuch Works, may not overlook the Works of the Place.

4. *The Flanked Angle, or Point of the Baſtion, ought to be at leaſt 70 Degrees*; that it may the better reſiſt the force of the Enemies *Batteries*, in caſe they deſigned to beat it down to lodge there.

5. *The Acute Flanked Angle near to a Right Angle, is preferable to all other.* It is certain, if the *Flanked Angle* be a *Right Angle*, it has all the Strength can be given it, having Solidity enough to withſtand the Enemies *Batteries*: but an *Angle* near to a Right, makes the *Tenaille* of the Place more compact, by the *Angle* of the *Shoulders* ſhortening and bettering the Defence, and by its not expoſing the *Face* ſo much to the Enemy. So that it follows of

consequence, that an *Obtuse Angle* is very deficient.

6. *The shortest Faces are the best*; because the longer they are the weaker, for the Enemy attacks them with a greater Front.

7. *The Flank must have some part under cover*; which signifies it must be cover'd by an *Orillon*, otherwise the Defence is presently ruined, and the Lodgment no sooner made on the *Counterscarp*, but the Place is obliged to capitulate, as has been often seen.

8. *There must be an Accord between these Maxims, to render the Fortification perfect.* For if the *Gorge* be too large, the *Face* suffers; the more the *Flank* is covered, the less it is subject to be ruined, but then the Defence is more oblique: in making a second *Flank*, the *Flanked Angle* is made too weak: in discovering the *Face*, the Defence is more easie, but it is more exposed to the Enemies *Batteries*. In a word, there is Advantages and Disadvantages over all; and the Secret consists in judging whether conforming with one *Maxim* be more Advantagious, than disagreeing with another.

MEASURE-ANGLE, is an Instrument of Brass for measuring *Angles*, either *Sailliant* or *Rentrant*, to know exactly the number of Degrees and Minutes, to lay them out upon Paper.

MECHANICKS, is a Science very useful in War, especially about an *Artillery*, as teaching a commodious way of moving heavy Bodies by the help of *Machines*.

MERLON, is that part of the *Parapet* which is terminated by two *Embrasures* of a *Battery*, so that its height and thickness is the same with that of the *Parapet*, but its breadth is ordinarily nine foot on the inside, and six on the outside. It serves to cover those on the Battery from the Enemy: and is better of Earth, well beat and close, than of Stone, because these fly about and wound those they should defend.

MILE, is a sort of Measure which is longer or shorter in different Countries; an *Italian* Mile is 1000 Geometrical Paces; an *English* Mile 1250, in *Scotland* and *Ireland* it is 1500, and in *Germany* 4000.

MILITARY-EXECUTION, is the ravaging and destroying a Country for Contribution.

MINE, is an Ouverture made in a Wall or other place, which is design'd to be blown up with Powder, it is composed of a *Gallery* and a *Chamber*. The *Gallery* is the first Passage made under ground, being no higher nor broader than to suffer a Man to work on his Knees. The *Chamber* is the small space at the end of the *Gallery*, like a small *Chamber* where the Barrels of Powder are deposited, for blowing up what is proposed to be sprung, When the *Chamber* is dry, the Powder is put in Sacks instead of Barrels, and the form of the *Chamber* is cubical; its height

and depth being about ſix foot. When the *Mine* is under the *Rampart* of an empty *Baſtion*, leaſt by the thinneſs of the Earth on the ſide next the Place, the Powder ſhould burſt forth that way; the top of the *Chamber* is cut into to Paſſages like *Chimneys*, to oblige the *Mine* to have its Effect upwards. If the *Chamber* be humide or damp, the Powder is put into *Barrels* or *Caiſſons*, and fired with a *Sauciſs*. If the Places to be blown up be rocky, or if there happen to be any thing elſe in the way to hinder the *Miners*, they make *Fourneaux*, *Arraignée*, or *Rameaux*, all which are the ſame thing, and ſignifie *Branches*, which terminate in ſmall *Mines*, and are fired all together by ſeveral *Sauciſſes*. The *Gallery* of a *Mine* goes turning and winding, the Earth is put in ſmall Baskets, and given out betwixt their Legs backwards from one to another, becauſe of the narrowneſs of the Paſſage. The Earth of the *Chamber* is to be ſupported with planks, and when the *Chamber* is damp, it muſt be floor'd with Boards.

MINER, is he that works in the *Mine*; he covers his Head with a *Hood*, to ſave his Eyes from the Earth that falls down, which by this *Hood* is thrown over his Shoulders.

MINION, is a Piece of *Cannon* carrying a Ball of 4 pound weight, the diameter of its Bore is 3 inches and 3 eighths, and the length of the Piece about ſix foot and a half; they are not much uſed in the Field now, becauſe the *Saker* which carries four pound and a quarter Ball is more eſteemed.

MINUTE, is the 60th part of a *Degree*, ſee *Degree*.

MOAT, *Ditch* or *Foſs*, is a Depth or Trench round the *Rampart* of a Place, to defend it, and prevent Surprizes; the brink of the *Moat* next the *Rampart*, is called the *Scarpe*, and that oppoſite on the other ſide, is called the *Counterſcarp*, which forms a *Reentring Angle* before the Centre of the *Curtin*. A dry *Moat* round a place that is large and has a ſtrong Garriſon, is preferable to one full of Water, becauſe the Paſſage may be diſputed inch by inch, and the Beſiegers when lodged in the *Moat* are continually expoſed to the *Bombs*, *Grenades*, and other *Fireworks* which are thrown inceſſantly over the *Rampart* on their Works. In the middle of dry *Moats*, is ſometimes made another ſmall *Moat*, called the *Cunette*, which is generally dug ſo deep till they find Water to fill it. The deepeſt and broadeſt *Foſſes* are counted the beſt, but a deep *Foſs* is preferable to a broad one; the ordinary breadth is about 20 fathom, and the depth about 16 foot. To drain a *Foſs* or *Moat* full of Water, is to dig a Trench deeper than the Level of the Water to let it run out; when it is drained, there are Hurdles thrown upon the Mud and Slime, and

cover'd

cover'd with Earth or bundles of Rushes, to make a sure and firm Passage.

MOINEAU, is a *French* Term for a little flat *Bastion* raised upon a *Re-entring Angle* before a *Curtin*, which is too long, between two other *Bastions*; it is commonly joined to the *Curtin*, but sometimes separated by a *Foss*, and is then called a *Detach'd Bastion*; they are not raised so high as the Works of the Place, because they must be exposed to the Fire of the Besieged, in case the Enemy should lodge themselves there. Their *Parapet*, as well as the *Parapet* of all Out-Works, ought to be *Cannon* Proof, that is to say, 18 foot thick.

MONT-PAGNOTE, *or Post of the Invulnerable*, is an Eminence chosen out of *Cannon-Shot* of the Place besieged, where curious Persons post themselves to see an Attack, and the manner of the Siege, without being exposed to any Danger. At the late Siege at *Landau*, there was a *Mont Pagnote* raised on a Rising Ground, half way betwixt the two Attacks, by the Orders of Prince *Lewis* of *Baden*, for the King of the *Romans* to see the Attacks.

MORTARPIECE, is a sort of a short piece of *Artillery*, reinforced, and of a wide Caliber, differing from a *Cannon*, both in Form and Use; the *Cannon* serving to throw *Ball*, and the *Mortar* to throw *Bombs*, *Carcasses*, *Fire Pots*, and several other sorts of *Fireworks*, as likewise *Stones*. *Mortars* are used both at Sea and Land, but they differ very much in Form. A Sea *Mortar* is generally 13 inches diameter of the Bore, is longer and more reinforced than a Land *Mortar*, because they are fired with a greater quantity of Powder, sometimes with 30 or 33 pound; some of them have their Beds or Stools of Metal cast in a piece with the *Mortars*, others have them of a thick square piece of Oak, which by the help of *Hand screws* or *Jacks*, is turned round upon a strong Axis of Iron to fire any way; they are always fixt at an Elevation of 45 Degrees; they carry *Bombs* of 200 pound, and generally weigh about 9 or 10000 weight.

Land Mortars are of different sorts; those used most in *England*, are 10, 13, 15, and 18 inches diameter; there are smaller *Mortars* of 6 and 8 inches; all but the 18 inch *Mortars* are mounted on a very thick plank of Oak, on which rise two *Cheeks* or *Brackets* on the sides of the *Mortar*, see *the Figure at Cheeks*. But the 18 inch is mounted on a *Low Dutch Carriage*, consisting of two strong planks of Wood bound with thick plates of Iron, and joined together with *Transums* of Wood. All *Land Mortars* may be elevated to any degree of the *Quadrant*. They have no Wheels, therefore on a March they are laid upon a *Block Carriage* made on purpose. They are never carried along with the *Army*, because of

of their great weight, except upon an Occasion of a Siege or Bombardment; but a sort of small *Mortars* called *Hobits*, mounted in *Gun-Carriages*, are always a part of the *Field Artillery*. See *Hobits*.

Hand Mortars, are likewise of several sorts, as *Tinkers Mortars*, which are fixed at the end of a staff of about four foot and a half long, the other end being shod with Iron to stick in the Ground, while a *Soldier* with his one Hand keeps it at an Elevation, and with the other Hand fires. *Firelock Mortars* are fixed in a Stock with a *Lock*, like a *Firelock*; they swing between two Arches of Iron, with Holes answering one another, by which the *Mortar* is elevated; these stand upon a sole or plank of Wood, and may be carried by one Man from one place to another. There are more sorts of *Hand Mortars*, but *Coehorne's* new Invention exceeds them all, so far as to deserve a

particular Description. They are made of hammer'd Iron of four inches diameter of the Bore, ten inches and a half long, and nine inches in the Chace, fixed upon a piece of Oak of 20 inches long, 10 and a half broad, and betwixt 3 and 4 thick; they stand fixed at 45 Degres of Elevation, and throw *Hand Grenades*, as all other *Hand Mortars* do; they are placed in the bottom of the *Trenches*, at 2 yards distance from one another, having each a *Soldier* to serve it, and an *Officer* to every 40 or 50, who lays them to what Elevation he thinks convenient, by raising or sinking the hind part of the *Bed*; three or four hundred of them are sometimes in Service at once, in different parts of the *Trenches*, 60, 70, or 80 in a place. Those in one place fire all at once, immediately after the *Batteries* have done, and are answered from another part of the *Trench*, which brings such a shower of *Hand Grenades* into the *Covert Way*, that those that defend it cannot prevent being in confusion.

MOTION, or *Movement of an Army*, is the several *Marches* and *Counter-marches* it makes, or changing of its *Post* for an Advantagious *Encampment*, either with a Design to engage the Enemy to a Battel, or to shun fighting.

Motion of a Bomb or Ball, is the Progress it makes in the Air after it is delivered, and is of three sorts: The *Violent Motion*, is the first Expulsion when the *Powder* has worked its Effect upon the *Ball*, or so far as the *Bomb* or *Ball* may be supposed to go in a right Line. The *Mixt Motion*, is when

when the weight of the *Ball* begins to overcome the force which was given by the *Powder*, and the *Natural Motion*, is when the *Ball* or *Bomb* is falling.

MOULDINGS *of a Gun or Mortar*, are all the eminent parts, as Squares or Rounds, which serve generally for Ornament, such as the *Breech-Mouldings* and *Muzzle Mouldings*; the *Rings of a Gun*, are likewise *Mouldings*.

MOUNT, see *Cavalier*.

To Mount the Guard, is to go upon Duty: *To Mount a Breach*, is to run up it or to attack: *To Mount the Trenches*, is to go upon Guard in the *Trenches*.

MUSQUET, is the most commodious and useful *Fire-Arm* used in the *Army*, either in attacking or defending a *Post*; it is easily managed, and is carried with small trouble, which makes its use the more common; the *Pikes* being laid aside in our *Army* of late, and *Musquets* brought in their stead, shows that tho' *Pikes* are useful, yet *Musquets* are much more, and can do better Service. They carry a *Ball* of 16 in the pound: the length of the *Line of Defence* is limited in *Fortification*, by the ordinary distance of a *Musquet-Shot*, which is about 120 fathom, and almost all the *Military Architecture* is regulated by this Rule for the length of the Defence, as the effect of *Cannon* gives a Rule for the thickness of the *Ramparts* and *Parapets*. See *Firelock*.

Musquet-Baskets, see *Baskets*.

MUSQUETEER, is a Foot *Soldier* arm'd with a *Musquet* or *Firelock*, *Sword*, *Bayonet*, &c. The *Grand Musqueteers* in *France*, are *Troopers* who fight sometimes on Foot, sometimes on Horseback; they are Gentlemen of good Families, and are divided into two *Troops*, the one called the *Grey Musqueteers*, because of the Colour of their Horses, the other the *Black Musqueteers*, for the same Reason.

MUSQUETOON, see *Blunderbuss*.

MUSTER, is a narrow Review of *Troops* under Arms, to see if they be compleat, and in good Condition; that their *Arms* and *Accoutrements* be in good Order; thereby to know the Strength of an *Army*: The *General* may order either *Muster* or *Review*, as often as he pleases.

Muster-Master, see *Commissary*.

Muster-Rolls, are the *Rolls* or *Lists* of the *Companies* or *Troops*, which are delivered to the *Commissary* by the *Captains*.

MUZZLE *of a Gun or Mortar*, is the extremity of the *Cylinder*, where the *Powder* and *Ball* is put in. The Metal which surrounds the Extremity of the *Cylinder*, is likewise called the *Muzzle*.

Muzzle-Mouldings is the Ornament round the *Muzzle*, see *Mouldings*.

N.

To nail Cannon, or as some say to *Cloy* them, is to drive an Iron Spike by main force into the Vent or Touch-hole, which renders the *Cannon* inserviceable, till the Spike be either got out, or a new Vent drilled. In all Sortees or Sallies of a place besieged, nothing is so glorious as the nailing of the Besiegers *Cannon*, nor so advantagious to the *Garrison*, for it takes the Enemy some time to repair it. Mr. *Dupas*, Governour of *Naerden* in *Holland*, to repair his Fault in delivering the Town to the *Hollanders*, threw himself into the Grave, which was then besieged by them in the year 1674, to serve as Volunteer, and in a Sally the *Garrison* made behaved himself with great Bravery in clearing the *Trenches*, and was killed nailing the Enemies *Cannon*. The late Sally of the Garrison of *Verue*, and their nailing up the Enemies *Cannon*, was no less disadvantagious to the Enemy, than it was glorious and advantagious to the Besieged.

NAVE *of a Wheel*, is that short thick piece in the centre of the *Wheel*, which receives the end of the *Axletree*, and in which the ends of the *Spokes* are fixed; it is bound at each end with Hoops of Iron, called the *Nave-bands*: It has likewise in each end of the Hole, through which the end of the *Axletree* goes, a Ring of Iron called the *Wisher*, which saves the Hole of the *Nave* from wearing too big.

NECK *of a Gun*, is that part betwixt the *Muzzle Mouldings* and the *Cornish-Ring*. *Neck of the Cascabel*, is the part betwixt the *Breech-Mouldings* and the *Cascabel*.

O.

OBLIQUE-FLANK, see *Flank*.

Oblique Defence, is that which is under too great an *Angle*, as is generally the Defence of a second *Flank*, which can never be so good as a Defence in *Front*, nor is it approved by *Engineers*.

OBTUSE-ANGLE, see *Angle*.

OCTOGON, is a Figure of 8 *Sides* or *Polygons*, forming the same number of *Angles*, and capable of being fortified with 8 *Bastions*.

OFFICER *in General*, is a Person employed in some Office. *Officer in the Army*, is a Person having a Command in the *Army*. Those having Commissions from the *Queen* or *General*, are called *Commission'd Officers*, which includes all from the *General* to an *Ensign*. Such as have no *Commission*, but only *Warrants* from their *Colonels*, are called *Warrant-Officers*, as *Quarter-Masters of Horse*, and *Surgeons*. Those that have neither *Commissions* nor *Warrants*, are called *Staff-Officers*, as *Serjeants*, *Corporals*, *Lanspesades*, &c.

An

An *Officer* ought to be endued with many good Qualifications, as *presence of Mind*, *Judgment* to execute what he is commanded to the best Advantage; *Conception*, to apprehend easily what he is to do; *Complexion*, to enable him to endure the Fatigues of War; *Integrity*, that whatsoever is entrusted to him, he may behave himself honourably; *Diligence* to execute every thing with dispatch; *Secrecy* in all Affairs of Consequence, with which he is entrusted; *Conduct* to gain the Esteem of his Superiours; No *Presumption* nor *Obstinacy*, as being the stumbling Blocks of such as have too much Spirit.

General Officers, are such as command a Body of *Troops* of several *Regiments*, as the *Captain-General*, *Lieutenant-General*, *Major-General*, *Brigadeer-General*, *Quarter-Master General*, and *Adjutant-General*, see each at their proper Letters.

Field-Officers, are those who have a command over a whole *Regiment*, as the *Colonel*, *Lieutenant-Colonel* and *Major*.

Subaltern Officers, are the *Lieutenants*, *Cornets*, and *Ensigns*.

TO OPEN *Trenches*, is the first breaking of Ground by the Besiegers, in order to carry on their *Approaches* towards a place: The difference betwixt opening and carrying on the *Trenches*, is that the first is only the beginning of the *Trench*, which is always turned towards the Besiegers: It is begun by a small *Foss*, which the *Pioneers* make in the night time on their knees, generally a *Musquet-Shot* from the place, or half a *Cannon-Shot*, and sometimes without the reach of *Cannon-Ball*; especially if there be no Hollows or Rising Grounds to favour them, or if the *Garrison* be strong, and their *Artillery* well served. This small *Foss* is afterwards enlarged by the next *Pioneers* which come behind them, who dig it deeper by degrees, till it be about 4 yards broad, and 4 or 5 foot deep, especially if they be near the place; to the end, the Earth which is taken out of it may be thrown before them to form a *Parapet*, to cover them from the Fire of the Besieged: The place where the *Trenches* are opened, is called the end of the *Trench*. See *Break Ground*.

OPEN, is a Word of Command, as *Open your Ranks backward to such a distance*, is when the Ranks fall back without changing Aspect, observing their *Right Hand Men* and their *Leaders*. *Open your Files from the Centre*, is when they face outwards from the Centre: if there be an odd *File* it stands, the rest take the distance commanded. *Open your Files to the Right or Left*.

ORDER, is a Word of Command, as *Order your Firelock*, is the planting the But end of the Piece against the middle of the outside of the Right Foot, with the *Lock* outwards. *Open or marching*

ing Order, and *close Order*, are likewise Words of Command in the Exercise of a *Battalion of Foot*.

Order of Battel, is a Disposition of *Battalions* and *Squadrons* of an *Army*, in one or more Lines, according to the Nature of the Ground, either to engage an *Army*, or to be reviewed by the *General*

Ord rs, are the Notice given every night by the *General* to the *Lieutenant-General* of the Day, who conveys them to the *Major-General*, and he to the *Brigade-Major*, who gives them to the *Adjutants*, and they to the *Serjeants*, that the *Army* may know when to march; what *Detachments*, *Convoys*, *Parties*, &c. are to go abroad next morning; when they are to Forrage or Graze; when they are to Muster or Review, and many other things; the Orders are generally given out in the evening at the *Head Quarters*, where all the *Generals* meet at that time. *Orders in general*, signifie all that is commanded by a Superiour *Officer*.

ORDNANCE, all sorts of *Guns*, *Mortars*, *Firelocks*, *Carabines*, *Pistols*, *Pikes*, *Swords*, &c. all sorts of *Arms* or *Stores*, belonging either to Offence or Defence. See *Cannon*.

ORGNES, are thick long pieces of Wood pointed and shod with Iron clear one of another, hanging each by a particular Rope or Cord over the Gate-way of a strong Place, perpendicular, to be let fall in case of an Enemy. Their Disposition is such, that

they stop the Passage of the Gate, and are preferable to *Herses* or *Portcullises*, because these may be either broke by a *Petard*, or they may be stop'd in their falling down, but a *Petard* is useless against an *Orgne*, for if it break one or two of the Pieces, they immediately fall down again, and fill up the vacancy; or if they stop one or two of the Pieces from falling, it is no hindrance to the rest, for being all separate, they have no dependance on one another. The Figure will explain them better.

ORILLON, is a Mass of Earth faced with Stone, built on the Shoulder of a *Cazemated Bastion*, to cover the *Cannon* of the retir'd *Flank*, and hinder its being dismounted by the Enemies *Cannon*. They are made sometimes round, and sometimes square; some maintain the round to be best, because they are not so easily beat down by the *Cannon* of the Besieged, for their roundness

neſs hinders the *Ball* very much from its Effect. Others like the ſquare *Orillons* better, becauſe they are leſs Charge, and can contain more Men to fire directly on the *Face* of the oppoſite *Baſtion*, than the round can do. *Orillon*, is likewiſe called the *Shoulder* and the *Epaulment*.

ORTHOGRAPHY or *Profile*, is the Repreſentation of a Work, ſhowing its breadth, thickneſs, height and depth, ſo as it would appear, it cut perpendicularly on the *Horizontal Line*, from the uppermoſt to the loweſt of its parts; as *Ichnography* ſuppoſes an Edifice or Work cut Horizontally, ſo *Orthography* ſuppoſes it cut Vertically, and never ſhows the length of any of its parts, as a *Plan* does, but then a *Plan* ſhows nothing of the height or depth of a Work. See the Figure at *Profile*.

OVAL, is a plain Figure bounded by its own Circumference, within which no Point can be taken, from which all Right Lines drawn to the Circumference can be equal.

OUT-WORKS, which are likewiſe *Advanced Works*, *Detached* and *Exterior-Works*, are Works of ſeveral ſorts, which cover the Body of the Place; as *Ravelins*, *Half-Moons*, *Tenailles*, *Horn works*, *Crown-works*, *Counter guards*, *Envelopes*, *Swallows-Tails*, *Lunettes*, *&c.* Theſe ſerve not only to cover the place, but likewiſe to keep an Enemy at diſtance, and to hinder his getting any Advantage of Hollows or Riſing Grounds that may happen near the *Counterſcarp* of the Place; for theſe Cavities and Eminences may ſerve for *Lodgments* to the Beſiegers, and facilitate the carrying on their *Approaches*, and raiſing their *Batteries* againſt the Town. When *Out-Works* are for ſome Reaſons placed one before another, as in the *general Plan*, you'll find a *Ravelin* before a *Curtin*, a *Horn work* before the *Ravelin*, and a ſmall *Ravelin* before the *Curtin* of the *Horn-work*; then the neareſt to the Body of the Place muſt be the higheſt, tho' lower than the Works of the Place, that they may command gradually thoſe which are without them, that the Enemy may be obliged to diſlodge, in caſe they had Poſſeſſion of them, as likewiſe leſt the Enemy, being Maſters of them, ſhould the eaſier cover themſelves; ſo that the firſt *Ravelin* ought to be lower than the *Tenaille* of the Place, and higher than the *Horn-work*, and the *Horn work* ought to be higher than the ſmall *Ravelin* which covers it. The *Gorges* of all *Out Works* muſt be plain, leſt the Beſiegers being Maſters of the Works, it ſhould ſerve to cover them from the Fire of the Place.

OXIGON, ſee *Triangle*.

P.

PACE, a Meaſure uſed in *Geometry* and *Fortification*, and much ſpoke of in *Military Diſcipline*. A *Geometrical* Pace is 5 Foot, an *Italian* Mile is 1000 *Geometrical* Paces; an *Engliſh* Mile is 1250; a common Pace is two Foot and a half.

PALISADES, are long pieces of Wood or Stakes, planted generally before *Poſts*, which might be taken by Surprize, or where the Acceſs is very eaſie, to ſecure them both from a ſudden and a regular Attack. They are generally 8 foot long, and 6 or 7 inches ſquare; the one end is pointed, and the other is let 3 foot perpendicularly into the Ground: ſometimes they are planted obliquely, pointing towards the Enemy, that in caſe the Beſiegers ſhould endeavour to pull them out with Cords, the Cords may ſlip off, having no hold. *Paliſades* are planted on the *Berm* at the Foot of *Baſtions* of Places ſurrounded with a wet *Foſs*, to prevent an *Eſcalade* or *Surprize*. They are likewiſe planted in the bottom of dry *Moats*, eſpecially if there be *Traverſes* made. Sometimes they are ſet in the *Gorges* of *Half-Moons* and other *Out-Works*. But above all, the *Parapet* of the *Covert Way* muſt be well paliſaded, either on the *Parapet*, or in the *Covert Way*. They are to ſtand ſo cloſe, that the Muzzle of a *Muſquet* can but juſt get betwixt them. The method of planting them, is by digging a *Trench* of about a foot, or a foot and a half wide, and three foot deep, which after the *Paliſades* are ſet in, as cloſe to one another as before ſaid, is filled with Earth, which is beat and ſet very hard about the *Paliſades* with *Rammers* or *Stampers*. Paliſades are very uſeful, and a good Defence in all ſorts of *Fortifications*, provided they be well planted and cloſe. They are likewiſe uſeful in Sieges, to plant on the out-ſide of the *Foſſees* of the *Batteries*, to prevent the Beſiegeds ſurprizing the *Batteries* in their Sallies, and their nailing the *Cannon*. *Paliſades* are either pull'd up by ſhaking them with *Ropes*, cut down by the *Grenadiers*, beaten down with *Cannon*, or burn'd down with ſmall *Faſcines* pitch'd over.

Turning-Paliſades, is an Invention of *Coehorne's*. To preſerve the *Paliſades* of the *Parapet* of the

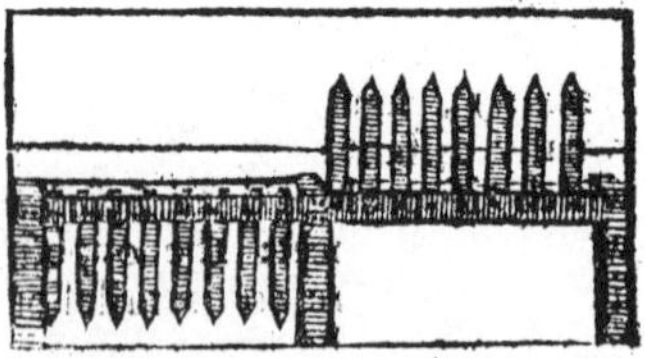

Redans from the Beſiegers *Shot*, he orders them ſo, that as many of them as ſtand in a rods length, turn up and down like a Trap, with all the facility imaginable. They are a good Defence, becauſe they are not in ſight of the Be-

Besiegers, but just when they bring on their Attack, and yet are always ready to do the proper Service of *Palisades*. They are likewise frugal, because they may be preserved in the *Magazines*, and need not be left on the *Parapet*; besides there may be square *Palisades* kept ready to supply the place of such as may be broke by the Besiegers *Cannon*. The Figure shows one set up, and another down.

PAN *of a Bastion*, is the same with the *Face* of a *Bastion*. See *Face*.

PARADE, is the place where *Troops* meet together to go upon Guard, or any other Service. In a *Garrison* where there are 2 or 3 or more *Regiments*, each has their parading Place appointed, where they are to meet upon all Occasions, especially upon any *Alarm*. In a *Camp*, all *Parties*, *Convoys*, or *Detachments* that are to go abroad, have a parading Place appointed them at the Head of some *Regiment*.

PARALLEL, is a Term in *Geometry* signifying two Lines equally distant from one another; which if prolong'd *ad Infinitum* would never meet: the opposite sides of a Square are parallel one to another. The *Ranks* of a *Battalion* are likewise parallel, so are the *Files* amongst themselves. The *Counterscarp* is generally drawn parallel to the *Faces* of the *Bastion*.

Parallels at a Siege, signifie the *Trenches* or *Lines* made parallel to the Defence of the Place besieged; they are likewise called *Lines of Communication* and *Boyau's*.

PARAPET, is an Elevation of Earth, designed for covering the *Soldiers* from the Enemies *Cannon* or *Small-Shot*, wherefore its thickness is from 18 to 20 foot; it is 6 foot high on the inside, and 4 or 5 on the side next the Country; it is raised on the *Rampart*, and has a Slope above called the *Superiour Talus*, and sometimes the *Glacis* of the *Parapet*, on which the *Soldiers* lay their *Musquets* for to fire over: This Pente or Slope makes it easie for the *Musqueteers* to fire into the *Ditch*, or at least on the *Counterscarp*. To fire razing the *Glacis* of the *Parapet*, is called *firing in Barbe*. The *Exterior Talus* of the *Parapet*, is the Slope facing the Country. The height of the *Parapet* being 6 foot on the inside, it has a *Banquett* or two for the *Soldiers* who defend it, to mount upon, that they may discover the Country the better, as likewise the *Foss* and *Counterscarp*, to Fire as they find Occasion.

Parapet of the Covert Way or *Coridor*, is what covers that Way from the sight of the Enemy, which renders it the most dangerous Place for the Besiegers, because of the Neighbourhood of the *Faces*, *Flanks*, and *Curtins* of the Place; it is the same with *Glacis*, which signifies that whole

Maſs of Earth that ſerves to cover the *Coridor*, and goes ſloping towards the Country.

PARK *of Artillery*, is the Place appointed for the Encampment of an *Artillery*, which is generally the *Rear* of both *Lines* of the *Army*. See *Artillery-Park*.

Park of Artillery at a Siege, is a *Poſt* fortified out of *Cannon-ſhot* of the Place beſieged, where are kept all the *Arms* and *Utenſils* neceſſary for a Siege; as *Bombs*, *Petards*, *Carcaſſes*, *Hand Grenades*, *Powder*, *Ball*, &c. with all ſorts of Inſtruments for removing the Earth, as *Spades*, *Shovels*, *Pickaxes*, *Bills Hows*, and *Wheelbarrows*, with a great many things more too tedious to be related here. Great Precaution is to be had about the *Park of Artillery*, for fear of Fire; therefore the *Centries* have either *Pikes* or *Staves* made on purpoſe, for they are not allowed to ſtand with *Firelocks*.

Park of Proviſions, is the Place where the *Sutlers* pitch their *Tents*, and ſell Proviſions to the *Soldiers*; which is in the *Rear* of each *Regiment*, but the chief of all is the Ground alotted at the *Head Quarters* for the *Sutlers*, where there is ſtill every thing requiſite to be had, and it is from thence for the moſt part that the *Sutlers* are furniſhed. But I think the place where the *Bread-Waggons* are drawn up, and where the *Soldiers* receive their *Ammunition-Bread*, being the Store of the *Army*, is properly the *Park of Proviſions*.

PARLEY, ſee *Chamade*.

PARTISAN, is a Perſon who is very dexterous in commanding a *Party*, and knows the Country very well; employed in ſurprizing the Enemies *Convoys*, or in getting Intelligence.

Partiſan-Party, is a ſmall *Body* of *Infantry* given to a *Partiſan*, to make an Incurſion upon the Enemy, to lurk about their *Camp*, to diſturb their *Forragers*, and to intercept their *Convoys*.

Partuiſan, is a Weapon ſometimes carried by *Lieutenants*, not unlike a *Halbert*.

PARTY, is a ſmall *Body* of *Horſe* or *Foot*, ſent into the Enemies Country, to pillage or take Priſoners, or to oblige the Country to come under Contribution, which is to pay a certain ſum of Money to redeem themſelves from Plunder. *Parties* are often ſent out by a *General* to view the Way and Roads, and to fetch Intelligence, to look for Forrage, or to amuſe the Enemy. Upon a March they are frequently ſent upon the *Flanks* of the *Army*, to diſcover the Enemy if near, and to prevent the *Army's* being ſurprized.

Party-Bleu, are a Company of Villains who infeſt the Roads in the *Netherlands*; they belong to neither Army, but rob both ſides, without any regard to Paſſes.

PAS DE SOURIS, is the ſame with *Liziere*, ſee *Liziere*.

PATEE a ſmall Work not unlike a *Horſe ſhoe*, that is to ſay an Elevation of Earth of an irregular Form, but for the moſt part oval

oval, having a *Parapet*; it is generally raised in marshy Grounds, to cover the Gate of a Place; it has only a fore-right Defence, and has nothing to flank it.

PATROUILLE, is a Night Watch, consisting of 5 or 6 Men commanded by a *Serjeant*, who are sent from the Guard to walk in the Streets and prevent Disorder.

To Patrouille, is to go over the Quarters of a Town to observe what is doing in the Streets, and to be careful of the Tranquillity and Surety of the Place.

PAVILION, an old Term for a *Tent*, see *Tent*.

PAY, is the Allowance or Wages which a *Soldier* has for his Maintenance in the Service, and is greater or lesser, according to the Custom of different Countries.

Pay-Master, is he who is entrusted with the paying of a *Regiment*.

PEDRERO, a small sort of *Guns* used on the *Quarter-Decks* of *Ships*; some of them have *Breeches* to screw out, so that they receive the Charge that way.

PELATOON, see *Platoon*.

PENTAGON, is a Figure bounded by five *Sides* or *Polygons*, which form so many *Angles*, capable of being fortified with the like number of *Bastions*.

PERPENDICULAR, is a straight Line raised upright upon another straight Line, without leaning to one side or another, but making the *Angles* on both sides equal.

PETARD, is an Engine of

Metal shaped like a *Sugar-Loaf* or *High Crown'd-Hat*, made for breaking open *Gates*, *Draw-Bridges*, *Barricades*, *Barriers*, &c its length is 7 or 8 inches, the diameter of the Mouth is 5 inches, and that at bottom one and a half; the thickness of Metal at the Neck is half an inch, and that of the Breech 12 or 15; its Charge of *Powder* is 5 pound or thereabouts, and it weighs about 55 or 60. There are much larger and stronger *Petards*, and there are likewise smaller: The first are employed in breaking open strong Reinforced *Gates*, and the last such as can make but small Resistance. When the *Petard* is loaded with *Powder*, it is put upon a strong piece of Plank, cover'd with a plate of Iron on the outside, which covers the *Ouverture*, being hollowed a little for the purpose; the place where they join, is done over with *Wax*, *Pitch*, *Rosin*, &c. to inforce the Effect. This being done, it is carried to the Place designed to be blown up, where

join-

joining the Plank exactly to the *Gate*, the *Petard* is ſtayed behind and fired by a *Fuſee*, that the *Petardeer* may have time to get off. They are ſometimes uſed in *Counter-mines* to break through into the Enemies *Galleries* to diſappoint their *Mines*.

PETARDEER, is he who loads, fixes and fires the *Petard*, and ought to be a Man of Courage, for he is often expoſed.

PICK-AXES, are a ſort of Hand-tools very uſeful in an *Army*, ſee *Ax*.

PICKET, is a ſmall pointed Staff ſhod with Iron, which ſerves to mark out the *Angles* of a *Fortification*, and the principal Parts, when the *Engineer* is tracing a *Plan* upon the Ground with a Line. There are likewiſe ſmall pointed Stakes, which ſerve to drive through *Faſcines* or *Gazons* to keep them faſt when the Earth is bad, or the Work raiſed in haſte

Pickets, is likewiſe the Stakes which the *Troopers* drive before their *Tents*, about 2 yards diſtance; from one to another of theſe *Pickets*, is ſtretched a Rope called the *Picket-Rope*, to which they tie their *Horſes*.

Picket, is likewiſe a Stake of nine or ten foot high, fixed in the Ground, and ſtanding upright; round the foot of it are ſmall Sticks with ſharp points: this is at the *Head* of each *Regiment* of *Horſe*, to puniſh Crimes that do not deſerve Death, by putting the Criminal with his Foot on one of theſe ſmall pointed Sticks, and tying up his Hand to a Ring above his Head, ſo that he neither ſtands nor hangs, nor can he ſhift his Foot, nor change Feet to eaſe himſelf.

Picket-Guard, ſee *Guard*.

PIECE *of Ordnance*, includes all ſorts of great *Guns* and *Mortars*. *Battering-Pieces*, are the large *Guns* uſed at Sieges for making the *Breaches*, ſuch as the 24 *Pounder* and the *Culverine*; the one carrying 24, and the other 18 pound *Ball*. *Field Pieces*, are 12 *Pounders*, *Demi-culverins*, 6 *Pounders*, *Sakers*, *Minions*, and 3 *Pounders*, which march with the *Army*, and encamp always behind the ſecond *Line*, but in Day of Battel are in the *Front*. A Soldiers *Firelock* is likewiſe called his *Piece*.

PIKES, are the Arms carried by *Pike-men*, who uſed formerly to be a third part of the *Company*, but they are now turned to *Muſqueteers*. The *Pike* is made of a point of Iron in form of the Leaf of an Apricock-Tree, called the *Spear*, about 4 inches long, and 2 broad in the middle, from whence it runs to a point: The *Spear* has two Branches or Plates of Iron to fix it to the Staff, of about 2 foot long, and ſtrong enough to reſiſt the ſtroak of a broad Sword. The Staff or Shaft of the *Pike*, is about 13 or 14 foot long, made of a ſlip of Aſh very ſtreight, about an inch and a

quarter thick at the greateſt end, which is ſhod with Braſs or Iron ſharpened to a point to ſtick in the Ground.

Half-Pike, is the Weapon carried by an Officer of Foot, and differs from a *Pike*, becauſe it is but 8 or 9 foot long, and the *Spear* is ſmaller and narrower.

PILE or *Pyramid* of *Bombs* or

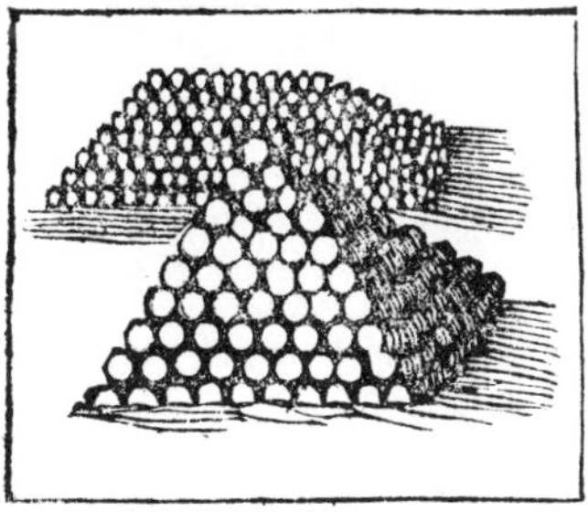

Ball, which is the way of diſpoſing them in *Magazines*, is the piling them up regularly in the Courts of the *Arſenal*, as may be ſeen at *Woolwich*; as ſuppoſe 385 *Bombs* to be made in a *Pile*, the firſt muſt be laid in a ſquare of 10 on each ſide, which makes 100 in the firſt *Bed*, and let half a foot in the Ground; to the end, the great weight which comes above them may not force them to ſlide out, for then the whole *Pile* falls: the ſecond *Bed* will be 81, which is 9 of a ſide, and muſt be laid on the vacant ſpace which happens between every 4 *Bombs* of the firſt *Bed*; and the third *Bed* being eight of a ſide, is 64 laid the ſame way, and ſo to the top of the *Pile*, which will terminate in one *Bomb* making a *Piramide*, whoſe *Baſis* is a Square.

PIONEERS, are ſuch as are commanded in from the Country, to march along with an *Army* for mending the Ways, for working on *Entrenchments* and *Fortifications*, and for making *Approaches*; but the *Soldiers* are moſt generally employed in all theſe things.

PISTOL, is a *Fire-Arm* uſed by the *Cavalry*; the length of it with the Stock, is about a foot and a half, carrying a *Ball* of half an ounce weight. Every *Trooper* has a pair of *Piſtols* before him.

PIVOT, is a piece of Iron or Braſs rounded at the point, that it may turn eaſily round in a piece or ſole of Iron or Braſs, hollowed to receive it.

PLACE, ſignifies a Fortreſs or Town fortified regularly or irregularly, and is often uſed; as when we ſay it is a ſtrong Place, &c.

Place of Arms in a Town, is a ſpace left near the Centre of the Place, where generally the *Grand Guard* is placed, and where the *Soldiers* of the *Garriſon* come to draw up in *Battalia* to mount the *Guard*, from whence they are marched to their particular Poſts. On an *Alarm*, the *Soldiers* who are not on the *Guard*, are to repair thither with their *Arms*. In Places regularly fortified, the *Place of Arms* ought to be in the Centre, and of a Figure like that of the

Poly-

Polygon; its greatneſs ought likewiſe to be proportioned to the *Polygon*.

Place of Arms of an *Attack*, or of a *Trench*, is a *Foſs* with a *Parapet*, or an *Epaulment* to cover a *Body* of *Horſe* or *Foot*, where they may be at their *Arms* to withſtand the Sallies of the Beſieged. The Places moſt convenient for making *Places of Arms*, are ſuch as can eaſily ſuccour one another, and are out of ſight of the Defences of the Place beſieged, as Hollows, or Hollow Ways, eſpecially if they croſs one another, for their depth ſerves as a *Parapet* to cover the *Infantry*: If they have not that natural Depth, they may ſupply that Defect with *Gabions*, *Sand-bags*, or whatever can hinder the Beſiegers from ſeeing into it. If there be a *Foſs* made round it, it is called a *Redoubt*. In carrying on the *Trenches*, there muſt be ſuch *Redoubts* raiſed at convenient diſtances to lodge the *Infantry*, which have the Guard of the *Trenches*.

Place of Arms of a Camp, is the Place choſen at the *Head* of the *Camp* for the *Army* to form themſelves in *Line of Battel*, for a *Review* or the like.

Place of Arms of a Troop of Horſe, or *Company of Foot*, is the Place where the *Troop* or *Company* are aſſembled.

PLAN, as a Term in *Geometry*, is a Superficies, whoſe parts are all equally diſpoſed betwixt its Extremities, ſo that one part is neither higher nor lower than another. *Horizontal-Plan*, is when is is parallel to the *Horizon*; and it is a *Vertical-Plan*, when it is perpendicular to the *Horizon*.

Plan, *Ground-Plot*, or *Ichnography* in *Fortification*, is the Repreſentation of the firſt or fundamental Tract of a Work, ſhewing the length of its *Lines*, the quantity of its *Angles*, the breadth of the *Ditches*, thickneſs of the *Ramparts* and *Parapets*, and the diſtance of one part from another, as the general *Plan* at the beginning ſhows. So that a *Plan* repreſents a Work, ſuch as it would appear if it were cut equal with the Level of the *Horizon*, or cut off at the Foundation. But it marks neither the Heights nor Depths of the ſeveral parts of the Works, which is properly *Profile*, and which expreſſes only the Heights, Breadths and Depths, without taking notice of the Lengths. As *Architects*, before they lay the Foundation of their *Edifice*, make their Deſign upon Paper, by which means they find out their Faults; ſo an *Engineer* before tracing his Works on the Ground, ſhould make *Plans* of his Deſigns upon Paper, to the end he may do nothing without ſerious Deliberation. *Plans* are very uſeful for *Generals* or *Governours*, in either attacking or defending a Place, in chuſing a *Camp*, determining *Attacks*, conducting the *Approaches*, or in examining the Strength and Weakneſs of a Place; eſpecially ſuch *Plans* as repreſent a Place with the Country about it

it, showing the *Rivers*, *Fountains*, *Marshes*, *Ditches*, *Valleys*, *Mountains*, *Woods*, *Houses*, *Churches*, and other particulars which happen about a Place.

PLANKS or *Madriers*, are pieces of Oak very thick and broad, see *Madrier*.

PLATES. The *Prise-Plates*, are two *Plates* of Iron on the *Cheeks* of a *Gun-Carriage* from the *Capesquare* to the *Centre*, through which the *Prise Bolts* go, and on which the *Hand-spike* rests, when it poises up the *Breech* of the Piece. *Breast-Plates* are the two *Plates* on the *Face* of the *Carriage*, one on each *Cheek*. *Train-Plates*, are the two *Plates* on the *Cheeks* at the *Train* of the *Carriage*. *Dulidge-Plates*, are the six *Plates* on the *Wheel* of a *Gun-Carriage*, where the *Fellows* are joined together, and serve to strengthen the *Dulidges*.

PLATFORM *in general*, is an Elevation of Earth on which *Cannon* is placed, such as the *Mounts* on the middle of *Curtins*: But it is likewise a sort of *Bastion* constructed on a *Re-entring Angle*, when its two *Faces* make a Right Line. *Platform of a Battery*, is a Floor of Boards nailed down upon Sleepers, sloping a little towards the *Embrasure*, for the *Guns* to run upon, and to keep the *Wheels* from sinking into the Ground. The Slope serves to diminish the Reverse of the Piece, and for the more easie running her up to her *Embrasure*. Each *Gun* has generally a *Platform* for herself. See *Battery*.

PLATOON, or rather *Peloton*, a small square *Body* of *Musqueteers*, such as is used to be drawn out of a *Battalion* of Foot, when they form the Hollow Square to strengthen the *Angles*; The *Grenadiers* are generally thus posted, *Peloton* is the *French* Word, only the Vulgar Corruption has brought it to be pronounced *Platoon*.

PONIARD, a sort of short Sword used in *Spain* and *Italy*.

POINT *Mathematick*, is what hath no parts; which is to say, neither Length, Breadth nor Thickness, and which consequently cannot be expressed or nconceived.

Point-Blank of a Gun, is the distance she throws a *Ball* in a supposed direct Line; the *Gun* being laid at no Elevation, but levell'd parallel to the *Horizon*. I say supposed direct Line, because it is certain and easily proved, that a *Ball* cannot fly any part of its Range in a right Line, but the swifter it flies, the nearer it approaches to a right Line; or the swifter it flies, the less crooked is his Range. This was the Opinion of *Nicolas Tartaglia*, and is since generally approved.

POLYGON, is a Figure of more than four Sides, and is either Regular or Irregular, Exterior or Interior.

Polygon Regular, is whose *Angles* and *Sides* are equal. It has an *Angle* of the *Centre*, and an *Angle* of the *Polygon*. The Centre of a Re-

Regular Polygon, is the Centre of a Circle which circumſcribes the *Polygon*, that is, whoſe Circumference paſſes through all the *Angles* of the *Figure*.

Irregular Polygon, is whoſe *Sides* and *Angles* are unequal.

Exterior Polygon, is the Lines drawn touching the Points of the *Flanked Angles*, when a Place is fortified inwards. *Errard* was the firſt that fortified after this Manner. Count *Pagan* improved it, and Monſieur *de Vauban* brought it to perfection.

Interior Polygon, is to fortifie outwards, which makes the *Angles* of the *Polygon* to be the *Angles* of the *Gorge*, ſo that the whole *Baſtion* is without the *Polygon*. *Maneſſon Mallet*, *De Ville*, and moſt *Engineers* follow this Method.

PONTON, is a late Invention, being a Boat of Latten of about 8 yards long and 2 broad; the form of it is a long Square, having a large Ring at each corner; it is laid upon a *Carriage* when the *Army* marches, and drawn by 5 Horſes. Each *Boat* has an *Anchor* and *Cable*, and *Baulks* and *Cheſts* belonging to it. The *Baulks* are about 5 or 6 inches ſquare, and about 7 yards long. The *Cheſts* are boards joined together by wooden bars, about a yard broad, and 4 yards long. When there is occaſion for uſing theſe *Boats*, they are ſlipp'd into the Water, and placed about 2 yards aſunder, each faſten'd with an Anchor, having beſides, a ſtrong Rope which runs through the Rings, and is faſten'd on each ſide the River to a Tree or a Stake made very faſt in the Ground. The *Baulks* are laid croſs the *Boats*, at ſome diſtance from one another, and the *Cheſts* upon them joined cloſe, which makes a Bridge in a very ſhort time, for *Horſe*, *Foot* or *Artillery* to march over.

PORT, ſee *Gate*.

PORTCULLISES, ſee *Herſe*.

PORT-FIRE, is a Compoſition of *Meal-powder*, *Sulphur* and *Saltpetre*, drove into a caſe of Paper, but not very hard; it is about 9 or 10 inches long, and is uſed to fire *Guns* or *Mortars* inſtead of Match, but then it is cut into pieces of about an inch long, and put in a *Linſtock*, or cleft Stick.

POST, is any ſort of Ground, fortified or not, where a Body of Men can fortifie themſelves, or be in a condition of fighting an Enemy. To relieve a *Poſt*, is to go upon Guard in a *Poſt*; to abandon or quit a *Poſt*; to gain a *Poſt* Sword in Hand, &c.

Poſt of Honour; the *Advance-Guard* is a *Poſt of Honour*; the Right of the two *Lines* is the *Poſt of Honour*, and is always given to the eldeſt *Regiments*; the Left is the next *Poſt*, and is given to the next eldeſt, and ſo on; the Centre of the *Lines* is the *Poſt* the leaſt honourable, and is given to the youngeſt *Regiments*.

Advanc'd-Poſt, is a ſpot of Ground ſeized by a Party to ſe-

cure

cure their *Front*, and to cover the *Posts* that are behind them.

POSTERN, is a ſmall Door in the *Flank* of a *Baſtion*, or other part of a *Garriſon*, to march in and out unperceived by the Enemy, either to relieve the Works, or to make Sallies.

POT, ſee *Fire-pot*.

POUCH; a *Grenadiers Pouch*, is a ſquare Caſe or Bag of Leather, with a flap over it, hanging in a ſtrap of about two inches broad over the Left Shoulder, in which he carries his *Grenades*.

POWDER, is a Compoſition of *Sulphur*, *Saltpetre*, and *Charcoal*. The *Sulphur* and *Charcoal* take fire, and the *Saltpetre* makes the Crack. Since the Invention of *Powder*, a great many Warlike *Machines* have been invented, which occaſions ſuch a conſumption of *Powder*, eſpecially at Sieges, that it cannot be determined; for the *Cannon*, *Bombs*, *Hand-Grenades*, *Muſquets*, *Mines*, and other extraordinary Diſtributions, beſides the Waſte, conſume a great deal.

POUNDER, as a 24 *Pounder*, is a *Gun* carrying a *Ball* of 24 pound, its diameter is ſix inches, the length is from ten to twelve foot long, it is a good battering *Gun*. *Twelve Pounder*, is a *Gun* carrying a *Ball* of twelve pound, the diameter is four inches and 5 eighths, its length from eight to ten foot. *Six Pounder* carries a *Ball* of ſix pouund, its diameter is three inches and ſix eights, its length from ſeven to eight foot; and *Three Pounder*, carries a *Ball* of three pound, the diameter of the Bore is three inches, and the length of the Piece about 6 foot or 6 and a half.

PRIESTS BONNET, ſee *Bonnet a Prêtre*.

PROVISIONS, is what a *General* ought to be very careful of, never ſuffering his *Army* to be in want of Subſiſtance, elſe they muſt needs periſh. A *Governour* of a *Garriſon* ought to be well provided of Proviſions of all ſorts, ſuch as *Wheat*, *Rye*, *Peaſe*, *Beans*, *Barley*, *Beef*, *Mutton*, *Veal*, *Bacon*, *Cheeſe*, *Butter*, *Salt*, *Pepper*, *Onions*, *Nutmegs*, *Beer*, *Wine*, *Brandy*, and many other things which are abſolutely neceſſary in a *Garriſon*.

PROVOST-MARSHAL *of an Army*, is one appointed to ſecure Deſerters and all other Criminals; he is to go often abroad round the *Army*, to hinder the *Soldiers* from pillaging; he indites Offenders, and executes the Sentence which is pronounced; he likewiſe regulates the Weights and Meaſures of the *Army*, and the Price of all ſorts of Proviſions.

PROFILE; *Engineers* to repreſent the Heights, Depths and Thickneſs of a Work, with the Depth and Breadth of the *Foſſees*, *&c.* do it by *Profile* or *Orthography*, which ſuppoſes the Work to be cut perpendicularly from top to bottom. The following Figure ſhows the *Profile* of a *Rampart*, *Fauſebray*, *Foſs*, *Covert Way*, and *Eſplanade*; every thing is ex-

An Explanation of the Figure.

1, 2, 3, 4. *The Solidity of the Rampart.*
1, 2. *The Basis of the Rampart.*
3, 4. *The Top of the Rampart.*
3, 5. *The Height of the Rampart.*
1, 3. *The Interior Talus or Slope of the Rampart.*
2, 4. *The Exterior Talus of the Rampart.*
3, 7. *The Terre-plein.*
7, 4. *The Basis of the Parapet.*
7, 15. *The Banquett.*
8, 13. *The Interior Talus of the Parapet.*
4, 12. *The Exterior Talus of the Parapet.*
12, 13. *The Superior Talus of the Parapet.*
9, 13. *The Interior Height of the Parapet.*
10, 12. *The Exterior Height of the Parapet.*
4. *The Cordon.*
11, 16, 17. *The Fausebray with its Banquett.*
17, 18. *The Liziere or Berm.*
18, 19, 20, 21. *The Foss.*
18, 20. *The Scarpe.*
19, 21. *The Counterscarp.*
22. *The Cunette.*
6, 21. *The depth of the Foss.*
19, 23. *The Covert Way.*
23, 24, 25. *The Glacis or Parapet of the Covert Way.*
23, 24. *The height of the Parapet of the Covert Way, with its Banquett.*
23, 25. *The Basis of the Glacis.*
24, 25. *The Pent or Slope of the Glacis.*

explained by Letters, and the pro-

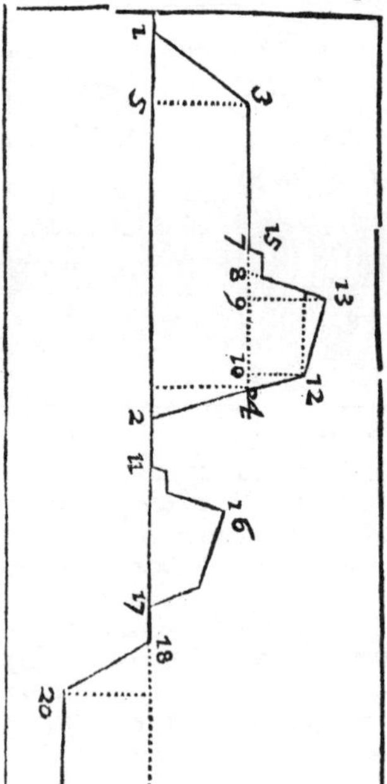

per Names set down. PUL-

PULLY, is a Wheel of Wood

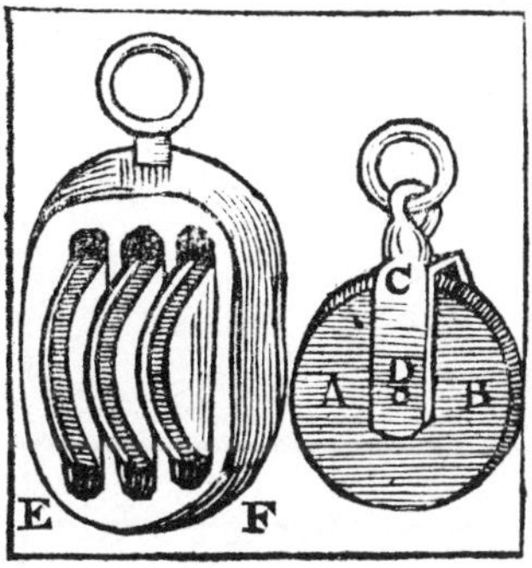

or Metal, *A*, *B*, put in a piece of Wood or Iron, as *C*, *D*, or into a Block, as *E*, F, where it has liberty to move in a Hole cut for the purpoſe,on an *Axis* or *Gudgeon* going through the Centre: Over the *Pully* goes a Cord which ſerves to raiſe Burthens. They are much uſed about an *Artillery*, in raiſing of *Guns* or *Mortars*, eſpecially in the *Gin*.

PYE, ſee *Gin*.

Q.

QUADRANT, or quarter of a Circle, is an Inſtrument of Braſs or Wood uſed by *Gunners*, in pointing their *Guns* to an Object, and by *Bombardeers*, in elevating their *Mortars*; it is made of two pieces of Wood joined at Right Angles, one of which is longer than the other, that it may enter the Muzzle of the Piece; they are joined by a quarter of a Circle, which is divided into ninety Degrees, the Centre of which is where the two Pieces join, from whence there hangs a Thread with a Plummet, which marks the different Elevations of Pieces, and the greatneſs of the Angles. The way of uſing it, is by putting the longeſt ſide into the Muzzle of the Piece; the Plummet falls perpendicularly, and marks the Angle on the *Quadrant*; when the *Gun* or *Mortar* is elevated to the Degree deſired, it is kept there by Coins of Wood put under the *Breech* of a *Gun*, or betwixt the *Bracket-Bolts* of a *Mortar*.

QUADRAT; to quadrat a Piece, is to ſee whether it is duly placed in its *Carriage*, and that the *Wheels* be of an equal height.

QUARTER, or *Quarters*, has ſeveral Significations in *Martial Affairs*.

Quarter, ſignifies the ſparing Mens Lives, and giving good Treatment to a vanquiſh'd Enemy; ſo it is ſaid the Conquerours offered good *Quarter*: The Enemy asked *Quarter*. We gave no *Quarter*.

Quarter in general, is the Ground on which a Body of *Troops* encamp, and ſignifies likewiſe the *Troops* encamped; as to beat up the Enemies *Quarter*, is to drive them from their Ground or Encampment; therefore it ought to be in the moſt convenient Place, as well for the Nature of the Ground, as for the Advantage of entrenching, and the conveniency of Forage and Water.

Quarter of an Aſſembly, is the Place where *Troops* meet for to march

march in a Body, and is the ſame with a Place of *Rendezvous*.

Quarter-Guard, ſee *Guard*.

Quarter at a Siege, is the Encampment upon one of the moſt principal Paſſages round about a Place beſieged, to prevent Relief and Convoys: When it is commanded by the *General*, it is called the *Head Quarters of the Army*: When the *Camp* is marked out about a Place beſieged, then the *Quarters* are ſaid to be diſpoſed: When great Detachments are made from a *Quarter* for Convoys, *&c.* ſuch a *Quarter* is ſaid to be weaken'd.

Head Quarters, is the Place where the *General* of an Army has his *Quarters*: it is generally near the Centre of the Army. The *Quarters* of the *Generals* of *Horſe*, is in the Villages that happen behind the *Right* and *Left Wings*; the *Generals* of *Foot*, are often in the ſame Village with the *General*.

Quarter entrench'd, is a Place fortified with a *Ditch* and *Parapet*, to ſecure a *Body* of *Troops*.

Winter-Quarters, is ſometimes taken for the ſpace of time included between the leaving the *Camp*, and taking the *Field*; but it is more properly the Places where the *Troops* are lodged during the Winter: According as the *Troops* are a long or ſhort time in *Garriſon*, the *Winter-Quarters* are ſaid to be ſhort or long.

Quarters of Refreſhment, is the place where the *Troops* that have been much fatigu'd are ſent to refreſh themſelves, during a part of the *Campaign*; that having refreſh'd and recover'd themſelves, they may be ready to take the Field again.

Quarter-Maſter of Horſe, is a Warrant-Officer appointed by the Colonel, he takes up the Ground for the *Troop*, and divides it, in giving ſo much for each Tent; he receives the Orders, keeps a Liſt of the *Troop*, viſits the Stables, and takes care of the Arms. He marches in the Rear of the *Troop*, but in Camp his Tent is pitch'd in the Front. In *Winter-quarters* he receives and diſtributes the Forage to the *Troop*: Each *Troop* has a *Quarter-maſter*.

Quarter-maſter of Foot, is an Officer who takes care of Encamping the Regiment, for there is but one to a Regiment of *Foot*: he attends the *Quartermaſter-General* upon a March, to know where the Ground is for the Regiment, which he divides among the Companies.

Quartermaſter-General, is a conſiderable Officer in an Army, and ought to be a judicious Man, and a Man of great Experience, and to underſtand Geography; and ſince his Function is to mark the Marches and Encampments of an Army, he ſhould know the Country perfectly well, all the Rivers, Plains, Marſhes, Woods, Mountains, Paſſages, Defiles, *&c.* even to the ſmalleſt Brook. The Evening before a March, he receives the Orders and Rout from the *General*, and appoints a place for the

Quarter-Masters of *Foot* and *Horse*, to meet him next Morning, with whom he marches to the next Camp, where being come, and having viewed the Ground, he marks out to the *Quartermasters* the Ground allow'd each Regiment for their Camp; he chuses the *Head-quarters*, and appoints the Villages for the *General-Officers* of the Army, where they shall quarter; he appoints a proper place for the Encampment of the *Train of Artillery*; he carries the Army a Foraging, and plants the Covering Party, for their Security at all the Passes round them, and assists in distributing the *Winterquarters* to the Army.

Quarter-Wheeling of a Body of Men, is turning the Front where the Flank was; which is done to the Right by the Man on the Right *Angle*, keeping his Ground, and facing about while the rest Wheel.

Queve d' Yronde, see *Swallows Tail*.

Quit your Arms, is a word of Command in the *Foot*, when they lay down their Arms, at which they stand up, till they are order'd to the Right about, at which they march clear of their Arms and disperse; but upon the Beat of Drum they run to their Arms with a *Huzza*, having their Swords drawn, and the Point upward.

R.

RABINET, a small sort of Ordnance between a *Falconette* and a *Base*, seldom used.

RACKET, is a case of Paper rolled very hard on a former of Wood, choaked at one end with a small Cord, and drove full of a Composition of *Meal-powder*, *Saltpetre*, *Sulphur*, and *Coal*, sometimes without *Saltpetre*; it has a small Rod tied to the side of it, called the Tail of the *Racket*; it is covered with a Cap of Paper running to a point, in which are placed Stars, which are a Composition of *Sulphur*, *Saltpetre*, *Meal-powder*, and *Camphire*, and other things that can make a clear Fire; when the *Racket* is fired, it cuts the Air with an admirable swiftness, leaving a stream of Fire behind it; when the *Racket* is burned, the Stars take fire, and make a very pleasant Sight.

To RAISE *a Siege*, is to give over the Attack of a Place, and to quit the Works thrown up against it, and the Posts taken about it. As all Enterprizes do not always succeed, so sometimes an Army is forced to raise a Siege, either because of Distempers in the *Camp*, or the unfitness of the Season, for the Rains, Snows, Winds, and cold Weather kills the Men; besides the Besieged may be strongly entrench'd, and receive Supplies of Men and Provisions. If there be no Ground to fear a Sally from the Place, then the Siege may be raised in the day time, by sending first the Sick and Wounded, the *Baggage*, the *Sutlers*, broken

broken *Cannon* and *Mortars*, and if possible, all the Instruments which have been used in the Siege. The *Artillery* and *Ammunition* may follow, and a strong *Rear-Guard* must face the Besiegers, in case they should offer to charge the *Rear*. But if there be any fear of an Enemy in *Front*, this Order must be altered according to the Prudence of the *General*, and according as the Nature of the Country will allow.

To raise a Plan of a Fortress, is the measuring with Cords and Geometrical Instruments, the length of the *Lines*, and the capacity of the *Angles*, that by knowing the length, breadth, and thickness of all the different parts of a *Fortification*, it may be represented in small upon Paper, so as to know the Advantages and Disadvantages of it.

RAMEAU, see *Mine*

RAMMER *of a Gun*, is a piece of Wood fitted to the diameter of the Bore, stuck upon a long Staff, and is used in setting home theCharge and theWadding.

Rammer, see *Beetle* and *Scourer*.

RAMPART, is an Elevation of Earth round a Place, capable of covering the Buildings from view, and of resisting the *Cannon* of an Enemy, as likewise of raising those that defend it, that they may discover the Country about it. A *Rampart* ought to be sloped on both sides, that is, the Mass of Earth which composes the *Rampart*, ought to be larger at bottom than at top, more or less, according to the nature of the Earth: It ought to be broad enough to allow the marching of *Waggons* and *Cannon*, besides the *Parapet* which is raised on it; its thickness is generally about 12 fathom, with the *Talus* or *Slope*. The Earth which makes the *Rampart*, is taken from the outside of it, because then the *Rampart* and *Foss* are made at the same time; from which it follows, that their Proportions depend on one another; for since the *Rampart* is made of a certain bigness, the *Foss* must be dug deep enough to afford Earth for the *Rampart*, the *Parapet*, and the *Esplanade*.

RENDEZVOUS, is the Place appointed by the *General*, where all the *Troops* which are to compose his Army are to meet at a day prefix'd, notwithstanding Wind, Rain, Snow, Cold, or other Chances of the Weather.

RANK, is the Order or streight Line made by the *Soldiers* of a *Battalion* or *Squadron*, drawn up side by side; this Order was established for the Marches, and for regulating the different Bodies of *Troops* and *Officers* which compose an *Army* or a *Battalion*. *Doubling of the Ranks*, is the putting two *Ranks* into one.

RATION, is a Portion of Ammunition, Bread or Forage, distributed to every Man in the *Army*. A *Foot Soldier* receives a *Ration* of Bread, which is a pound and a half for each day; and a *Trooper* a *Ration* of Bread, and another of Forage.

RA-

RAVELINS, are Works raised on the *Counterscarp* before the *Curtin* of a Place, and serve to cover the Gates of a Town and the Bridges. They consist of two *Faces* forming a *Sailliant Angle*, and are defended by the *Faces* of the Neighbouring *Bastions*. The *Half-Moons* which cover the points of the *Bastions*, have their Defences from the *Ravelins*. They are the most in use of all Out-Works, and are by the *Soldiers* most commonly called *Half-Moons*. They ought to be lower than the Works of the Place, that they may be under the Fire of the Besieged, in case the Enemy should endeavour to lodge themselves there. Their *Parapets*, as those of all Out-Works, ought to be *Cannon-Proof*; that is to say, about 18 foot thick; their *Ramparts* ought to be the half or third of one of the *Flanks* of the Place, and the breadth of their *Moats* half the breadth of the *Moat* of the Place.

RAZANT, *Line of Defence Razant*, see *Line*.

REAR *of an Army*, or of a *Battalion*, signifies generally, either the hindmost part of the *Army*, or *Battalion*, or the Ground behind it.

Rear-Guard, is that *Body* of the *Army* which marches after the *Main-Body*; for the March of an *Army* is always composed of an *Advanc'd-Guard*, a *Main-Body*, and a *Rear-Guard*; the first and last commanded by a *General* Person. The Old *Grand Guards* of the *Camp*, are always the *Rear-Guard* of the *Army*, and are to see that every thing come safe up to the New *Camp*.

Rear-Half-Files, are the three hindmost Ranks of a *Battalion*, when it is drawn up six deep.

Rear-Line of an Army encamp'd, or *second Line*, is always 400 or 500 yards distant from the first *Line*, which is likewise called the *Front Line*; these two *Lines* run parallel, and have sometimes a third, which is called the *Reserve*.

Rear-Rank, is the last *Rank* of a *Battalion*, when drawn up.

RECOILE, or *Reverse of a Gun*, is its running back when fired, which is occasion'd by the strugling of the Powder in the Chamber, and its seeking every way to fly out. *Guns*, whose Vents are a little forward in the *Chace*, recoile most. To lessen the *Recoile* of a *Gun*, the *Platforms* are generally made sloping towards the *Embrasures* of the *Battery*.

RECRUITS, are new Men raised to supply the Places of such as have lost their Lives in the Service, or are rendred unserviceable by Age or Wounds. *Recruit-Horses*, are the *Horses* brought up for compleating the *Regiments* of *Horse* or *Dragoons* every year.

RECTANGLE, see *Angle*.

REDANS, or *Indented Works*, are *Lines* or *Faces* forming Sallying and Re-entring *Angles* flanking one another, and are generally used on the sides of a River which runs through a *Garrison* Town.

Town. They were used before *Bastions* were, and are by some thought preferable to them.

REDOUBTS, are square Works of Stone raised without the *Glacis* of a Place, about *Musquet-Shot* from the Town, with a *Foss* round 'em, having *Loop holes* for the *Musquettetrs* to fire thro; sometimes they are of Earth, having only a Defence in Front, surrounded with a *Parapet* and *Foss*: both the one and the other serve for *Detached Guards* to interrupt the Enemies Works. They are sometimes made for the *Angles* of the *Trenches*, for covering the Work-men against the Sallies of the *Garrison*. The length of their Sides may be from ten to twenty fathom; their *Parapet*, having two or three *Banquetts*, must be about nine or ten foot thick, and their *Foss* the same, both in breadth and deepness; they contain a Body of Men for the Guard of the *Trenches*, and are likewise called *Places of Arms*.

REDUCE *a Place*, is to oblige the *Governour* to surrender it to the Besiegers, by Capitulation.

REDUIT, *Castle* or *Donjon*, is a Place more particularly entrench'd and separated from the rest by a *Foss*. There is generally in each of them a high Tower, from whence the Country round the Place may be discovered.

REFORM; *to reform*, is to reduce a Body of Men, by either disbanding the whole, or only breaking a part, and retaiing the rest; or sometimes by incorporating them into other *Regiments*.

REFORM'D-OFFICER, is one whose *Troop* or *Company* is broke, and he continued in whole or half Pay, doing Duty in the *Regiment*; he preserves his Right of Seniority, and continues in the way of Preferment.

REGIMENT, is a Body of Men either Horse or Foot, commanded by a *Colonel*, *Lieutenant-Colonel*, and *Major*; each *Regiment* of *Foot* is divided into *Companies*, but the number of *Companies* differs; though in *England* our *Regiments* are generally thirteen *Companies*, one of which is always *Grenadiers*. *Regiments* of *Horse* are most commonly six *Troops*, but some of nine. *Dragoon-Regiments*, are generally in time of War eight *Troops*, and in time of Peace but six. Each *Regiment* has a *Chaplain*, and a *Surgeon*. Some *German Regiments* consist of 2000 *Foot*, and the *Regiment* of *Piccardy* in *France* of 6000, being 120 *Companies*, at 50 in a *Company*.

Regiment of Guards, see *Guards*.

REGULAR-ATTACKS, are such as are made in form, that is, by *Regular-Approaches*.

REINFORCED-RING *of a Gun*, is that next the *Trunions*, betwixt them and the Vent. The reinforced part of a *Gun*, is from the *Base-Ring* to the *Reinforced-Ring*, being much thicker of Metal than any other part of the Piece, because of the great force of the *Powder*.

REINFORCEMENT *to an Army*, is an Addition of fresh *Troops* to strengthen an *Army*, and to enable them to go on with an Enterprize.

RELAIS, see *Liziere*.

RELIEVE; *to Relieve the Guard*, is to put fresh Men upon the *Guard*. *To relieve the Trenches*, is to relieve the *Guard* of the *Trenches*, by sending off those that have been there upon Duty before.

REMOUNT; *to remount the Cavalry or Dragoons*, is to furnish them with Horses in the room of those which have been either killed or disabled.

RESERVE, is a Body of *Troops* sometimes drawn out of the *Army*, and encamped by themselves in a *Line* behind the other two *Lines*. See *Camp*.

RETIRADE, is a *Trench* with a *Parapet*. But *Retirade* or *Coupure*, is most ordinarily taken for a *Retrenchment* formed by the two *Faces* of a *Re-entring Angle* in the Body of a Place, after the first Defence is ruined, and the Besieged obliged to abandon the Head of the Work, without quitting it entirely; therefore while some are making *Head* to the Enemy, others ought to be busie in making the *Retirade*, which is only a simple *Barricade* or *Retrenchment* thrown up in haste, with a sort of *Foss* before it; it depends upon the Knowledge of the *Engineer* to direct, and the Honour of the *Officers* and *Soldiers* to work at such a time, since they do it for the Defence of their Liberty; so that no *Officer* ought to think it below him to carry *Fascines*, or to throw up Earth to cover himself. The *Retirade* ought to be raised as high as possible; and some *Fourneaus* or *Fougades* made under it, to blow up the Enemies *Lodgments*.

RETRAITE, or *Berm*, see *Liziere*.

RETREAT, or *Tattou*, is a Beat of the *Drum* in the evening at the firing of a Piece, called the *Warning-Piece*, at which the *Drum-Major*, with all the *Drums* of the *Battalion*, except such as are upon Duty, beats round the *Regiment*; the *Drums* of the *Quarter-Guards*, of the *General's Guards*, and all other small *Guards*, do likewise beat, the *Trumpets* at the same time sounding at the Head of their respective *Troops*. This is to warn the *Soldiers* to forbear firing, and the *Centries* to challenge till break of Day, that the *Reveille* is beat. The *Retreat* is likewise called setting the *Watch*.

RETRENCHMENT, is any Work raised to cover a *Post*, and fortifie it against an Enemy, such as *Fascines* loaded with *Earth*, *Gabions*, *Barrels* of *Earth*, *Sand-bags*, and generally all things that can cover the Men, and give a stop to the Enemy. But it is more particularly applicable to a *Foss* bordered with a *Parapet*; and a *Post* fortified thus, is called

Post

Post retrench'd, or *strong Post*. *Retrenchments* are either general or particular.

General Retrenchments, are new *Fortifications* made in a Place besieged, for to cover themselves when the Enemy becomes Masters of a *Lodgment* on the *Fortification*, that they may be in a Condition of disputing the Ground inch by inch, and of putting a stop to the Enemy's Progress, in expectation of Relief. As, if the Besieged attack a *Tenaille* of the Place which they judge the weakest, either by its being ill flanked, or being commanded by some Neighbouring Ground; then the Besiegers make a great *Retrenchment*, inclosing all that part which they judge in most Danger, as you may see in the general *Plan*. These ought to be fortified with *Bastions* and *Demi bastions*, with a good *Foss*, and ought to be higher than the Works of the Place, that they may command the old Works, and put the Besiegers to a great trouble in covering themselves; they ought likewise to be Countermined.

Particular Retrenchments, are such as are made in the *Bastions*, when the Enemy are Masters of the *Breach*. They can never be made but in full *Bastions*, for in empty or hollow *Bastions* there can be made only *Retirades*. These particular *Retrenchments* are made several ways, according to the time they have to cover themselves; sometimes they are made before hand, which is certainly the best; as Count *Pagan* who makes a double *Parapet* in all his *Bastions*; and a *Retrenchment* made before hand, requires no more Men for its Defence, than if it were not made, because they never defend it till the principal Work be lost. The *Parapets* of such *Retrenchments* ought to be 5 or 6 foot thick, and 5 foot high, with a large and deep *Foss*, from whence ought to run out small *Fougades* and *Countermines*.

RETURNS *of a Mine*, are the Turnings and Windings of the *Gallery*; see *Gallery* and *Mine*.

Returns of a Trench, are the Turnings and Windings which form the Lines of the *Trench*, and are as near as they can be made parallel to the Place attacked, to shun being enfiladed. These *Returns* when followed, make a long way from the end of the *Trench* to the Head, which going the streight way is very short, but then the Men are exposed; yet upon a Sally, the best Men never consider the Danger, but getting over the *Trench* with such as will follow them, take the shortest way to repulse the Enemy, and to cut off their Retreat if possible.

REVEILLE, is a Beat of the *Drum* about break of day, to advertise the *Army* that it is day light, and that the *Centries* forbear challenging.

REVERSE, signifies on the back, or behind. So we say a

Reverse View, a *Reverse commanding Ground*, a *Reverse Battery*, &c.

Reverse of a Gun, see *Recoile*.

REVIEW, is the drawing out of the *Army*, or part of the *Army* in Line of Battel, to be viewed by the *General*, that he may know the Condition of the *Troops*.

RHILAND-ROD, is a Measure of two fathom or twelve foot, used by the *Dutch Engineers*.

RHOMB, is a four sided Figure, whose *Sides* are equal, but the *Angles* unequal.

RHOMBOIDE, is a four sided Figure, whose *Angles* and opposite *Sides* are equal, but all its four sides are not equal.

RIDEAU, is a Rising Ground or Eminence commanding a Plain, which is sometimes near parallel to the Works of a Place. It is a great Disadvantage to have *Rideaus* near a *Fortification*, especially when they shoot from far, and terminate on the *Counterscarp*; for they not only command the Place, but likewise facilitate the Enemies Approaches.

Rideau, is likewise a *Trench* covered with Earth, in form of a *Parapet* to cover the *Soldiers*.

ROLL; *Muster-Roll*, is a scroll of Parchment, which each *Captain* gives the *Muster-Master*, on which are writ the Names of the *Soldiers* of his *Company*.

To roll in Duty, is when Officers of the same Rank take their turns upon Duty, as *Captains* with *Captains*, and *Subalterns* with *Subalterns*, and command according to the Seniority of their *Commissions*.

ROLLERS, are round pieces of Wood of about 9 inches diameter, and four foot long, which serve in moving *Mortars* from one place to another when it is near, by raising the fore part of the *Bed* so high that one of these *Rollers* may be laid under it, then pushing the *Bed* forwards, and laying another in its way, and another before that, and so on: Thus the *Mortar* is with little trouble brought to another place.

ROPE; for *Gin-Rope*, see *Gin*; and for *Draught-Ropes*, see *Draught-Hooks*.

ROUND, is a Night *Watch* commanded by an *Officer* that goes round the *Rampart* of a *Garrison*, to listen if any thing be stirring without the Works, to see that the *Centries* be watchful and diligent upon their Duty, and that every thing be in order: In strict *Garrisons*, the *Rounds* go every quarter of an hour, that the *Rampart* may still be furnished: The *Centries* ought to challenge at a distance, and are to rest their *Arms* as the *Round* passes, letting no Man come near them. When the *Round* is near the *Corps de Garde*, the *Centry* calls aloud, *Who comes there*; when it is answered, *the Round*; he says *stand*, and calls for the *Corporal* of the *Guard*, who drawing his Sword calls, *who comes there*, and is answered *the Round*; then, *let him*

who has the Word advance. The *Corporal* receives the Word with his Sword drawn, and pointed at the Heart of him who gives it. When the *Major* goes the *Round*, the *Officers* of the *Guard* receive him with two *Musqueteers*, and give him the Word only once, which is when he goes his *Round-Major*. When the *Governour* goes his *Round*, the *Officers* draw out the *Guard* without *Arms*, and send four *Musqueteers* to receive him at ten Paces distance, and give him the Word as often as he pleases to demand it: All other *Rounds*, without exception, are obliged to give the Word to the *Corporal* of the *Guard*.

To RUN *the Gauntlet*, is a punishment for considerable Offences; when a *Soldier* is sentenced to run the *Gauntlet*, the *Regiment* is drawn out and make a Lane, each *Soldier* having a Switch in his hand; the *Criminal's* Shoulders and Back are naked, and as he runs along, every one has a stroak at him; while he runs, the *Drums* beat at each end of the Lane; sometimes he runs 3 times, sometimes 5, and sometimes 7 times, according to the Nature of the Offence.

S.

SAFE-GUARD, is a Protection granted by a *Prince* or his *General* to some of the Enemies Lands, to preserve 'em from being plunder'd. It signifies likewise a *Trooper* who stays at the entry of a Place protected, to hinder *Soldiers* which straggle off from the *Army*, from committing any Disorder.

SAKER, is a Piece of Ordnance, carrying a *Ball* of 5 pound and a quarter weight; the diameter of the *Bore* is 3 inches and 9 sixteenths of an inch; the length of the *Gun* about 8 or 9 foot; it is a very good Field Piece, and is always a part of the marching *Artillery*.

SALLY, is when the Besieged march out a part of the *Garrison* in the night time, to attack the Besiegers in their Works, to nail their *Guns*, and to hinder the Progress of their Approaches. When a Place besieged is weak in Men, they make few *Sallies*; but when the *Garrison* is strong, and the *Inhabitants* numerous, the *Governour* ought to disturb the Enemy by frequent *Sallies*, which ought to be as secret as possible. Those who make the Sally, are to be armed with short *Arms*, and are to have *Grenades*, *Fire-pots*, *Gouderons* and *Pioneers* to destroy and level the Enemies Works.

To SALUTE a Prince or Person of extraordinary Quality at his coming into a *Garrison*, is the firing of the *Cannon* round the Place: likewise in the Field when a *Regiment* is to be viewed by a *King* or his *General*, the *Drums* beat a March as he approaches, and the *Officers* salute one after another as he passes by, stepping back with the right Foot and

 Hand,

Hand, and bowing the ſpears of their *Half Pikes* to the Ground, and afterwards recovering them gently, and bringing up the Foot and Hand and planting them. As ſoon as they have ſaluted, they are to pull off their Hats without bowing, but ſtanding upright. The *Enſigns* ſalute all together, bringing down their *Colours* near the Ground directly before them at one Motion, and having taken them up again gently, lift their Hats. If it be a Review of the *Army*, every *Battalion* is to ſalute with *Pikes* and *Muſquets* charg'd.

SAND-BAGS, are *Bags* con-

taining about a cubical foot of Earth; they are uſed for raiſing *Parapets* in haſte, or to repair what is beaten down; they are of uſe when the Ground is rocky, and affords no Earth to carry on their Approaches, becauſe they can be eaſily brought from far off, and removed at will. The ſmaller *Sand-bags* hold about half a cubical foot of Earth, and ſerve to be placed upon the *Superior Talus* of the *Parapet*, to cover thoſe that are behind, who fire through the *Embraſures* or *Intervals* which are left betwixt them.

SAP, is the digging deep under the Earth, in ſinking lower by degrees, to paſs under the *Glacis*, and open a way to come under Cover to the Paſſage of the *Moat*. After they have overcome all the Obſtacles which the Beſieged have oppoſed to hinder the Advancement of their Approaches, and that notwithſtanding their frequent Sallies, they are at laſt got near the Foot of the *Glacis*, the *Trench* is carried directly forwards, the Work-men covering themſelves the beſt way they can, with *Blinds*, *Woolpacks*, *Sand-bags*, or *Mantelets* upon *Wheels*; when they are got to the Foot of the *Glacis*, they make *Epaulments* or *Traverſes* on each ſide to lodge a good Body of Men. The *Sap* is made five or ſix fathom from the *Salliant Angle* of the *Glacis*, where the Men are only cover'd ſide ways, wherefore they lay Planks over head, with Hurdles and Earth above them. Having by this means obliged the Enemy to quit the *Covert-Way*, the *Pioneers*, with *Mantelets*, *Wool-packs*, or *Sand-bags*, make immediately a *Lodgment*, covering themſelves the moſt advantagiouſly they can, from the Fire of the oppoſite *Baſtion*.

SARRAZINE, is the ſame with *Herſe* or *Portcullis*, ſee *Herſe*.

SAUCISSE, is a long train of *Powder* ſewed up in a Roll of

pitch'd

pitch'd Cloath, of about two inches diameter; the uſe of it is to fire *Mines*, or *Caiſſons*; the length of it muſt reach from the *Mine* to the place where the *Engineer* is to fire it, to ſpring the *Mine*.

SAUCISSONS, are *Faggots* or *Faſcines*, uſed in covering of Men, or making *Epaulments*. They differ from the ordinary *Faſcines*, becauſe they are made of thicker Wood or Branches of Trees, and tied at both ends and in the middle, and are about a foot and a half or two foot thick, and four foot long. They are good to ſtop Paſſages, and being mixed with *Earth* and *Faſcines* to make *Traverſes* over a wet *Ditch*.

STALADE, or *Eſcalade*, is a furious Attack upon a *Wall* or *Rampart*, contrary to Form, and with no Precaution, carried on with *Ladders*, to inſult the *Wall* by open Force.

SCALE, is a right Line divided into equal Parts, repreſenting Miles, Fathoms, Paces, Feet, Inches, or any other Meaſure; it is uſed in making *Plans* upon Paper, in giving each Line its true Length.

SCALENE, ſee *Triangle*.

SCARP, or *Eſcarpe*, is the *Interior Talus* or *Slope* of the *Ditch* next the Place, at the Foot of the *Rampart* or *Liziere*.

SCHENOGRAPHY, which is likewiſe called *Profile* or *View*, is the natural Repreſentation of a Place, ſuch as it appears to us, when we look upon it, from without, conſidering its Situation, the Form of its Walls, the Number and Figure of its Steeples, and the Tops of its Buildings, both Publick and Private.

To SCOUR *a Line*, is to flank it ſo as to ſee directly along it, that a *Muſquet-Ball* entring at one end, may fly to the other, leaving no place of Security.

SCOURER or *Rammer*, is that wherewith a *Soldier* rams down the *Powder* and *Ball* into his Piece.

SENIORITY, is the difference of time betwixt the raiſing of two *Regiments*, whereby the one is ſaid to be ſo much Senior than the other; all *Regiments* take place according to Seniority. The difference of time betwixt the Date of two *Commiſſions* makes the one Senior to the other; and all *Officers* of the ſame Rank, roll by the Seniority of their *Commiſſions*.

SENTINEL, ſee *Centinel*.

SERJEANTS, are *Staff-Officers* in a *Company* of *Foot*, and ought to be vigilant and active in their Buſineſs; they ought to Read and Write, becauſe they are obliged to keep a Liſt of the *Soldiers* and their *Lodgings*, and to viſit them often; they are to teach the *Company* the Exerciſe of their *Arms*, and how they are to keep their *Ranks* and *Files*; their Poſt on a March is on the *Flanks*, to cauſe the *Company* to march in good Order. A *Serjeant* of each *Company* is to be on the *Parade*

at night, to receive the Orders and the Word from the *Adjutant*, which he is to carry to his *Captains* and *Subalterns*: When the *Adjutant* comes, the *Serjeants* place themſelves in a Ring with him, according to the precedency of their *Companies*, with their Hats on the Spears of their *Halbards*; and after he has given them the Orders, he whiſpers the Word to the firſt *Serjeant*, who gives it to the next, and ſo on, till it come to the youngeſt, who gives it to the *Adjutant*. They acquaint the *Officers* that are to go next upon Duty; they viſit the Mens *Arms*, and diſtribute Ammunition to them.

SHAFTS *of Limbers*, ſee *Limbers*.

SHOT; all ſorts of *Ball*, either for *Cannon*, or for *Muſquets*, *Carabines* and *Piſtols*, ſee *Ball*.

Chain-Shot, is two whole or half

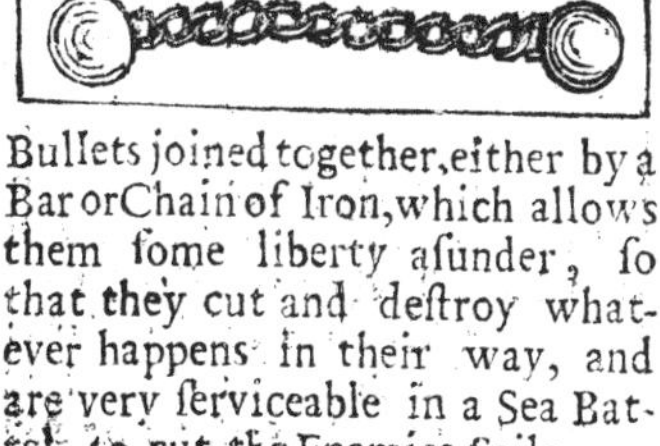

Bullets joined together, either by a Bar or Chain of Iron, which allows them ſome liberty aſunder, ſo that they cut and deſtroy whatever happens in their way, and are very ſerviceable in a Sea Battel, to cut the Enemies Sails.

SHOVEL, is an Inſtrument carried along with an *Artillery*, to be delivered out to the *Pioneers* who are to mend the Ways, or to work in the Approaches; it ſerves to gather and throw up the Earth which the *Pick-ax* and *Mattock* raiſe; it is made of a Shaft of about three foot and a half long; the Head of it which is thin and ſhod with Iron, is about fifteen inches deep, and eight broad.

SHOULDER *of a Baſtion*, is where the *Face* and the *Flank* meet.

SIDES *of Horn-works*, *Tenailles*, *Crown-works*, &c. are thoſe parts of their *Ramparts* which reach from the Border of the *Foſs* of the Place, to the Head of the Work, which in *Horn-works* and *Tenails* are parallel; ſometimes theſe Sides are no longer than the reach of a *Muſquet-Shot*, and are then defended from the *Faces* of the Place; but when they are longer, they have either *Flanks* made in the long Sides, which are then ſaid to have *Shoulders*; or elſe they are indented or made with *Redans*, or with *Traverſes* or croſs *Entrenchments* in the *Ditch*.

SIEGE, is the Encampment of an *Army* entrenched and fortified round a place which is attacked, with a Deſign to take it. When a *General* deſigns to beſiege a Place, he muſt firſt order it to be inveſted by a Body of *Horſe*, under the Command of a *Lieute-*

nant-

nant-General, to hinder any Succours getting into the Place, till the rest of the *Army* arrive. The method of encamping, is quite otherwise in a *Siege* than in a *March*; for in a *Siege* the *Army* environs the Place that nothing may enter, and lies without *Cannon-Shot* of the Town. If the Place be situated on a River, a *Detachment* is made of a part of the *Army* to the other side; and there are *Bridges* of *Communication* made both above and below the Town, with *Redoubts* guarded by a *Body* of *Foot* to secure them. If the Place be environ'd with Mountains, they possess all the Heights from whence they can any way annoy the Enemy. At a *Siege* the *Army* encamps with their Backs to the Place; *Battalions* and *Squadrons* interlin'd. The *Engineers* trace the *Lines* of *Circumvallation* and *Contrevallation*, with *Redoubts* and *Angles*, at convenient distances, and every *Regiment* works at the Place appointed them. The *Line* of *Circumvallation* is without the *Camp* to hinder Succours. The *Line* of *Contrevallation*, is that betwixt the Army and the Place, which covers the Besiegers from the Sallies of the *Garrison*. When the *General* has disposed his *Camps*, placed his *Guards*, as well towards the Place, as towards the Country, and established the *Lieutenant-Generals* to command in the particular *Quarters*, with Orders what they are to do: He goes with the *Engineers* to view the Place, and orders the Attack in the Place he judges the weakest; but because it is difficult to find two places situated after the same manner, so it is hard to make two Sieges after the same way; for there are some Towns, where without opening or carrying on of *Trenches*, the Besiegers come at once and lodge themselves on the *Counterscarp*, by the help of some Hollow Ways, Ruins or Cavities, or by some ill fortified Suburbs. And there are others, where the Ground is better managed, where within *Cannon-Shot* of the *Out-Works*, there is nothing which can facilitate the Enemies Approaches. To such sort of places, which are not the worst, there must be *Trenches* and *Approaches* to gain the Ground foot by foot, which renders such *Sieges* dangerous and very long, because of many Accidents which happen daily in the *Attacks*, *Sallies*, and *Mines*, and other Accidents of *War*.

To make or form a Siege, there must be an *Army* sufficient to furnish five or six Reliefs for the *Trenches*, *Pioneers*, *Guards*, *Convoys*, *Escorts*, and what else may happen: An *Artillery* with *Magazines* furnished with a sufficient quantity of Warlike Ammunition, and Provisions of all sorts: And an *Hospital* with *Physicians*, *Chirurgeons*, &c. and Med'cines of all sorts.

To raiſe a Siege, is to give over attacking a Place, in abandoning the Works, and levelling the Works or Poſts which they were in poſſeſſion of before the place.

To turn a Siege into a Blockade, is to give over the Attacks, and to poſſeſs themſelves of all the Avenues leading to the Place, to hinder any *Succours* or *Convoys* getting into it, with a Deſign to take it by Famine.

SILLON, is a Work raiſed in the middle of a *Foſs*, to defend it when it is too wide: It has no particular Form, ſometimes being made with little *Baſtions*, *Half-Moons*, and *Redans*, which are lower than the Works of the Place, but higher than the *Covert-Way*. The Towns of *Doway* and *Bruges* in *Flanders*, are both fortified this way. The Word *Sillon*, is wearing out of Uſe, being now called *Envelope*. See *Envelope*.

SINGLE *Tenaille*, ſee *Tenaille*.

SIXAIN, an antient Order of Battel for ſix *Battalions*, which, ſuppoſing them all in a Line, is formed thus. The ſecond and fifth *Battalions* advance and make the *Van*; the firſt and ſixth fall to the *Rear*, leaving the third and fourth to form the Body. Each *Battalion* ought to have a *Squadron* on its Right, and another on its Left. Any number of *Battalions* produced of the number ſix, may be drawn up by this Order; ſo 12 *Battalions* may be put in two *Sixains*, and 18 *Battalions* in three *Sixains*.

SKIRMISH, a ſudden Encounter between two ſmall Bodies of Men, without Order.

SOLDIER, is he who is liſted in the Service, and receives Pay; he who ſerves on Foot, is commonly called a *Soldier*; and he who ſerves on Horſe-back, a *Trooper*.

SOUND, ſee *Trumpet*.

SPADE, is an Inſtrument for digging up the Ground; the Handle or Shaft is about three foot long; the Head of it is all of Iron, the upper part being flat for the *Pioneer* to ſet his Foot on to force it into the Ground; the length of the Head is a foot or fifteen inches, and the breadth ſix or eight.

To SPIN *Hay* is to twiſt it it up in Ropes very hard for an Expedition in the Winter time; each *Trooper* carrying as much as he can behind him.

SPOKES *of a Wheel of a Cannon*, are thoſe ſhort pieces of Wood, being twelve in number; which by having one end fixed in the *Fellows*, and the other in the *Nave*, keep the *Nave* in the Centre, and make the *Wheel*.

SPUNGE *of a Gun*, is a long Staff put into a roll of Wood, which is covered over with a Sheep-skin, the Wool outwards, to ſpunge and clean the *Gun*. As ſoon as the *Gun* has fired, a *Matroſs* is ready with the *Spunge*, while another claps his Finger on

on the Vent to ſtop the Air, and ſtifle what Fire may remain in the *Chamber*. The *Spunge*, *Rammer*, and *Ladle*, after the *Gun* is loaded, are laid under her betwixt the *Wheels*.

SQUADRON, is a ſmall *Body* of *Horſe*, compoſed of 3 *Troops*, each 50 *Troopers*, making 150, and ſometimes 200, when the *Troops* are larger, but never above that; becauſe a greater number can never be advantagiouſly poſted, nor have room to act in narrow Grounds, as *Woods*, *Marſhes*, and *Defiles*. The eldeſt *Troop* takes always the Right of the *Squadron*; the ſecond the Left, and the youngeſt the Centre. A *Squadron* is always drawn up three deep, that is to ſay, in three *Ranks*; having the length of a *Horſe*, or rather more between *Rank* and *Rank*. The *Standard* is always in the Centre of the firſt *Rank*. When the *Army* is encamped, a *Squadron* of *Horſe* is allowed 30 Paces for their Front, and 30 paces Interval between one *Squadron* and another; on a March the *Squadrons* of the ſame *Column* ought to keep a convenient diſtance.

SQUARE, is a Figure compoſed of four equal *Sides*, and four right *Angles*.

Long Square, is a Figure compoſed of four Sides, whereof the two oppoſite are equal, and all the Angles Right Angles.

Square Battalion of Men, is that which is compoſed of an equal number of Men in *Rank* or *File* or when the number of Men in each *File*, is equal to the number of Men in each *Rank*. *Square Battalion of Ground*, is when the Ground of the *Flanks* is of the ſame Extent, as the Ground of the *Front* and *Rear*. To make a *ſquare Battalion* of Men, whoſe number is known, as 50, take the neareſt Radix or ſquare Root, which is ſeven, for the number of Men in *Rank* and *File*. To make a *ſquare Battalion* of Ground, the number being likewiſe determin'd, as 60, *Maneſſon Mallet* ſays that Number muſt be multiplied by 3, which is the number of feet that every Man takes in *Front*, and the Product 180 divided by 7, which is the number of feet that each Man taketh up in deepneſs, or the diſtance of the *Ranks*; the Quotient is 25; the ſquare Root of which is 5, which is the number of Men in each *File*; and if by this Radix 5, you divide 60, the Quotient is 12 for the number of Men in each *Rank*.

Hollow Square, is a *Body* of *Foot* drawn up with an empty ſpace in the middle for the *Colours*, *Drums* and *Baggage*, facing, and covered by the *Pikes* every way, to oppoſe the *Horſe*.

STAFF-OFFICER, ſee *Officer*.

STANDARD, is a piece of Silk or Damask, about a foot and a half ſquare; on which is embroidered, the Arms, Device, or Cypher of the *Prince*, or of the *Colonel*,

Colonel: It is fixed on a Launce of about 8 or 9 foot long, and carried in the Centre of the first *Rank* of the *Squadron*; in Rainy or bad weather, it has a Case of Leather over it to preserve it.

STAR-REDOUBTS are now out of use, both because their *Re-entring Angle* is not well flanked, and because the square *Redoubts* are sooner raised, and equally serviceable. They were made with *Salliant* and *Re-entring Angles*, and had from five to eight Points; and each of their Sides or Faces was from 12 to 25 fathom long.

STORM, see *Assault*.

STRAIKS, are strong Plates of Iron six in number, fixed with large *Nails*, called *Straik-Nails*, on the Circumference of a a *Cannon Wheel*, over the Joints of the *Fellows*, both to strengthen the *Wheel*, and to save the *Fellows* from wearing out on hard Ways or Streets.

SUBALTERN-OFFICERS, see *Officers*

SUB-BRIGADEER, is a Post in the *Troops* of *Guards*, next under a *Brigadeer*.

Sub Lieutenant, is an *Officer* in *Regiments* of *Fuseleers* where there are no *Ensigns*, having a Commission as youngest *Lieutenant*, and Pay only as *Ensign*; but takes place of all *Ensigns*, except the *Guards*.

Subsistance, is the Money paid to the *Soldiers* weekly, not amounting to their full Pay, because their Cloaths, Acoutrements, Tents, Bread, *&c.* are to be paid; it is likewise the Money paid the *Officers* upon Accompt, till their Accompts be made up, which is generally once a year, and then they are paid their Arrears, which signifies the Money they were behind.

Sub-divisions, are the lesser Parcels, into which a *Regiment* is divided in marching, being half the greater Divisions.

SUCCOUR, is the Enterprize made to relieve a Place, that is, to raise the Siege, and force the Enemy from it.

SURFACE or *Superficies*, is an Extent, having Length and Breadth, but no Thickness: It is evident, that the Extremities of a *Surface* are *Lines*.

Surface, as a Term in *Fortification*, is that part of the *Exterior Side*, which is terminated by the *Flank* prolong'd or extended, and the *Angle* of the nearest *Bastion*. The Double of this *Line* with the *Curtin*, is equal to the *Exterior Side*.

SUTLER, is he who follows the *Army* to sell all sorts of Provisions to the *Troops*. They pitch in the *Rear* of each *Regiment*, and about the *Generals Quarters*; their Beer, Wine, Bread, *&c.* they either buy from the Boors, or fetch from the nearest Towns.

SWALLOWS-TAIL, is an *Out-Work* differing only from a single *Tenaille*, in that its Sides are not parallel, like those of a *Tenaille*;

but

but if prolong'd, would meet and form an *Angle* on the middle of the *Curtin*; its *Head* or *Front* is compos'd of two *Faces* forming a *Re entring Angle.* This Work is extraordinary well flanked and defended by the Works of the Place, which discover all the Length of its long Sides. But their great Fault is their not covering sufficiently the *Flanks* of the opposite *Bastions.*

T.

TAIL *of the Trenches*, or *Opening of the Trenches*, is the Post where the Besiegers begin to break Ground to cover themselves from the Fire of the Place, in advancing the *Lines* of *Approach*; see *Opening of the Trenches.*

TALUS or *Epatement*, is the Slope given to the *Rampart* or *Wall*, that it may stand the faster, and is more or less according as the Earth is looser or more binding. All *Ramparts* ought to have a *Slope* or *Talus* on each side; that is, they ought to be broader at the *Basis*, than at the *Top*: the one is called the *Exterior Talus*, the other the *Interior Talus.* And there is likewise a *Superior Talus.* See *Profile.*

Exterior Talus, is the Slope given to a Work, one the side towards the Country, and ought to be as small as possible, that the Enemy may not find it easie to be mounted, either by *Scalade* or otherwise. But if the Earth be not good, the *Talus* must be large, that it may keep it up the better. In such a case it were good to support the Earth with a Wall, which the *French* call *Chemise*, when it is not thick, and otherwise *Revetement*, which signifies cloathing or fencing, to make the Earth last longer, and to save the making too large a *Talus.* This *Wall* ought to have a small *Talus* of a fifth or sixth part of its height, and for a Reinforcement it is generally supported in the inside by *Counter-forts*, or a sort of *Buttresses.*

Interior Talus, is the Slope of the inside of the Work next the Town, which is much larger than that of the outside; and has at the *Angles* of the *Gorge*, and sometimes in the middle of the *Curtin*, *Ramps*, or *sloping Roads*, to mount upon the *Terre-plein* of the *Rampart.* The *Interior Talus* of the *Parapet*, ought to be very small, that the Men may with more ease fire over it. See *Profile.*

Superior Talus of the *Parapet*, is the Slope on the top of the *Parapet*, marked 13, 12, in the Figure at *Profile.* This Slope allows the *Soldiers* to defend the *Covert Way* with small *Shot*, which they could not do were it level.

Tarpaulins, are pitched Cloths, with which the Decks of Ships where there are Stores are covered, to save them from Rain; or to throw over Stores in open

Boats, or upon *Battery*, or in *Magazines*.

TATTOU, see *Retreat*.

TEMOINS, is a *French* Term for the pieces of Earth left standing as Marks or Witnesses, in the *Fossees* of Places they are emptying, to the end they may know exactly, how many cubical fathoms or feet of Earth has been carried away, thereby to pay the Work-men. who are always sure to leave some of the highest spots of Ground for *Temoins*, to have more deepness to measure. But the *Engineers* are generally careful to mark out indifferent Places, some high, some low, to measure as exact as they can.

TENAILLE, is an *Out-Work* longer than broad, whose long Sides are parallel; and is either single or double. There are likewise *Tenailles* in the *Foss*.

Single Tenaille, is a Work whose *Front* is advanced towards the Country, having two *Faces* forming a *Re entring Angle*; its two long Sides terminate on the *Counterscarp*, opposite to the *Angle* of the *Shoulder*.

Double Tenaille, is a Work whose *Front* having four *Faces*, forms two *Re entring*, and three *Salliant Angles*; its long Sides are likewise parallel, and terminate on the *Counterscarp*, opposite to the *Angle* of the *Shoulder*. Both the single and double *Tenailles* have this fault, that they are not flanked or defended at the *Re-entring Angle*, because the heighth of the *Parapet* hinders the *Soldiers* from discovering before that *Angle*. Therefore *Tenailles* are only made when there is not time enough to make *Horn works*. The *Ramparts*, *Parapets*, *Fosses*, *Covert-Way*, and *Glacis* of *Tenailles*, are the same with other *Out Works*.

Tenaille in the Foss, is a low Work raised before the *Curtin* in the middle of the *Foss*, and is of three different sorts, as may be seen in the *Foss* of the *Cittadel*, in the *Grand Plan*. The first is composed of a *Curtin*, two *Flanks*, and two *Faces*. The *Rampart* of the *Curtin*, including the *Parapet* and *Talus*, is but five fathom thick, but the *Rampart* of the *Flanks* and *Faces* is seven. The second, which *Vauban* has by Experience found to be of a very good Defence, is composed only of two *Faces*, made on the Lines of Defence, whose *Rampart* and *Faces* are parallel. The third differs from the last, only in having its *Rampart* parallel to the *Curtin* of the Place. All these sorts are very good Defences for the *Foss*, and lye so low, that they cannot be hurt by the Besiegers *Cannon*, till it be on the *Covert-Way*.

Tenaille of a Place, or *Front of a Place*, is what is comprehended between the Points of two Neighbouring *Bastions*, as the *Faces*, the *Flanks*, and the *Curtin*. So it is said, The Enemy attacked the whole *Tenaille* of a Place, when they made two Attacks on the *Faces* of the two *Bastions*.

TENT, is a ſort of Pavilion of Cloth which is pitched upon Poles, with Cords and Pegs, and pulled down when the *Army* moves; it ſerves to keep an *Officer* under Cover, as the ſmall *Tents* do the *Soldiers*.

TERRE-PLEIN *of a Rampart*, is the *Horizontal Superficies* of the *Rampart*, between the *Interior Talus* and the *Banquett*; 'tis on the *Terre-plein*, that the Defendants go and come; it is likewiſe the Paſſage of the Rounds. Trees on the *Terre-plein* of a *Rampart*, ſerve to bind it, but in a Siege they are inconvenient; for the noiſe made by the Wind amongſt the Leaves, hinders the Beſieged from hearing the Work-men in the *Approaches*.

To TERTIATE *a Piece*, is to examine it, whether it has the due thickneſs of Metal in every place, and whether it be true bored.

THUNDERING-BARRELS, ſee *Barrels*.

TOISE, is a Meaſure uſed by the *French Engineers* in all their *Fortifications*, and is ſix foot; a ſquare *Toiſe* is 36 ſquare feet, and a cubical *Toiſe* is 216 cubical feet.

TOMPION, is a ſtopple of Wood or Cork, which is uſed in loading a *Mortar*; it is exactly fitted for the mouth of the *Chamber*, and is drove hard in after the *Powder*, and the *Bomb* is placed above it; it ſerves by confining the *Powder*, to make it burſt out with the more Violence. *Tompion* is likewiſe a ſtopple of Wood for the mouth of the *Mortar* or *Gun*, to keep out the Rain.

TOUCH-HOLE or *Vent*, is the ſmall Hole at the end of the *Cylinder* of a *Gun* or *Muſquet*, by which the Fire is conveyed to the *Powder* in the *Chamber*. In a *Fire-lock*, *Carabine*, or *Piſtol*, it is called the *Touch-hole*, but in a Piece of *Cannon* it is more properly called the *Vent*.

TOWN-MAJOR, ſee *Major*.

TRANSUM, is a piece of Wood which goes acroſs betwixt the *Cheeks* of a *Gun-Carriage*, or of a *Gin*, to keep them fixed together; each *Tranſum* in a *Carriage* is ſtrengthened by a Bolt of Iron. See *Carriage*.

TRAPEZE, a four ſided Figure, having only two of its four Sides parallel.

TRAPEZOIDE, is a Figure of four Sides, all unequal, and whoſe *Angles* are likewiſe unequal, and none of its Sides parallel.

TRAVERSE, is a *Trench* with a little *Parapet*, ſometimes two, one on each ſide; to ſerve as a cover from the Enemy that might come on their *Flank*: ſometimes it is covered over head with Planks, and loaded with Earth. They are very advantagious in ſtopping an Enemies Way, and to prevent being enfiladed. They are likewiſe a good Defence in a dry *Foſs*, in making the

the *Parapet* on the ſide next the oppoſite *Flank*.

Traverſe in a wet Foſs, is made by throwing into the *Foſs* over againſt the place where the *Miner* is to be put to the Foot of the *Wall*, abundance of *Sauciſſons*, *Joyſts*, and other pieces of Wood, with *Faſcines*, *Stones*, *Earth*, and all other things which can help to fill up the *Foſs*, and be capable of carrying a *Gallery* for ſuch as uſe it.

Traverſe is likewiſe a Wall of Earth or Stone croſs a Work which is commanded, to cover the Men; as at *Coehorne*'s Work at *Namure*, which lies on the ſide of a high Ground, and is open to the other ſide the *Sambre*; there are two high *Traverſes* croſs the Work one behind another.

To Traverſe a Gun or Mortar, is to bring her about with *Hand-ſpikes* to the Right or Left, till ſhe is pointed exactly at the object.

Trench, which is likewiſe called *Lines of Approach*, and *Lines of Attack*, is a way hollowed in the Earth, in form of a *Foſs*, having a *Parapet* towards the place beſieged, when the Earth can be removed; or elſe it is an Elevation of *Faſcines*, *Gabions*, *Wool-packs*, and ſuch other things that can cover the Men, and that does not fly in pieces or ſplinters to hurt them: This is to be done when the Ground is rocky, but when the Earth is good, the *Trench* is carried on with leſs trouble; and the *Engineer* is to demand only Proviſion of *Spades*, *Shovels*, and *Pickaxes*, to enlarge the *Trench* to five foot deep, and two fathoms wide. The greateſt Fault a *Trench* can have, is to be enfiladed; to prevent which they are ordinarily carried on with Turnings and Elbows. As the *Trenches* are never carried on but in the night time, therefore the Ground ought to be viewed and obſerved very nicely in the day. On the *Angles* or *Sides* of the *Trench*, there ought to be *Lodgments* or *Epaulments* in form of *Traverſes*, the better to hinder the Sallies of the *Garriſon*, to favour the Advancement of the *Trenches*, and to ſuſtain the Work-men. Theſe *Lodgments* are ſmall *Trenches* fronting the Place beſieged, and joining the *Trench* at one end. The *Platforms* for the *Batteries* are made behind the *Trenches*, the firſt at a good diſtance, to be uſed only againſt Sallies of the *Garriſon*; as the *Approaches* advance, the *Batteries* are brought nearer, to ruin the *Defences* of the Place, and diſmount the *Artillery* of the beſieged: The *Batteries* for the *Breaches*, are made when the *Trenches* are advanc'd near the *Covert-Way*. If there be two *Attacks*, there muſt be *Lines of Communication* or *Boyaus* between the two, with *Places of Arms*, at convenient diſtances. The *Trenches* ought to be ſix or ſeven foot high with the *Parapet*, which ought to be five foot thick, and

have

have *Banquetts* for the *Soldiers* to mount upon.

Returns of a Trench, are the Elbows and Turnings which form the *Lines of Approach*, and are made as near as can be parallel to the Defences of the Place, to prevent their being enfiladed.

To mount the Trenches, is to mount *Guard* in the *Trenches*. *To relieve the Trenches*, is to relieve the *Guards* of the *Trenches*. *To dismount the Trenches*, is to come off the *Guard* of the *Trenches*. *To cleanse or scour the Trenches*, is to make a vigorous Sally upon the *Guard* of the *Trenches*; to force them to give way and quit their Ground, to drive away the Work-men, break down the *Parapet*, fill up the *Trench*, and to nail their *Cannon*.

Counter-Trenches, are *Trenches* made against the *Besiegers*, which consequently have their *Parapet* turned against the Enemies *Approaches*, and are enfiladed from several parts of the Place, on purpose to render them useless to the Enemy, if they chance to be Masters of them; but they ought not to be enfiladed or commanded by any Height in the Enemies Possession.

TRIANGLE, is a Figure comprehended betwixt 3 Sides, and is either *Rectilineal* or *Spherical*. A *Rectilineal* or *plain Triangle*, is a Figure consisting of 3 straight Sides. A *Spherical Triangle*, is a Figure formed by three Arches, of three great Circles, cutting one another on the Surface of a Sphere.

A *Rectilineal Triangle*, considered according to the Sides, may be either *Equilateral*, *Isosceles*, or *Scalene*; and considered according to its Angles, may be either *Rectangle*, *Ambligon*, or *Oxigon*.

Triangle Equilateral, is what has the three Sides equal; it is evident, the three Angles must likewise be equal, each being sixty Degrees. *Triangle Isosceles*, is what hath two Sides equal; whence it follows, that all *Equilateral Triangles* are *Isosceles*; tho' all *Isosceles Triangles* are not *Equilateral*. *Triangle Scalene*, is what hath three unequal Sides. *Triangle Rectangle*, is what hath one Right Angle. *Triangle Ambligon*, is what hath one Obtuse Angle. *Triangle Oxigon*, is whose Angles are all Acute.

TROOP *of Horse or Dragoons*, is a small Body of about 50 or 60; sometimes more, sometimes less; commanded by a *Captain*. Each *Troop* has, besides a *Captain*, a *Lieutenant*, *Cornet*, *Quarter-Master*, and three *Corporals* who are the lowest Officers in a *Troop*. A *Regiment* of *Light Horse* in *England*, consists of six *Troops*, and sometimes nine.

Troop; *to Beat the Troop*, is the same with *Assembly*; see *Assembly*.

TROOPER, is a private Man in a *Troop of Horse*.

TRUCKS, are small Wheels of one piece of Wood, about a foot

foot and a half or two foot diameter, uſed for *Sea Carriages*, and likewiſe for the *Truck-Carriages* by Land, and ſometimes for *Garriſon Guns*.

TRUMPET, is an Inſtrument of Wind Muſick, uſed in publick Rejoicings, but eſpecially in the War; each *Troop* of *Horſe* has two *Trumpets*. It is made of Metal, moſt commonly of Braſs, but ſometimes of Silver. The Mouth of the *Trumpet* is always of Braſs, and is to take out and in at Pleaſure. He who blows the *Trumpet*, is called the *Trumpeter*, who ought to be a Man fit for Fatigue and Vigilant. The firſt Sound of the *Trumpet* before a March, is to *Boot and Saddle*, at which the *Troopers* get themſelves ready to mount; this is ſounded when the *Drums* beat the *General*. When the Aſſembly is beat, then the *Trumpet* ſound *to Horſe*, and they all mount; the third is *to March*. The *Trumpets* likewiſe ſound a *Charge* in day of Battel, and the *Retreat* at night.

TRUNIONS *of a Gun*, are the two pieces of Metal ſticking out on the ſides of a Piece, by which it ſwings in its *Carriage*. They are generally the diameter of the *Ball* of the Piece in length, and their diameter is the ſame with the diameter of the *Ball*. The *Axis* of the *Trunions*, is equal with the lowermoſt ſide of the *Chace* of the *Gun*.

Trunion-Ring, is that Ornament or Jutting out a little before the *Trunions*.

TURNPIKE, ſee *Chevaux de Friſe*.

V.

VAN, or *Van-Guard*, is that part of the *Army* which marches in the *Front*. See *Guard*.

VEDETTE, is a *Centry* on *Horſe-back*, or a *Trooper* upon a *Centry Poſt*. His Horſe Head is towards the Place from whence any Danger is feared, and his *Carabine* is advanced with the Butt End againſt his right Thigh; when the *Army* lies encamped, there are *Vedettes* poſted at all Avenues, and on all Riſing Grounds, to watch for its Security.

VENT or *Touch-hole*, ſee *Touch-hole*.

To VIEW a Place in order to beſiege it, which the *French* call *Reconnoitre*, is when the *General* accompanied by the *Engineers* rides round the Place, obſerving the Situation of it, with the Nature of the Country about it; as *Hills*, *Valleys*, *Rivers*, *Marſhes*, *Woods*, *Hedges*, &c. thereby to judge of the moſt convenient place for opening the *Trenches*, and carrying on the *Approaches*; to find out proper places for encamping the *Army*, for the *Lines of Circumvallation* and *Contrevallation*, and for the *Park* of *Artillery*.

To View or Reconnoitre an Enemy, is to get as near their *Camp* as poſſible, to ſee the Nature of the

the Ground, and the Avenues to it, to find out the Strength and Weakness of their Encampment, where they may be best attacked, or whether it be proper to hazard bringing them to Action. *Parties of Horse*, are generally sent out to view the Enemies March, to know whether it tends, thereby to guess at their Designs, and to regulate the Motions of the *Army* accordingly.

To View or Reconnoitre, is likewise when the *Quarter-Master-General*, with a strong Party of *Horse*, goes to view the Ways for the March of the *Army*, and to find the most convenient place for an Encampment, to wit where there is Water and Forage, where the *Army* may not be too much exposed to the Insults of the Enemy, but covered by Rivers, Marshes, or strong Grounds, where they cannot easily be forced.

VOLUNTEERS, are Persons of Quality, who of their own accord, either for the Service of their *Prince*, or out of the Esteem they have for the *General*, serve in the *Army*, without being engaged to any *Captain*; but upon their own Expence are ready upon all Occasions to gain Honour and Preferment, by exposing themselves in the Service.

UTENSILS, are the Necessaries which are to be furnished by a *Landlord* to the *Soldiers* quarter'd upon him, such as *Beds*, *Sheets*, *Pots*, *Dishes*, *Spoons*, *Cups*, *Fire*, *Candle*, &c. Sometimes the *Landlords* make an Agreement fo these things, and allow Money in lieu of 'em.

Utensils, are likewise all sorts of Hand-Tools used in an *Army* or at a *Siege*, such as *Spades*, *Shovels*, *Pick-axes*, *Hatchets*, *Bills*, &c. or the Instruments used about a *Gun*, as the *Ladle*, *Rammer*, *Spunge*, *Wad-hook*, *Linstock*, *Coins*, *Hand-spike*, *Priming-Iron*, &c.

W.

WADD, is a Stopper of Hay or Straw forced into a *Gun* upon the *Powder*, to keep it close in the *Chamber*; when it is home at the *Powder*, the *Gunner* gives it generally three thumps with the *Rammer* Head.

WAD-HOOK or *Worm*, is a small Iron turned Serpent-wise like a Screw, and put upon the end of a long Staff, to draw out the Wad of a *Gun* when she is to be unloaded.

WAGGONS, see *Baggage-Waggons*.

Waggon-Master-General, is he who has the ordering and marching of the Baggage of the *Army*. On a day of March, he meets the Baggage at the Place appointed in the Orders, and marshals it according to the *Rank* of the *Brigade* or *Regiment* each *Waggon* belongs to, and marches it according to the Route given him, which is sometimes in one *Column*, sometimes in two; sometimes after the *Artillery*, and some-

ſometimes the Baggage of each *Column* follows their reſpective *Column*.

WARNING-PIECE, is the *Gun* which fires every night about Sun-ſet, to give Notice to the *Drums* and *Trumpets* of the *Army*, to beat and ſound the *Retreat* or *Tattou*, which is likewiſe called ſetting the Watch.

WARRANT OFFICER, ſee *Officer*.

WELL, is a Depth ſunk in the Ground by the *Miner*, from whence he runs out Branches or Galleries in ſearch of the Enemies *Mine* to diſappoint it, or to make a *Mine*.

WHEEL, is a Word of Command, when a *Battalion* is to alter their *Front*, either one way or other. When a *Battalion* is to *Wheel to the Right*, every Man moves and wheels from the Left to the Right, only the Man on the *Right Angle* turns very ſlowly, being as it were the Hinge on which the reſt move. When a *Battalion* is commanded to *Wheel to the Left*, the Soldier on the *Left Angle* turns ſlowly, while the Right wheels from the Right to the Left. When the Word of Command *Wheel*, is given to a Diviſion of Men upon a March, if it be *to the Right*, the Right Hand Man keeps his Ground, turning only on the Heel, while the Left Hand Man moves about quick, till he makes an even Line with the Right Hand Man. If it be *Wheel to the Left*, the contrary is to be obſerved. *To wheel by ſingle Ranks*, if it be *to the Right*, the Right Hand Man of each *Rank* turns on his Heel, while the Left Hand Men move round, and the whole are formed into one *Rank*, fronting as their *Flank* was before. To reduce them into *Ranks* again, the Left Hand Men turn on their Heels, while the Right Hand Men move round. The Motion of wheeling is of great uſe, if a *Battalion* be threaten'd with an Attack upon its *Flank*, or if there be a Deſign of falling upon the Enemies *Flanks*. *Squadrons* of *Horſe* wheel after the ſame manner.

WHEEL-BARROW, is one of the moſt neceſſary Utenſils about a *Fortification*, for rolling the Earth from one place to another; it is ſo well known, it needs no Deſcription. See the Figure at *Doſſer*.

WHEELS *of a Gun Carriage*, which ſerve for marching the *Gun* with more eaſe, are two large Circles of Wood compoſed of the *Fellows*, the *Spoaks*, and the *Nave*. The *Fellows* are ſix pieces of Wood, their thickneſs being that of the diameter of the *Ball*, their breadth ſomething more, forming each an Arch of ſixty Degrees, ſo that being joined, they make one entire Circle; they are joined where their ends meet by a peg of Wood, called the *Duledge*, and the Joint is ſtrengthened on the outſide of the

VVheel

VVheel by a ſtrong plate of Iron, called the *Duledge-Plate*, and on the Circumference, by a plate of Iron of the length and breadth of one of the *Fellows*, called the *Straiks*, fixed on it with ſtrong Nails, called the *Straik-Nails*; the *Straiks* cover the Joints of the *Fellows*. The *Spoaks* are twelve in number, being ſhort pieces of VVood let into the inner Circumference of the *Fellows*, and into the *Nave*, and appear like ſo many Semi-diameters or Rayons; they keep the *Nave* in the Centre, and eaſe the *Fellows*. The *Nave* is a ſhort thick piece in the Centre of the *Wheel*, through which the *Axletree* goes, and is fixed on the other ſide by a *Linſpin*. The length and diameter of the *Nave*, is proportioned to the nature of the *Gun*. It is bound with two ſtrong Hoops of Iron, called the *Nave-Bands*. The two *Wheels* are one on each end of an *Axletree*, which keeps them at a fix'd diſtance, and upon which the fore part of the *Carriage* is fixed by ſtrong Bands of Iron, called the *Axletree Bands*.

VVICKET, is a ſmall Door in a Gate of a fortified Place, at which a Man on Foot may get in, which is ſometimes opened when the Gate is ordered to be kept ſhut. The heighth of it is about three foot and a half, and the breadth two.

VVINDAGE *of a Gun*, is the difference between the diameter of the *Bore*, and the diameter of the *Ball*; for ſince the *Ball* are rough, if they were not ſomewhat leſs than the *Bore*, they might jamme in the Piece; ſo the *Windage* of a *Demi Culverin* is a quarter of an inch.

VVINDLACE, is a Roller

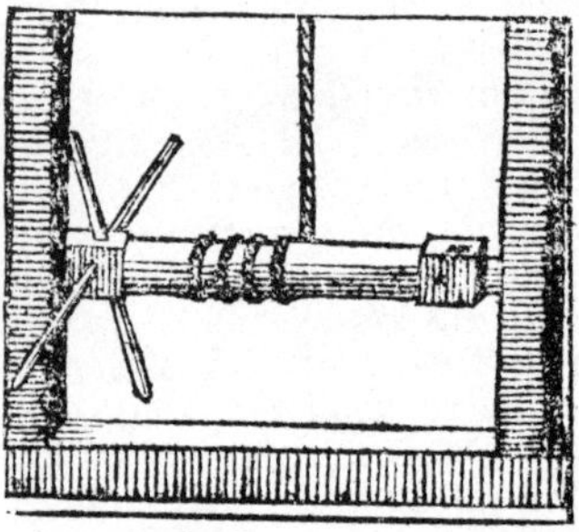

of VVood ſquare at each end, through which is either croſs Holes for Hand-ſpikes, or Staves acroſs, to turn it round; by this means it draws a Cord, one end of which is faſtened to ſome weight which it raiſes up. They are uſed in *Gins*, and about *Dutch Mortars* to help to elevate them.

VVINGS *of an Army*, ſee *Army*.

Wings of a Battalion, are the Right and Left Hand *Files*; when a *Battalion* is drawn up, the Diviſions on the Right and Left are called the *Wings*.

VVINTER-QUARTERS, are ordinarily the Place where *Troops* are lodged during the VVinter. It is likewiſe the time comprehended betwixt the end of the Campaign, and the beginning of the next.

VVIS-

WISCHER, see *Nave.*

VVITNESSES, see *Temoins.*

VVORD in an *Army* or *Garrison*, is a Token or Mark of Distinction, by which Spies or Treacherous Persons are known; it serves likewise to prevent Surprizes. It is given out every night by the *General* to the *Lieutenant-General*, or *Major-General* of the Day, who gives it to the *Majors* of *Brigades*, they to the *Adjutants*, who give it first to their *Field Officers*, and afterward to a *Serjeant* of each *Company*, who carry it to the *Subalterns.* In a *Garrison*, it is given by the *Governour* after the Gates are shut, to the *Town-Major*, who gives it to the *Adjutants*, and they to the *Serjeants*.

Words of Command, are the Terms used by Officers in exercising *Battalions* or *Squadrons*, or when they are upon Action.

VVORKS; all the *Fortifications* about a Place, are called the *Works of the Place*; and more particularly all detach'd VVorks, are called the *Out-Works*.

VVORM, is a Screw of Iron to be fixed on the end of a Rammer, to pull out the VVad of a *Firelock*, *Carabine* or *Pistol*; it is the same with *Wad-hook*, only the one is properer for small *Fire-Arms*, and the other for *Cannon.*

Y.

YOUNGER *Officer*, is he whose Commission is of a later Date.

Younger Regiment, is that which is of a shorter standing, in respect of another.

THE

A Deſcription of a Ship, with all Her Tackling.

Mizan Maſt: And Running Rigging.

1 *Mizan Maſt and Top Maſt.*
2 *Mizan Yard & Sail.*
3 *The Croſs-tree Yard.*
4 *Mizan Topſail Yard and Sail.*
5 *Mizan Top and Cap.*
6 *Mizan Topſail Lifts.*
7 *Mizan Topſail Sheets.*
8 *Mizan Shrowds.*
9 *Mizan Topmaſt Shrowds.*
10 *Mizan Chains.*
11 *The Gallery.*
12 *Tafferail.*
13 *The Cap Enſign-ſtaff and Enſign.*
14 *The Quarter Deck.*
15 *The Half Deck.*
16 *The Mizan Sheet.*
17 *Mizan Topſail Braces.*
18 *Mizan Topſail Bowlines.*
19 *Mizan Topſail Clewlins.*
20 *Mizan Brailes.*
21 *Mizan Topmaſts Stay.*
22 *Mizan Stay.*
23 *Mizan Sheet.*
24 *Yards & Oars for the Boat.*
25 *The Ships Boat.*
26 *A Hoiſting-line for Pennant.*
27 *Mizan Bowlin.*
28 *A Mizan Crowfoot.*
29 *Croſs-tree Braces.*
30 *Mizan Tack.*
31 *The Lanyards.*
32 *The Mizan Topmaſt Crowfoot.*

Main Maſt: And Running Rigging.

1 *Main-maſt & MainTopmaſt.*
2 *Main Yard and Main Sail.*
3 *MainTopſail Yard and Sail.*
4 *Topgallant Yard and Sail.*
5 *Topgallant Croſs-trees and Cap.*
6 *Main Lifts.*
7 *Main Topſail Lifts.*
8 *Topgallant Lifts.*
9 *Main Braces.*
10 *Main Sheets.*
11 *Main Bowlings.*
12 *Main Clewgarnets.*
13 *Main Stay.*
14 *Main Shrowds.*
15 *Main Chains and Chain-Plaits.*
16 *Back-ſtays for the main Top Maſt.*
17 *Main Topmaſt Stay.*
18 *Main Topſail Braces.*
19 *Main Topſail Clewlines.*
20 *Main Topſail Bowlines.*
21 *Main Topmaſt Shrowds.*
22 *Main Topgallant Shrowds.*
23 *Topgallant Clewlines.*
24 *Topgallant Braces.*
25 *Topgallant Bowlines.*
26 *The Horſe for the main Top ſail Yard.*
27 *Main Topſail Leathlines.*
28 *Main Topſail Buntlines.*
29 *A Crowfoot from the Top to the Stay.*
30 *Main Tacks.*
31 *Main Sheets.*
32 *The main Tye and fall of the Garnet.*
33 *The Main Horſe and Tackle.*
34 *The Tye of the main Topſail Jeers.*
35 *The main Topgallant Sails Tye and Jeers.*

Foremaſt, and Rigging.

1 *Foremaſt and Topmaſt.*
2 *Fore yard and Sail.*
3 *Fore Topſail yard and Sail.*
4 *Topgallant yard and Sail.*
5 *Fore Lifts.*
6 *Fore Topſail Lifts.*
7 *Fore Topgallant Lifts.*
8 *Croſs-trees and Cap.*
9 *Fore Braces.*
10 *Fore Topſail Braces.*
11 *Fore Topgallant Braces.*
12 *Fore Sheets and Tacks.*
13 *Fore Bowlines.*
14 *Fore Topſail Bowlines.*
15 *Fore Topgallant Bowlines.*
16 *Fore Clewgarnets.*
17 *Fore Topſail Clewlines.*
18 *Fore Shrowds.*
19 *Fore Topmaſt Shrowds.*
20 *Fore Topgallant Shrowds.*
21 *Fore Stay.*
22 *Fore Topmaſt Stay.*
23 *Fore Topgallant Stay.*
24 *Horſe for the fore Topſail Yard.*
25 *Fore Topgallant Clewlines.*
26 *Fore Topmaſt Backſtaies.*
27 *Fore Chains and Plaits.*
28 *Fore Topſail Leathlines.*
29 *Fore Topſail Buntlines.*
30 *Fore Leathlines.*
31 *Fore Buntlines.*
32 *The Fore Topſails Tye and Jeer.*
33 *The Yard of the fore Topſail.*
34 *The Fore Horſe and Tackle.*
35 *The foretop Gallantſails Tye and Jeer.*
36 *The Fore Tacks.*

The Bowſprit, and Rigging.

1 *Bowſprit.*
2 *Spritſail Yard and Sail.*
3 *Spritſail Top and Croſs-trees*
4 *Spritſail Topſail Yard and Sail.*
5 *Spritſail Topmaſt.*
6 *Spritſail Sheets.*
7 *Standing Lifts for the Spirt ſail yard.*
8 *Spritſail Braces.*
9 *Spritſail Sheets.*
10 *Spritſail Clewlines.*
11 *Spritſail Topſail Braces.*
12 *Spritſail Topſail Lifts.*
13 *Spritſail Topmaſt Shrowds.*
14 *Jack Staff and Jack.*
15 *The Ships Head.*
16 *The Cat Head.*
17 *The Harſſes.*
18 *The Fore Caſtle.*
19 *The Crain Line.*
20 *The Horſe on the Bowſprit.*
21 *Steps on the ſide.*
22 *The main Spritſail.*
23 *The main Spritſail Tye and Jeers.*
24 *The Spritſail Topſail Clewlines.*
25 *The Spritſail Topſail Jeers.*
26 *The Spritſail Topſail Crowfoot.*
27 *The Trents Bowſprit.*

Mathematical Recreations, done from *Ozanam*, *Oughtred*, and all other the beſt Authors; with Explanations thereon, and many Cuts to illuſtrate it. To which are added, ſeveral curious Tracts; As the Laws of Chance, the Art of making Rivers Navigable, and other Water-Works, done from the *French*.

PROPOSALS for printing a compleat Hiſtory of *England*, from the Earlieſt Account of Time, to the Death of his late Majeſty King *William* III. Containing a faithful Relation of all Affairs of State, Eccleſiaſtical and Civil, with the Effigies of all the Kings and Queens taken from the Originals, and curiouſly Engraven by the beſt Maſters; wherein you have the following Authors, *viz.* *Milton's* Hiſtory of *England* to the Conqueſt; from the Conqueſt to the end of the Reign of *Edward* III. by Mr. *Daniel*; —The Reigns of *Richard* II. *Henry* IV, V, and VI. are new writ in Mr. *Daniel's* Method;—*Edward* IV. by Mr. *Habington*; —*Edward* V. and *Richard* III. by Sir *Thomas Moor*, Mr. *Buck*, &c. *Henry* VII. by Sir *Francis Bancan*, Sir *James Weare*, &c. *Henry* VIII. by the Lord *Herbert*; Queen *Elizabeth*, by Mr. *Camden*; King *James*, by *Wilſon*; and the Continuation to this time, writ by a Learned and Impartial Hand, in 3 Vol. *Fol.* containing about 600 Sheets. Price to Subſcribers 3 Guineas, a ſeventh Gratis.

There will be ſhortly Publiſh'd,

ALL the Critical Works of Monſieur *Rapin*, containing his Compariſons of *Plato* and *Ariſtotle*, of *Homer* and *Virgil*, *Demoſthenes* and *Cicero*, *Thucydides* and *Livy*, &c. —His Reflections on *Ariſtotle*'s Treatiſe of Poetry. —On Eloquence, particularly the Barr and Pulpit. —On Moral Philoſophy, &c. In 2 Vol. in 8*vo*. Tranſlated into *Engliſh* by Mr. *Baſil Kennet*, and others.

A new Three Years Voyage to *China*: Or Travels Over-land from *Muſcow*, thro' great *Uſtiga*, *Sirania*, *Permia*, *Sibiria*, *Daour*, and *Great Tartary*, to *China*. Containing an exact and particular Deſcription of the Extent and Limits of theſe Countries, and the Cuſtoms of the barbarous Inhabitants; as their Religion, Government, Marriages, daily Employments, Habits, Habitations, Diet, Death and Funerals. By his Excellency *E. Yſbrantz Ides*, Ambaſſadour from the Czar of *Muſcovy* to the Emperour of *China*. Illuſtrated with a Map of the Countries, drawn by the Ambaſſadour upon his Journey, and about 30 very fine Cuts. To which is annex'd, An accurate Deſcription of *China*, done originally by a *Chineſe* Author; with ſeveral Remarks, by way of Commentary, ſhewing the Falſhood of what our *European* Authors have writ of *China*. Printed in *Dutch*, by the Direction of *Burgomaſter Witſen*, and now faithfully done into *Engliſh*, in 4*to*.

Ma-

THE
Gentleman's Dictionary.

PART III.

CONTAINING

The Art of NAVIGATION: With the *Explanation* of all the *Terms* relating to *Naval-Affairs*; as in the *Building*, *Rigging*, *Working*, and *Fighting* of SHIPS: Of *Officers*, and their *Duty*, &c.

The Art of Navigation *Teaches how to Conduct or Guide a Ship from one Place to another, by the help of* Sea-Charts, Magnetic-Compass, Sounding-Line, Log-Line, *and Due Observations of the Height of the* Sun *and* Stars; *As also, a thorow-Knowlege in the Working of a Ship, upon all Occasions whatsoever: And therefore the* Art of Navigation *has Two Parts,* (viz.) *The* Piloting, *and the* Working-part *of* Navigation.

The Names of each Part, and Material belonging to a Ship compleatly Rigg'd, are here rank'd in an Alphabetical Order, and fully Described, with their several Uses; as also the principal Sea-Terms, and Phrases in the Working of a Ship, in all Circumstances, amply Explain'd; with variety of Instances, shewing when properly to apply such Expressions: And all the Terms of Art, as well in the *Piloting*, as in the *Working-part* of *Navigation*, are here largely insisted on.

ABAFT, or *Aft*, is that part of the Ship towards the *Stern*, or the Hind-part of the Ship.

The Mast hangs Aft; that is, The *Mast* hangs towards the *Stern*.

How Chear ye Fore and Aft? that is, How fares all your Ship's Company?

ABASE, signifies to Lower or Take in; as, *to Abase the Flag*, is, to Take in the *Flag*.

ADMIRAL, is the Chief of a Fleet, or He that has the Com-

mand of a whole Fleet, or is Intrusted with the Naval Affairs of a Kingdom. See *Flags*, and *Officers*.

AFTWARD, the Hinder-part of the Ship.

ALOOF, or, *Keep your Loof*, is a word of Command from him that Cons the Ship, to the Man at Helm, when the Ship Sails upon a Wind, or on a Quarter-Wind, directing him to keep the Ship near the Wind. See *Cond.*

ALTITUDE *of the Sun*, or *Star*. See *Height*.

Meridian Altitude. See *Meridian*.

AMAIN, is a word sometimes used by a Ship to his Enemy, by way of Defiance, or Commanding him to Strike his *Top-Sails*, that is, to Yield.

Strike Amain; that is, Lower your *Top-Sails*.

AMPLITUDE *of the Sun*, is an Arc of the *Horizon*, intercepted between the *Prime Vertical* and the *Sun*, in his *Rising* or *Setting*.

Magnetical Amplitude, is an Arc of the *Horizon*, intercepted between the *Sun* in his Rising or Setting, and the *East* or *West* Point of the *Compass*.

ANCHOR, is an Instrument to stop Ships on the Sea or River: It consists of a *Shank* or *Beam*, with *Flooks* at one end, which fasten to to the Ground, their length being one third of the *Shank*; and at the other end is fasten'd the *Stock*, serving to guide the *Flook*, that it may fix in the Ground; as also a *Ring*, to which the *Cable* is fasten'd. There are several sorts of *Anchors*, differing only in Weight, which is according to the Burden of the Ship.

Sheet-Anchor, is the greatest of all that belong to a Ship, and is never used but in great Necessity, as being their utmost Refuge, when forced to Ride on a Lee-shore.

Bow-Anchors, or *First* and *Second Bower*, or *Best* and *Small Bower*, are such as the Ship, in Fair Weather, may Ride by.

Kedge-Anchor, or *Kedger*, used in Calm Weather, in a low Stream; or to *Kedge* up and down a narrow River, lest the Wind or Tide should drive the Ship ashore.

Grapples, or *Grapplings*, are the smallest of all the *Anchors*; they have four *Flooks*, but no *Stock*: Their Use is, for a Boat to ride by, or to throw into an Enemy's Ship in a Close Engagement; to catch hold of the *Gratings*, *Rails*, &c. in order for *Boarding* her.

To Ride at Anchor, is said of a Ship which is held so fast by her *Anchors*, as not to be driven with either Wind or Tide. The best *Riding at Anchor*, is when the Ship is *Land-lock'd*, and out of the Tide.

To Cast, or *Drop Anchor*, is to let it fall, to stop the Course of the Ship.

To Weigh Anchor, is to get it up, in order to Set Sail.

To Shooe an Anchor, is to put Boards fitted for that purpose on the *Flooks*, that the *Anchor* may the better hold in Soft Ground.

The Anchor is Foul; that is, the *Cable* is hitch'd about the *Flook*, by

by the turning about of the Ship.

The Anchor is come Home, is, when it cannot hold the Ship, but that ſhe drives away by the violence of the Wind or Tide.

The Anchor is a Peek; that is, the *Anchor* is right under the *Hawze* (or *Hole*) through which the *Cable* belonging to that *Anchor* runs out.

The Anchor is a Cock-bell; that is, the *Anchor* hangs up and down by the Side of the Ship.

Pudding of the Anchor. See *Pudding*.

ANCHORAGE, or *Anchoring*, a Ground fit to hold a Ship's *Anchor*, that ſhe may ride it out ſafely. The beſt *Anchoring* Ground is ſtiff Clays, or hard Sand.

ARMED: A *Croſs-Bar-Shot* is ſaid to be *Armed*, when ſome Rope-yarn, or the like, is rolled round about the End of the Iron-Bar which runneth through the Shot. This is done, that the Shot may be the better Ramm'd into the Gun; as alſo, leſt its Ends ſhould catch into the *Honey-combs* in the Piece.

ARMINGS, or *Waſte-Cloths*, are a ſort of Red Cloths hung about the Out-ſides of the Ship's Upper Works, *fore* and *aft*, and before the *Cubbridge-heads*: Its Uſe is for Show, and to Grace the Ship; as alſo, to cover the Men, in time of an Engagement, from being ſeen by the Enemy.

Top-Armings, thoſe about the *Round-Tops*, for the ſame Purpoſe.

ARM-MAST. See *Maſt*.

AVAST, ſignifies to Stay, Hold, or Stop.

AWNING, is a Sail, or the like, hung over any part of the Ship above the Decks, to keep off the Sun's ſcorching Heat, in Hot Climates; and ſometimes, as a defence from Rain, or Wind. In a Boat, an *Awning* is made, by bringing the *Sail* over the *Yard* and *Stay*, and booming it out with the *Boat-hook*.

AZIMUTH *of the Sun*, is an Arc of the *Horizon*, intercepted between the *Meridian*, and the *Vertical Circle* which paſſeth the Centre of the *Sun*.

Magnetical Azimuth, is an Arc of the *Horizon*, intercepted between the *Magnetical Meridian* and the Sun's *Azimuth Circle*: 'Tis found by an *Azimuth Compaſs*, by obſerving the *Sun*, when he is about 10 or 15 Degrees above the *Horizon*.

AZIMUTH-*COMPASS*. See *Compaſs*.

B

BACK-STAFF, is an Inſtrument of excellent Uſe in taking the Sun's Altitude at Sea, Invented by one Captain *Davis* a *Welſhman*, and is call'd, by ſome, *Davis*'s *Quadrant*, or the *Sea-Quadrant*; but by the *French*, 'tis call'd, the *Engliſh Quadrant*.

BACK-STAYS. See *Stays*.

BAIL, or to *Baile*, is to throw the Water, by Hand, out of the Boat.

BAILS, are the Hoops that bear up the Title of a *Boat*.

BALLAST *of a Ship*, is either Gravel, Stones, *&c.* ſtow'd in the Hold, to keep her ſtiff in the Sea, that ſhe may bear the more Sail.

Trench the Ballaſt; that is, (in ſeeking a Leak) to part the *Ballaſt*.

The Ballaſt Shoots; that is, it runs over from one ſide to the other.

BARE-PUMP. See *Pump*.

BARGES, are Boats of State and Pleaſure, which *Men of War* have to carry Generals, Admirals, and Chief-Commanders; they are generally finely built, and adorn'd with various Ornaments, having Bales and Tilts, and Seats furniſh'd with Cuſhions and Carpets, *&c.* and Benches for many Oars: Theſe are alſo uſed in Navigable Rivers that lead to Rich Cities. Alſo, thoſe great Flat-bottom'd Veſſels, employ'd in Navigable Rivers, for carrying of Goods, *&c.* are call'd *Barges*.

BARQUE, or *Bark*, is a Veſſel with Three Maſts, (*viz.*) Main-Maſt, Fore-Maſt, and Miſſon-Maſt: They carry about 200 Tuns.

BARQUE-*Longue*, is a ſmall, low, and ſharp-built, but very long Veſſel, without a Deck; they go with Sails and Oars, and are common in *Spain*.

BARRS, *Capſtan-Barrs*. See *Capſtan*.

BEAK, or *Beak-Head of a Ship*, is that Part without the Ship, before the *Fore-Caſtle*, which is faſten'd to the *Stem*, ſupported by the *Main-Knee*: 'Tis commonly Carv'd and Painted, and is a great Ornament to the Ship, beſides other neceſſary Uſes.

BEAKONS, or *Beacons*, are Fires kept on the Sea-Coaſts, to prevent Shipwracks, and to give Notice of Invaſions, *&c.*

BEAMS *of a Ship*, are the great Main Croſs-Timbers which keep the Sides of the Ship aſunder, and which ſupport the *Decks* and *Orlops*: They are reckon'd by *Firſt*, *Second*, and *Third Beam*, from the

Main-Beam, or that which is next the *Main-Maſt*.

Midſhip-Beam, is the Greateſt *Beam* of all.

BEAR, is a word uſed in theſe ſeveral ſenſes:

Bear up; that is, Let the Ship ſail more before the Wind.

Bear up round; that is, Put her right before the Wind.

Bear in with the Land, or *Harbour*; that is, when ſhe ſails towards the Shore, with the Wind Large, or before the Wind.

The Ship Bears off; that is ſaid of a Ship that keeps off from the Land.

The Ship Bears a good Sail; that is, (having all her Sails abroad, in a Gale of Wind) ſails upright in the Water.

To

To bring the Guns to Bear; that is, to lie right with the Mark.

BEARING, the Point of the *Compass*, that one Place *Bears* from another.

BECALM'D, is said of a Ship that has not a Breath of Wind: And 'tis usual to say, *Our Ships lay so close together, that we Becalm'd one another*; that is, The *Wind* had no Power over us. See *Rain*.

BED *of the Carriage of a Gun*, is the Plank which lies under the Piece.

To BELAY, is to make fast the Ropes in their proper Places.

BENDS, *Wails*, or *Wales*, the Outmost Timbers of a Ship's-side, on which Men set their Feet, in climbing up: They are reckon'd from the Water, and are call'd the *First*, *Second*, or *Third Bend*, or *Wail*: They are the chief strength of the Ship's-Sides, and have the *Beams*, *Knees* and *Foot-hooks* bolted to them.

To Bend the Cable; that is, to make it fast to the Ring of the Anchor.

To Unbend the Cable; that is, to loosen it from the Ring of the Anchor, as it is often done, when the Ship is long at Sea.

To Bend the Main-Sail, is to make it fast to the Yard.

BENDING, is Tying two Ropes, or Cables, together.

BENEAP'D: *A Ship is Beneap'd*, that is, when the Water does not flow high enough to bring the Ship off the Ground, out of the Dock, or over the Barr.

BIGHT; any part of a Rope between the Ends.

Holding by the Bight, is to hold by that part of the Rope which is *Quoyl'd*, or Roll'd up.

BILBOWS, a Sea-Punishment, answering to *Stocks* at Land. See *Duckings*.

BILGE *of a Ship*, is the bottom of her Floor, or the breadth of the Place the Ship rests on, when she is a ground: Therefore,

Bilge-water, is that which lies on her Floor, and cannot go to the Well of the Pump.

Bilge Pumps, or *Burr-Pumps*, are those which carry off the *Bilge-Water*.

The Ship is Bilged; that is, she has some of her Timber struck off, on a Rock or Anchor, and Springs a Leak.

BILLOWS, or *Surges* of the Sea; that is, the Waves raised by the Wind.

BIRTH, is a convenient Place to *Moor* a Ship in; as also a due Distance observed between Ships lying at Anchor, or under Sail. And a proper Place aboard for a *Mess* to put their Chests, *&c.* is call'd the *Birth* of that Mess.

BITTACLE, is a Square Box standing before him that Steers the Ship, with the *Compass* placed therein, to keep, and direct the Ship in her Course.

BITTER, is a Turn of the Cable about the *Bits*.

She is brought up to a Bitter, is said of a Ship that is to stop by a *Cable*.

BITTER's-*End*, is that part of the Cable which stays within Board.

BITS, *in a Ship*, are two great Pieces of Timber (usually placed abaft the Manger in the Ship's Loof) thorow which the *Cross-piece* goes; their lower-parts are fasten'd to the *Riders*, and their middle parts (in great Ships) are bolted to two great Beams cross the *Bows*: Their Use is to Belay the Cable thereto, when the Ship rides at Anchor. In great Storms, to strengthen the *Bits*, and secure the *Bows*, the Cable is fasten'd to the *Main-Mast*.

Fore-jeer-Bits, to which the *Fore-jeer* is fasten'd and Belay'd.

Fore-top-sail-sheet-Bits, to which the *Fore-top-sail-sheet* is belay'd.

BLOCKS, (at Sea, is the usual Name for what we call *Pulleys* at Land,) are thick Pieces of Wood, some with three, four or five Shivers in them, thorow which all the *running Ropes* runs.

Blocks, whether *Single* or *Double*, are distinguish'd and call'd by the Names of the Ropes they carry, and the Uses they serve for.

Double Blocks, are used when there is occasion for much strength; because they'll Purchase with more ease than *Single Blocks*, tho' much slower.

Block and Block, a Phrase signifying, that two *Blocks* meet, in Haling any *Tackle*, or *Halliard*, having such Blocks belonging to them.

Fish-Block, is hung in a Notch at the End of the David; it serves to hale up the Flooks of the Anchor to the Ship's Bow.

Snatch-Block, is a great Block with a Shiver in it, and a Notch cut through one of its Cheeks, for the more ready receiving in of any Rope; since by this Notch the middle part of a Rope may be reev'd into the Block, without passing it end-wise. 'Tis commonly fasten'd with a Strap about the Main-Mast, close to the Upper-Deck; and is chiefly used for the Fall of the Winding-Tackle, which is reev'd into this Block, and then brought to the Capstan.

BLUFF, or *Bluff-headed*, is said of a Ship with an upright Stern.

BOARD, is a word variously used at Sea: As,

To go Aboard, signifies to go into the Ship.

To heave Over-board, is to throw a thing out the Ship, into the Sea.

To slip by the Board, is to slip down by the Ship's side.

Board and Board, is when two Ships come so near as to touch one another, or when they lie side by side.

To make a Board, is to turn to Windward: And the longer your *Boards* are, the more you work into the Wind.

To Board it up, is to beat it up, some-

ſometimes upon one Tack, and ſometimes upon another, or bolt it to and again, in ſailing againſt the Wind.

She makes a good Board; that is, the Ship advances much at one Tack.

The Weather-Board, is that ſide of the Ship which is to Windward.

To Board a Ship, is to Enter an Enemy's Ship in a Fight. In *Boarding a Ship*, 'tis beſt to Bear directly up with him, and to cauſe all your Ports to Leeward to be beat open, and bring as many Guns from your Weather-ſide thither, as you have Ports for, and then Lay the Enemy's Ship on Board, Loof for Loof; and order your Tops and Yards to be Mann'd, and furniſh'd with Neceſſaries; & let all your Small-ſhot be in a readineſs; then Charge at once with both Small and Great, and at the ſame time Enter your Men in the Smoak, either on the Bow of your Enemies Ship, or bring your Midſhip cloſe up with her Quarter, and ſo enter your Men by the Shrouds. Or if you would uſe your Ordnance, 'tis beſt to Board your Enemy's Ship *a-thwart her Hawſe*; for then you may uſe moſt of your Great-Guns, and ſhe only them of the *Prow*. Let ſome of your Men endeavour to cut down the Enemy's Yards, and Tackle, whilſt others clear the Decks, and beat the Enemy from aloft: Then let the Scuttles, and Hatches be broke open with all poſſible ſpeed, to avoid Trains, and the Danger of being Blown-up, by Barrels of Powder placed under the Decks. Thus your Men being in Poſſeſſion of the Sails and Helm, and the Enemy every way ſtow'd below the Decks, the Ship is taken, and all lies at your Diſcretion.

LOGBOARD. See *Log*.

BOATS, of which there are ſeveral ſorts belonging to a Ship; whoſe Uſes you have under their Names.

Long-Boat, is the largeſt and ſtrongeſt Boat belonging to a Ship, that can be hois'd aboard of her: It hath a Maſt, Sail, and Oars, as other Boats, as alſo a Tiller to the Rudder, which anſwers to the Helm of a Ship: It's Uſe is, to Weigh the Anchor, bring Goods, Proviſion, *&c.* to, or from the Ship; and other Services, as Occaſion requires.

To Trim the Boat, is to keep her Even.

To Wind the Boat; is, to bring her Head the other way.

BOAT-HOOK. See *Hook*.

BOATROPE. See *Rope*.

BOATSWAIN, is a Ship-Officer, to whom is committed the Charge of all the Tackling, Sails, and Rigging, Ropes, Cables, Anchors, Flags, Pendants, *&c.* He alſo calls out the ſeveral Gangs and Companies aboard, to the due Execution of their Watches, Works, and Spells, *&c.*

BOATSWAIN's-MATE, has the peculiar Command of the

Long-Boat, for the ſetting forth of Anchors, Weighing or fetching Home an Anchor, Warping, Towing, or Mooring ; and to give an Account of his Store.

BOLD-BOW. See *Bow*.

BOLD-HAWSE. See *Hawſe*.

BOLTS, in a Ship, are Iron-Pins, of which there are ſeveral ſorts, according to their different Make, and Uſes.

Drive-Bolts, are long Pieces of Iron, ſerving to drive out other Bolts, Tree nails, or any ſuch thing.

Rag-Bolts, are ſuch as have Jags or Barbs on each ſide, to keep them flying out of the Hole, wherein they are driven.

Clench-Bolts, are thoſe that are Clench'd with a Rivetting-Hammer, at the end where they come through.

Forelock-Bolts, are thoſe, which have at the end a Forelock of Iron driven in, to keep it from ſtarting back.

Set-Bolts, are thoſe uſed for forcing of the Planks, and the other Works, and bringing them cloſe together.

Fend, or *Fender-Bolts*, are made with long and thick heads, and ſtruck into the uttermoſt Bends or Wales of the Ship, to ſave her Sides from Bruiſes, Gallings, eud Hurts.

Ring-Bolts, are thoſe uſed for *bringing to* of the Planks, and thoſe Parts whereto are faſten'd the Breeches and Tackles of the Ordnance.

BOLT-ROPE. See *Rope*.

BOLTSPRIT. See *Bow-Sprit*

BONAVENTURE-*Miſen*. See *Maſts*.

BONNETS, are ſmall Sails to put to the Main or Fore-Courſe, when they are too narrow or ſhallow to cloath the Maſt.

Lace on the Bonnet; that is, faſten it to the Courſe.

Shake off the Bonnet; that is, Take it off the Courſe.

BOOM, is a long Piece of Timber, with which the Clew of the Studding-Sail is ſpread out ; and ſometimes the *Boom* is uſed to ſpread or Boom out the Clew of the Main-Maſt. Alſo, a Cable ſtretch'd a-thwart the Mouth of a River or Harbour, with Yards, Top-maſts, Battling or Spars of Wood, *&c.* laſh'd to it, to prevent an Enemy's entring in, is call'd a *Boom*: And ſuch Monſieur *Chateau-Renault* had with Diligence and Art prepared at *Vigo* in *Gallicia*, for the Defence of the Plate-Fleet lying there : But how ſtrong ſoever, 'twas Forc'd by Sir *Thomas Hobſon*, who, by that means, made way for the reſt of the *Engliſh* Ships to come in, to partake of the Taking, or Deſtroying all that Fleet, in 1702.

A Ship comes Booming; that is, ſhe makes all the Sail ſhe can.

BORES-*Tackles*. See *Tackles*.

BOW of a Ship, is that Part of her, which begins at the Loof and

and compassing Ends of the Stem, and ends at the Sternmost-part of the Fore-castle.

Bold-Bow, that is, a Broad *Bow*.

Lean-Bow, that is, a narrow thin *Bow*.

Bow-Pieces, are the Pieces of Ordnance at the *Bow*.

BOWER, *Small*, or *Best*. See *Anchor*.

BOW-LINE, is a Rope made fast to the Leech or Middle-part of the out-side of the Sail; 'tis fastened by two, three, or four Ropes, like a Crow's-Foot, to as many parts of the Sail; only the Misen-Bow-line is fastened to the lower-end of the Yard: This Rope belongs to all Sails except the Sprit-Sail, and Sprit-Top-sail; therefore those Sails cannot be used close by a Wind, for want of room to hale the Bow-Line forward by; since the Use of the Bow-Line, is to make the Sails stand sharp, or close, or by a Wind.

Sharp the Bow-Line; that is, *Hale it taught*; or, Pull it hard.

Hale up the Bow-Line; that is, Pull it harder forward on.

Check, or *Ease*, or *Run up the Bow-Line*; that is, Let it be more Slack.

BOW-LINE-BRIDLES, are the Ropes by which the Bow-Line is fastened to the Leech of the Sail.

BOW-LINE-KNOT, is a Knot that will not slip, by which the Bow-Line-Bridles is fastened to the Crengles.

BOWSE, signifies as much as to Hale, or Pull.

Bowsing upon the Tack; that is, Haleing upon a Tack.

Bowse away! that is, Pull away all together!

BOW-SPRIT, or *Bolt-Sprit*, is a kind of Mast resting slopewise on the head of the Main-Stem, and having its lower-end fasten'd to the Partners of the Fore-mast, and farther supported by the Fore-stay: It carries the Sprit-sail, Sprit-top-sail, and Jack-staff: And its Length is usually the same with the Fore-mast.

BOW-SPRIT-LADDER. See *Ladder*.

BRACES, are Ropes belonging to all the Yards of a Ship, except the Misen, two to each Yard, reev'd thorow Blocks that are fasten'd to Pennants seiz'd to the Yard-arms: Their Use is either to Square, or to Traverse the Yards: And all these *Braces* come aftward on, as the Main-Brace comes to the Poop; the Main-top-sail Brace comes to the Misen-top, and and thence to the Main-shrowds: The Fore and Fore-top-sail-Braces come down by the Main and Main-top-sail-stays, and so of the rest. But the Misen bow-line serves to *Brace to* the Yard; and the Cross-jack-Braces are brought forwards to the Main-shrowds, when a Ship sails close by a Wind.

BRACKETS, are small Knees serving to support the Galleries;

Galleries; and are commonly Carved: Alſo, the Timbers that ſupport the Gratings in the Head, are call'd *Brackets*.

BRAILES, are ſmall Ropes made uſe of to Furl the Sails a-croſs; they belong only to the two Courſes, and the Miſen-ſail: They are reev'd through Blocks ſeiz'd on each ſide the Ties, and comes down before the Sail, being at the very skirt thereof faſten'd to the Crengles.

Hale up the Brails, or, *Brail up the Sail*; that is, Hale up the Sail, in order to be Furl'd, or bound cloſe to the Yard.

BRAKE, is the Handle of a Pump.

BREAD-ROOM See *Room*.

BREAMING *of a Ship*. See *Brooming*.

BREAST-CASKETS. See *Caskets*.

BREAST-FAST, is a Rope faſten'd to ſome part of a Ship forward on, to keep her Head faſt to a Wharf, or the like.

BREAST-HOOKS, are the Compaſſing Timbers before, in a Ship, which help to ſtrengthen her Stem, and all her Fore-part.

BREAST-ROPE. See *Rope*.

BREECHINGS, are thoſe Ropes with which the Great-Guns are laſh'd, or faſten'd to the Ship's-ſide.

BREEZE, a ſhifting Wind that blows from Sea, or Land, for ſome certain Hours of the Day, or Night; common in *Africa*, and ſome Parts of the *Eaſt*, and *Weſt-Indies*.

BRIGANTINE, is a ſmall light Veſſel, which can both Row and Sail well, and is either for Fighting, or giving Chaſe.

BROOMING *of a Ship*, is to waſh and burn off, all the Filth that ſhe has contracted on her Sides, with Weeds, Straw, Broom, or the like, when ſhe is on the Careen, or on the Ground.

BUCKET-ROPE. See *Rope*.

BULK *of a Ship*, is her whole Content in the Hold, for the Stowage of Goods.

To Break the Bulk; that is, to open the Hold, and take out Goods thence.

BULK-HEADS, are Partitions made a-thwart the Ship with Boards, whereby one part is divided from the other, as the Great-Cabbin, Gun-Room, Bread-Room, and ſeveral other Diviſions. The *Bulk-Head* afore, is the Partition between the Fore-caſtle and Gratings in the Head.

BUNT *of a Sail*, is the Middle-part thereof, which is deſignedly form'd into a Bag or Cavity, that the Sail may receive the more Wind. It is uſed moſtly in Top-ſails; becauſe Courſes are generally cut ſquare, or with but ſmall allowance for *Bunt* or Compaſs.

The Bunt holds much Leeward Wind; that is, The Bunt hangs too much to Leeward.

BUNT-

BUNT-LINES, are ſmall Lines made faſt to the bottom of the Sails, in the middle-part of the Bolt-rope to a Crengle, and ſo are reev'd through a ſmall Block ſeiz'd to the Yard: Their Uſe is, to trice or draw up the Bunt of the Sail, for the better furling of it up.

BUOY, is a ſhort piece of Wood, or cloſe-hoop'd Barrel, faſten'd ſo as to float directly over the Anchor; and by that means, 'tis always known where about it lies.

Stream the Buoy; that is, Let the Anchor fall, while the Ship has way.

To Buoy up a Cable, is to faſten ſome Pieces of Wood, Barrels, *&c.* to the Cable, near the Anchor; that the Cable may not touch the Ground, in caſe it be foul or Rocky, leſt it ſhould be fretted, and cut off.

BUOYANT, ſignifies any thing that is floating, or floatable.

CAN-BUOYS, are of a larger ſize; and uſed to diſcover dangerous Rocks and Shelves, by being faſten'd over them.

BUOY-ROPE. See *Rope*.

BURTON, is a ſmall Tackle, conſiſting of two ſingle Blocks, and may be made faſt any where, at pleaſure, for hoiſting of ſmall things in or out; and will purchaſe more than a ſingle Tackle with two Blocks.

BUT-ENDS, are the Fore-ends of all Planks under water, as they riſe, and are join'd one End to another; which, in great Ships are moſt carefully bolted; for if any one of theſe Ends ſhould ſpring, or give way, the Leak would be very dangerous, and difficult to ſtop.

BUTTOCK *of a Ship*, is that part of her, which is her Breadth right a-ſtern, from the Tack upwards: And a Ship is ſaid to have a Broad or Narrow *Buttock*, according as ſhe is built Broad or Narrow at the Tranſom.

C

CABINS, are the little Rooms, or Appartments, wherein the Officers lie, ſuch as are on the Quarter-deck, and on each ſide of the Steerage, *&c.*

Great-Cabin, is the chief of all, and that which properly belongs to the Captain, or Chief-Commander.

CABLE *of a Ship*, is a great Rope faſten'd to the Anchors, and proportional to them; ſerving to keep the Ship faſt, whilſt ſhe rides at Anchor.

Sheet-Anchor-Cable, is the greateſt of all, belonging to a Ship.

Serve, or *Plat the Cable*; that is, Bind it about with Ropes, Clouts, *&c.* to keep it from galling in the Hawſe.

To Splice a Cable, is to make two Pieces faſt together, by working the ſeveral Threads of the Rope, the one into the other.

Pay

Pay more Cable; that is, Let it more out from the Ship, that the Boat which carries the Anchor may the more easily drop it into the Sea.

Pay cheap the Cable; that is, Put, or, Hand it out a-pace.

Veer more Cable; that is, Let more of it run out.

The Cable is well-Laid; that is, well-Wrought, or well-Made.

A Shot of Cable; when the Cable is double in length, to make the Ship ride with more ease.

To Quoil a Cable, is to roll it up in a Ring, or sack one above another.

Cable-Tire, is where these several Rolls of Cables are quoil'd.

To Bend, or *Unbend the Cable*. See *Bend*.

Pointing the Cable. See *Pointing*.

CABURNS, are small Lines made of spun Yarn, to bind Cables, seize Tackles, or the like.

CALKING, the same with *Caulking*; which see.

CALM, is when there is no Wind stirring. See *Rains*.

Calm Sea; that is, when the Sea appears very smooth.

CAMBER'D-*Deck*. See *Deck*.

CAN, *a Pump's-Can*, is a great *Can* wherewith Seamen pour Water into Pumps, to make them Pump.

CAN-BUOYS. See *Buoys*.

CAN-HOOK. See *Hook*.

CANOW, is a little Boat, chiefly for the Service of great Ships: The *Indians* make these *Canows* of the Trunk of a great Tree, hollow'd or scoop'd; or of the Barks of Trees sew'd together.

CANTING-COINS. See *Coins*.

CAP, *in a Ship*, is a square Piece of Timber put over the Head, or upper-end of any Mast, having a round Hole to receive the Mast. By means of these *Caps*, the Top-masts, and Top-gallant-masts are kept steady, and firm, in the Tressel-trees, where their Feet stand; as those of the lower Masts do, in the Steps; lest they should be *Born by the board*, in a stiff Gale: So that every Mast that has a Top, has a *Cap*.

CAP *of a Gun*, is a Piece of Lead, which is put over the Touch-hole of a Gun, to keep the Priming from being wasted, or spilt.

To CAP, is said of a Ship, in the Trials of the Running or Setting of Currents.

CAPE, is a High Land running out with a Point into the Sea; as *Cape St. Vincent*, *Cape de Verde*, *Cape of Good-Hope*, *Cape-Horn*, &c.

CAPSTAN, *Main-Capstan*, is a great Piece of Timber in the nature of a Windlass, placed next behind the Main-mast; its foot standing in a step on the Lower-deck, and its head between

tween the two Upper-decks; form'd into ſeveral ſquares, with holes in them: It's Uſe is, to weigh the Anchors, to hoiſe up, or ſtrike down Top-maſts, to heave any weighty thing, or to ſtrain any Rope that requireth a main force.

Jear-Capſtan, is placed between the Main-maſt and the Fore-maſt; and ſerves to ſtrain any Rope, heave upon the Jear-rope, or upon the Viol, or hold-off by, at the weighing of an Anchor.

Capſtan-Bars, are the Bars or Pieces of Wood, that are put into the Capſtan-holes, to heave up any thing of weight into the Ship, by the help of as many Men, as can well ſtand at them.

Man the Capſtan! that is, Have all the Hands neceſſary to heave at the Capſtan-Bars.

Spindle of the Capſtan, is the main Body thereof.

Whelps of a Capſtan, are ſhort Pieces of Wood made faſt to it, to keep the Cable from coming too nigh, in turning it about.

Pawl of a Capſtan, is a ſhort piece of Iron made faſt to the Deck, and reſting upon the Whelps, to keep the Capſtan from recoiling, which is of dangerous conſequence.

Pawling the Capſtan; that is, ſtopping it from turning, by means of the Pawl.

Come up Capſtan! or, *Launch out the Capſtan!* that is, Slack the Cable which you heave by.

CAPTAIN *of a Ship*, is he that Commands a Ship of War in Chief, in a Fight; And is a Charge, as great, as that of a Colonel at Land: Beſides his being Accountable for the Ship, if by his Miſ-conduct Loſt, or Taken. See *Officers*.

CARDINAL-*Winds*, or *Points*, are the North, Weſt, South, and Eaſt Points of the Compaſs.

CAREEN *a Ship*, is to bring her to lie down on one ſide, in order to Trim, and Caulk the other ſide. Or a Ship is ſaid to be brought to the *Careen*, when the moſt part of her Lading, &c. being taken out, ſhe is hall'd down on one ſide by a ſmall Veſſel, as low as neceſſary, and there kept by the weight of the Ballaſt, Ordnance, &c. as well as by Ropes, leſt her Maſts ſhould be ſtrain'd too much; and this is done, that her Sides or Bottom may be Trimm'd, Seams Caulk'd, or any thing that's faulty under Water, mended.

A Ship ſails on the Careen; that is ſaid of a Ship that lies much on one ſide, when ſhe ſails.

CARGO: By a *Ship's Cargo*, is meant, the Quantity of Goods that a Merchant-ſhip is laden with. See *Burden*.

CARLINGS, are Timbers in a Ship, lying Fore and Aft along from Beam to Beam, whereon the Ledges reſt,

on which the Planks of the Ship are faſten'd : And all the *Carlings* have their Ends let into the Beams Culvertail-wiſe.

CARLING-KNEES, are thoſe Timbers which go athwart the Ship, from her Sides to the Hatch-way, and which bear up the Deck on both ſides.

CARNELS; ſo ſome call thoſe Veſſels which go with Miſen-ſails inſtead of Main-ſails.

CARNEL-WORK: In building of Ships, firſt with their Timbers and Beams, and after bringing on their Planks, is call'd *Carnel-work*, to diſtinguiſh it from *Clinch-work*.

CARPENTER *of a Ship*, his Office, when at Sea, is to have all things ready that relate to his buſineſs; that is, of keeping the Ship in Repair, as Stopping of Leaks, Fiſhing the Maſts or Yards, Caulking, Breaming, and the like. He hath a *Mate* under him, and a *Crew*, or *Gang*, to Command on neceſſary Occaſions. See *Officers*.

CARRIAGE *of a Gun*, is the Frame of Timber on which a Piece of Ordnance is laid, fix'd, and mounted.

CARTOUCHE, the ſame with Cartridges.

CARTRIDGES, are Caſes of Paper, or, as now they are uſually made for Ships of War, (to prevent Danger from fire in a Gun not well ſpung'd) or Parchment, exactly fitted to the Bore of a Piece of Ordnance, and containing its due Charge of Powder. There are alſo *Tin-Cartridges*, in which the Paper or Parchment ones, are both form'd and carried.

CASE-SHOT, is when Musket-Bullets, Stones, old pieces of Iron, &c. are put up into Caſes, and ſo ſhot out of Great-Guns: They are generally uſed at Sea, to clear the Enemy's Decks, when they are full of Men.

CASKETS, are ſmall Ropes made of Sinnet, and faſten'd to *Gromets*, or little Rings upon the Yards: Their Uſe is, to make faſt the Sail to the Yard, when it is to be furl'd.

Breaſt-Caskets, are the longeſt or biggeſt of theſe, or thoſe in the midſt of the Yard, betwixt the Ties.

CAST *a Point of Traverſe*, ſignifies, to prick down on a Chart the Point of the Compaſs any Place bears from you; or to find what point of the Compaſs the Ship bears at any Inſtant, or what Way the Ship has made.

CATARACT, is a Precipice in the Chanel of a River, cauſed by Rocks, or other Obſtacles, ſtopping the Courſe of its Stream, from whence the Water falls with a great noiſe and impetuoſity; as the *Cataracts* of *Nile*, *Danube*, *Rhine*, &c.

CAT-HARPINGS, are ſmall Ropes, running in little Blocks from one ſide of the Shrowds to the other, near the Deck: Their Uſe is to force the Shrowds, and make them taught, for the more ſecurity and ſafety of the Maſts.

CAT, or *Cat-head*, is a ſhort Piece of Timber in a Ship, lying aloft right over the Hawſe, having at one end two Shivers, wherein is reev'd a Rope, with a great Iron-hook faſten'd to it call'd *Cat-Hook*: It's Uſe is to trice up the Anchor from the Hawſe to the top of the Fore-caſtle.

CAT-HOLES *in a Ship*, are over the Ports, as right with the Capſtan as they can be: Their Uſe is, to heave the Ship a-ſtern, upon occaſion, by a Cable, or a Hawſer, call'd *Stern-faſt*.

CAT-HOOKS. See *Cat*.

CAT-ROPE. See *Rope*.

CAULKING *of a Ship*, is driving Okum, or the like, into all the Seams of the Planks of the Ship, to prevent Leaking, or keep out the Water.

CAULKING-IRONS, are Iron-Chiſſels for that purpoſe.

CHAFE, or *Chafing of a Rope*, is ſaid of a Rope that is gall'd or fretted, or when a Rope rubs againſt any thing.

The Cable is Chafed in the Hawſe; that is, fretted, or begun to be worn out there.

CHAIN-PUMP. See *Pump*.

CHAINS *in a Ship*, are thoſe Irons to which the Shrowds of the Maſts are made faſt to the *Chain-Wails*.

CHAIN-SHOT, is two Bullets, or rather Half-Bullets with a Chain between them; they are uſed at Sea, to ſhoot down Yards or Maſts, to cut the Shrowds, or any Rigging of a Ship.

CHAIN-WAILS *in a Ship*, are the broad Timbers which are made jetting out of her Sides; to which, with Chains, the Shrouds are faſten'd, and by them ſpread out, the better to ſecure the Maſts.

CHANEL, is a narrow Paſſage of Sea between two Lands, joining one part of the Sea or Ocean to another part; as the *Britiſh Chanel* between *England* and *France*, and the like.

CHARGE: *A Ship of Charge*, is ſuch as draws much Water, or ſwims deep in the Sea: Though ſometimes an unweildy Ship, that will not ware nor ſteer, is call'd *a Ship of Charge*.

CHART, a *Nautical*, *Marine*, or *Sea-Chart*, is a Deſcription or Draught of any Place, projected on Paper or Parchment, for the benefit of Seamen; diſcovering the Sea-Coaſts, Sands, Rocks, Depth of Water, *&c.* as alſo the Latitude, Longitude, Diſtances and Bearings of Places.

Plain-Chart, is a Plat or Chart, uſed by Seamen, having the

Degrees

Degrees of Longitude thereon, made of equal Length with thoſe of Latitude. This *Chart* is very eaſie and uſeful in ſhort Voyages: But in long Voyages, except under the Equinoctial, or near a Meridian, 'tis impoſſible to ſail by them, any thing correctly: For few or no Places, but ſuch as lie under the ſame Meridian, or under the Equinoctial, can be expreſs'd therein, according to their true Diſtance and Situation one from the other. See *Mercator*.

Mercator's, or *Wright*'s *Chart*, is a Projection of the Surface of the Earth in *Plano*; wherein the Degrees upon the Meridian Encreaſe towards the Poles, in the ſame Proportion that the Parallel Circles Decreaſe towards them. This Projection was hinted by *Ptolemy*; and a General *Chart* accordingly was made, and publiſh'd by *Mercator*, and therefore have ever ſince bin call'd by his Name: But the Thing *Demonſtrated*, and a ready Way ſhew'd for Deſcribing it, was not, till Mr. *Wright* taught to Enlarge the Meridian-Line, by the continual Addition of Secants: So that all the Points of Latitude in each Parallel might be protracted in like Proportion with thoſe of Longitude; and therefore will truly ſhew, in any aſſign'd Courſe, the Ship's Motion on an Imaginary Plane, where the Parallels of Latitude are ſtrait Lines, and the Meridians alſo parallel to one another, by a Method that is eaſie, certain, and demonſtrable both in Longitude, Latitude, and Diſtance, and is the only true way of Sailing that is Practicable: And the moſt Curious and Correct *Chart* of this kind, is that done by the Excellent Mathematician Captain *Halley*; where you have not only the Places juſtly and accurately laid down, but alſo the Degrees of the Variation of the Magnetic-Needle, or Sea-Compaſs, and that by Inſpection only; as Deſign'd by his own Obſervations, in a Voyage purpoſely made to the Weſtern and Southern Ocean, at the Public Charge, in the Year of Our Lord, 1700. And theſe *Charts* are to be Sold by *R. Mount*, and *T. Page*, at the *Poſtern* on *Tower Hill*, *London*.

CHASE, ſignifies, the Ship Chaſed or Purſued.

To Chaſe, or *Giving Chaſe*, is to Purſue a Ship at Sea.

CHASER, is the Ship in Purſuit of the Chaſe.

Stern-Chaſe, is when the Chaſe is right a-head with the Chaſer.

To lie with a Ship's Fore-foot in a Chaſe, is to ſail the neareſt way to meet her, and ſo to croſs her in her way, or *to come a-croſs her Fore-foot*.

A Ship of a good Forward, or *Stern-Chaſe*, is a Ship that is ſo built forward on, or a-ſtern, that ſhe can carry many Guns, to ſhoot right forwards, or backwards.

In *Chaſing*, theſe Rules are generally obſerved: If the *Chaſed* be

be found any thing to the Windward, the *Chaser* is to bring all his *Tacks* aboard, and to shape his Course to meet her at the nearest Angle. If the *Chased* be to the Leeward, then the *Chaser* may come in with her; except she bear upright before the Wind, and so out-sail her; or that she bring her self close by a Wind, and the *Chaser* prove the more Leeward Ship, and so lose her that way. If the *Chase* be found right-a-head, and so the *Chaser* be put to a *stern-Chase*; then the best Sailer shall carry it, if there be Sea-room, and Day-light.

Being come up close with the *Chase*, endeavour to cross her Fore-foot; and, by that means, you'll both hinder her Way, and avoid the Fury of her Ordnance, (except those in her Chase;) and use your own, if required; to more Advantage; and that as well your Chase-pieces, at your first getting up within reach, as your Broad-side, and Quarter-pieces, as you pass *thwart her Hawse*; and scoure her Decks from Stem to Stern.

If she makes away from you, ply your Guns, as many as possible, at her Sails, Yards, Masts, and general Tackling: And being near, spare not your Case-shot, or Cross-bar-shot, to make the greater Damage.

CHASE-GUNS, are such, whose Ports are either in the Head, (and then they are used in Chasing others;) or in the Stern, (which are only useful, when they are Chased, or or Pursued by another Ship, or Ships.

CHASE *of a Gun*, is its whole Length.

CHECK *the Bow-line*. See *Bow-line*.

CHECQUE, *Clerk of the Cheque*. See *Clerk*, and *Officers*.

CHEEKS, are two Pieces of Timber, fitted on each side of the Mast, at the Top, serving to strengthen the Masts there. The uppermost Rail, or Piece of Timber, in the Beak of a Ship, is call'd the *Cheek*. The Knees, which fasten the Beak-head to the Bow of the Ship, are call'd *Cheeks*. And the Sides of any Block; or the Sides of a Ship's Carriage of a Gun, are call'd *Cheeks*.

Upper and Lower Cheeks, are those Pieces of Timber on each side of the *Trail-board*.

CHESSE-TREES, are two small Pieces of Timber, with a Hole in them, on each side of the Ship, a little before her Loof; their Use is, for the Main-Tack to run thorow, and to hale it down to.

CHEST-ROPES. See *Ropes*.

CHEVILS, the same with *Knevels*.

CHIRURGEON, the same with *Surgeon*; which see.

CLAMPS, *in a Ship*, are those thick Planks, which lie Fore and Aft, under the Beams

 of

of the First *Orlop*, or Second-Deck; bearing them up at either End; and are the same that the *Rising-Timbers* are to the *Deck*.

CLEAR *the Hawse*. See *Hawse*.

CLEAT, is a Piece of Timber, fasten'd on the Yard-Arm of a Ship, to keep the Ropes from slipping off the Yard.

CLERKS *of the Cheque*, are Officers belonging to the Principal *Dock-Yards*; they are employ'd in Mustering the Workmen in the Yards; the Ship's Company, within Reach; and *Chequing* them out of Wages, when Absent. See *Officers*.

CLERKS *of the Survey*. See *Officers*.

CLEW *of the Sail*, is the lower-corner of it, to which are made fast the Sheats and Tacks: A square Sail hath no *Clew*.

A Sail with a great Clew; that is, with a great goaring, or sloping down.

To Spread a great Clew; that is said of a Ship that has a very long Yard, and therefore has much Canvas in her Sail.

CLEW-GARNET, is a Rope made fast to the *Clew* of the Sail, and running from thence to a Block seiz'd to the middle of the Main, and Fore-Yard; which, in Furling, does hale up the *Clew* of the Sail close to the Middle of the *Yard*.

CLEW-LINE, is the same to the Top-sails, Top-gallant-sails, and Sprit-sails, that the *Clew-garnet* is to the Main-sail and Fore-sail, and has the same Use. In a Gust of Wind, when a Top-sail is to be taken in, 'tis usual first, to hale home the *Lee-Clew-Line*; then 'twill be easier to take in the Sail.

CLINCH-BOLTS. See *Bolts*.

CLINCH *of a Cable*, is that part of it which is bended about the Ring of the Anchor, and then seized, or made fast.

CLINCHING, is a kind of slight *Caulking*, used at Sea, in a prospect of Foul-weather, about the Ports; that is, to drive a little Okum into their Seams, to prevent the Water's coming in at them.

CLOATHED: A Mast is said to be *Cloathed*, when the Sail is so long, as to reach down to the Gratings of the Hatches, so that no Wind can blow below the Sail.

A Ship spreads much Cloth; that is, she has Broad Sails.

CLOSE-QUARTERS. See *Quarters*.

CLOYED: *The Touch-hole is Cloyed*; that is, something is got into the *Touch-hole*; so that with the *Priming-Iron*, way cannot be made, for the Powder to be put in, to *Prime* her.

COACH, is the *Council-Chamber* on-board a *Flag-Ship*.

COAMINGS, or *Coomings*, are those Planks, or that Frame, which lie upon the

Carling-

Carling-knees, and bear up the Hatches higher than the reſt of the Decks, to keep the Water from running down at the Hatches; and in which alſo Loop-holes, for Muskets to ſhoot out at, are uſually made; in order to Clear the Decks of the Enemy, when a Ship is Boarded.

COASTING, is that Part of *Navigation*, where the Places aſſign'd, are not far diſtant, ſo that a Ship may ſail in ſight of Land, or within Sonndings between them. In this, there is only required a good Knowlege of the *Land*, the Uſe of the *Compaſs*, and *Lead*, or *Sounding-Line*: Such are the Voyages in the *Britiſh* Seas, between *England*, *Holland*, and *France*, in the *Mediterranean*, and *Baltic-Seas*.

COATS, are Pieces of Tarr'd-Canvaſe which are put about the Maſts, at the *Partners*, to keep out Water: They are alſo put about the Pumps at the Decks, that no Water may go down there; and ſuch are alſo uſed at the *Rudder's-head*.

COCK-BOATS, are Small Boats uſed in Rivers, or near the Shoare: They are of no ſervice at Sea, becauſe too tender, weak, and ſmall.

COCKS, are little ſquare Pieces of Braſs, with Holes in them, put into Wooden-ſhivers, to keep them from ſplitting, and galling, by the Pin of the Block.

COCK-PIT, is a Place on the Lower Floor, or Deck, abaft the Main-Capſtan, lying between the Platform, or *Orlop*, and the Steward's-Room; where are Partitions, for the Purſer, the Surgeon, and his Mates.

COCKSWAIN, or *Coxſon*, is an Officer a-board a Man of War, who has the care of the *Barge*, or *Shallop*, and all things belonging to it; to be always ready with his Boat's-Gang, or Crew, and to Man the Boat, on all Occaſions: He ſits at the Stern of the Boat, and ſteers; and has a Whiſtle, to Call and Encourage his Men.

COILE, the ſame with *Quoil*; which ſee.

COINS, or *Quoins*, are Wedges of Wood, ſerving to raiſe, or lower the Breech of a Gun, upon Occaſion.

Canting-Coins, are little ſhort Pieces of Wood, or Billets, cut Wedge-like, to lie betwixt the Casks.

Standing-Coins, are Billets, or Pipe-ſtaves, to keep the Casks from ſtirring, or giving way.

COLLAR *of a Ship*, is a Rope faſten'd about her Beak-head, unto which the *Dead-man's-eye* is ſeiz'd, that holds her Main-ſtay. And that Rope which is wound about the Main-maſt-head, to ſave the Shrowds from galling, is called alſo a *Collar*.

COMB, in a *Ship*, is a little Piece of Timber, ſet under the lower-part of the Beak-head, near the middle: It has two

Holes in it; and supplies to the *Fore-tacks*, what the *Chest-trees* do to the *Main-tacks*; that is, to bring the *Fore-tacks* a-board.

COMPASS, in *Navigation*, is an Instrument to direct the Ship's Course by; being a Round Box, with a graduated Circle, drawn on a Card, or Pastboard, hanging Horizontally therein, by means of a Brass Centre, or Cap, on an Erect Pin: The *Card* is divided into Four Quarters, representing the Four *Cardinal-Winds*, or *Principal Points*, *East*, *West*, *North*, and *South*; and each Quarter subdivided into Eight other Equal Parts, making in all Two and thirty *Points*, or *Rhumbs*. And under the *Card*, from *North* to *South*, is placed a *Magnetic-Needle*, or Wire; whose Ends, *touch'd* by a *Loadstone*, always stand towards the *North*, and *South*; tho' in different Parts of the World, with different *Variations* from those *Points*. This Instrument is kept in the *Bittacle*; hanging so in Brass-Rings, as to give such Way to the Motion of the Ship, that the Box will stand Horizontally steady. And by Steering by this *Compass*, (well Made, and duly Rectified,) is known how, or which Way the Ship sails, at all Times; and how to keep in, and direct her to, her true Course.

Fly of the Compass, is the same with the *Card*, or the round piece of Pastbord which has the *Points* drawn thereon; and the *Magnetical-Needle*, or *Touch'd*-Wire, underneath.

Variation of the Compass, is the Deflection of the *Magnetical-Needle* from the true *Meridian*; or that *Arc* of the *Horizon*, (either *Eastward*, or *Westward*,) intercepted between the *True*, and *Magnetical-Meridian*: And is found, either by the *Sun's Amplitude*, or *Azimuth*; as is seen in most Books of *Navigation*. And this *Variation* is not always the same, in the same Place; but varies, in process of Time, from what it was.

Who it was that Discover'd this Useful Property of the *Loadstone*, is unknown; tho' tis conjectured, by very Eminent Persons, that the *Sea*, or *Magnetic-Compass*, was first made by an *English-man*. But we are certain, that the *Phænomena* of the *Variation of the Compass*, have bin Accounted for, only by our Learned Captain *Halley*, (now *Savilian* Professor of *Geometry* in *Oxford*; to whom the World is entirely Obliged, for his Excellent *Theory*, and Wonderful Discoveries in this Subject: as also for his most-Curious *Chart of the World*, after Mr. *Wright's* Projection; wherein the *Variation of the Needle* is found at any Place by *Inspection*: The want of which, render'd one of the Noblest Inventions, in a manner, useless.

AZIMUTH-COMPASS, is an Instrument made of Brass; resembling the Common *Sea-Compasses*; but has a Broad Limb,

Limb, graduated Diagonally, with an Index, and Thread; and is fitted up in a square Wooden-Box, with Jambols, and other Contrivances, for a free liberty of hanging Horizontally: It chiefly serves for taking the *Sun*'s *Azimuth*, or *Amplitude*, in order to find the Difference between the *Magnetical Meridian*, and the *True Meridian*, which shews *the Variation of the Compass*. These Instruments, if truly wrought, are very useful for that Purpose; and are excellently made, as also all Things of this Kind, by Mr. *Richard Glynne*, a very Skilful and Accurate Mathematical Instrument-Maker, next Door to the *Latin Coffee-House* in *Ave-Maria-Lane*, near St. *Paul*'s, *London*.

CON'D, *Con*, or *Cun*, is to Guide, or Direct the Ship in her right Course, or to give Words of Direction to the Man at Helm, how to Steer. And he that *Con*'s the Ship, uses these Terms to him at Helm:

Starboard, or *Port*, *the Helm*; that is, Put the Helm a *Starboard*, or to the *Larboard*; that is, to the Right, or Left of the Ship; and then the Ship will go to the *Larboard*, or *Starboard*: For the Ship always sails contrary to the *Helm*.

Right the Helm! or, *Helm a Midship!* that is, Keep it right up, or in the Midships, when 'tis required the Ship should go right before the Wind.

Aloof! or, *Luff!* *Keep your Loof!* *Fall not off!* *Veer no more!* *Keep her to!* *Touch the Wind!* *Have a care of the Lee-latch!* are Directions much to the same purpose; implying only, that the *Steers-man* shou'd keep the Ship near the Wind.

Ease the Helm! *No Near!* *Bear up!* That is, Let her fall to Leeward; or sail more Large, or more Before the Wind.

Steady! *As you go!* That is, Keep her upon the same Point; or, Keep her from going in and out, or making *Yaws*; whether she sails Large, or Before the Wind.

Keep her Thus! *Thus!* That is, Let her go just as she does.

CONTINENT, is a vast Continued Space of Land, containing many Countries, and Kingdoms; and consequently, not easily discernible to be surrounded with the Sea.

CONVOYS, are *Ships of War*, employed for the Security and Safety of particular Traders; to prevent their being Insulted, in Time of War.

COOK, *Ship's Cook*; whose Business is, to Dress, and Deliver out the Victuals. He has a *Mate*, under him.

COOK-ROOM, is the Place where the Victuals are Dress'd: In some Ships, 'tis seated in the *Hold*; but generally in the *Forecastle*, where there are Furnaces contriv'd, and other Necessaries, for the Purpose.

COOPER, *Ship's Cooper*, is he that looks to the Casks, and all other Veſſels, for Beer, Water, or any other Liquor. He has a *Mate*, under him.

CORDAGE, ſignifies in general, all the Ropes belonging to the Rigging of a Ship.

CORPORAL *of a Ship*, is an Officer that has the Charge of Setting and Relieving the Watches, and Centryes; and ſees all the Soldiers and Sailers keep their Arms Clean, and Neat; and teaches them their Uſe. He has a *Mate* under him.

COUNTERS in a *Ship*, are Parts of her: As,

Upper-Counter, is the Hollow Arching from the *Gallery*, to the lower part of the ſtrait Piece of the *Stern*.

Lower-Counter, is between the *Tranſom* and the lower part of the Gallery.

COURSE of a *Ship*, is that Point of the Compaſs on which ſhe ſails; and is indeed, the Angle, that the *Rhumb-Line* ſteer'd upon, makes with the *Meridian*.

COURSES, in a *Ship*, are her *Main-ſail*, and *Fore-ſail*; which are call'd, the *Main*, and *Fore-Courſe*.

To Sail under a Main-Courſe, and Bonnet; is, to ſail under *Main-ſail*, and *Bonnet*.

To go under a Pair of Courſes; that is, to ſail under *Main*, and *Fore-ſail*, without lacing on any *Bonnets*.

COXSWAIN, the ſame with *Cockſwain*.

CRABB, is an Engine of Wood with three Claws, placed on the Ground, for Launching of Ships, and to heave them into the Dock.

CRADLE, is a Timber-Frame, made along the Out-ſide of a Ship, or Galley, by the *Bildge*; ſerving for the more ſecure and commodious Launching of her.

CRAFT, at *Sea*, ſignifies, all manner of Lines, Nets, Hooks, *&c.* which ſerve for Fiſhing.

Small-Craft, are all ſuch little Veſſels; as *Ketches*, *Hoys*, and *Smacks*; made uſe of, in the Fiſhing-Trade.

CRANE-LINES, in a *Ship*, are Lines going from the upper-end of the *Sprit-ſail-top-maſt*, to the middle of the *Fore-ſtays*; they ſerve to keep the *Sprit-ſail-top-maſt* upright and ſteady in its Place, and to ſtrengthen it; in order that it might bear its Yard and Sail the better.

CRANK-*Sided*, is ſaid of a Ship that cannot bear her Sails; or can bear but ſmall Sail, for fear of Over-ſetting.

CRANK *by the Ground*; that is ſaid of a Ship that cannot be brought on the Ground, without danger of Overthrowing her.

CREEK, is a ſmall Bay, or Nook, by the Shore.

CRENCLES, in a *Ship*, are ſmall Ropes, ſpliced into the Bolt-Ropes of the Sails of the *Main maſt*, add *Fore-maſt*: They are faſten'd to the *Bow-line-Bridles*;

Bridles; and are alſo to hold by, when a *Bonnet* is ſhaken off.

CREW, *Ship's Crew*, or *Boats's Crew*; that is, the Seamen belonging to a *Ship*, or *Boat*. There are alſo in a Ship, ſeveral particular *Crews*, or *Gangs*; as the *Gun-Room Crew*, belonging to the *Gunner*; the *Carpenters Crew*, &c.

CROSS-*Barr-Shot*, is a great *Shot*, or *Bullet*, with a Barr of Iron put thorow it.

CROSS-*Jack-Yard*, is a ſmall Yard, ſlung at the upper-end of the *Miſen-maſt*, under the *Top*. It has no *Halliards*, nor *Ties*, belonging to it. Its Uſe is, to ſpread and hale out the *Miſen-top-ſail Sheats*.

CROSS-*Piece*, is a great Piece of Timber, going a-croſs the *Bolts* of a Ship: To this the Cable is *belay'd*, when the Ship Rides an Anchor.

CROSS-*Staff*, or *Fore-Staff*, is an Inſtrument, made commonly of Box, or Pear-tree, with Sliding Vanes: 'Tis uſed at Sea, for taking the *Altitude* of the *Sun*, or *Star*, &c. in order to find the *Latitude*.

CROSS-*Trees*, in a *Ship*, are four Pieces of Timber, bolted, and let-in to one another a-croſs, at the Head of the Maſt: Their Uſe is, to keep and bear the *Top-maſts* up; for the Foot of the *Top-maſt* is always faſten'd into them: But ſome call only thoſe two of theſe Timbers which go a-thwart Ships, the *Croſs-Trees*; and then, they call the others, the *Treſſel-Trees*.

CROSS; *To Ride a-croſs*. See *Ride*.

CROSS-*Tree-Yard*, is a Yard, ſtanding ſquare, juſt under the Miſen-top; and to it the *Miſen Top-Sail* is faſten'd below.

CROW-FEET, are ſmall Lines, or Ropes, ſometimes ſix, eight, or ten, reev'd thorow the *Dead-mens-Eyes*: They are ſcarce of any more Uſe, than to make a Shew of ſmall Rigging. They are placed at the bottom of the *Back-ſtays* of the *Fore-top-maſt*, *Miſen-top-maſt*, and *Gallant-top-maſt*.

The Spritſail-Topſail Crow-Foot, is a Rope, divided into two, or four Portions, from the upper-end of the *Spritſail-Topmaſt*, to the the *Fore-top-maſt-Stay*.

CRUISERS, are *ſmall Men of War*, made uſe of, to and fro, in the Chanel, and elſewhere, to ſecure our Merchant-Ships, and Veſſels, from the Enemies ſmall Frigats, and Privateers: They are generally thoſe that Sail well, and therefore are commonly well Mann'd: And indeed, the Safety of the Trade in the *Chanel*, and up and down the *Soundings*, and other Places, does abſolutely require the conſtant keeping out of ſuch Ships at Sea.

CUBBRIDGE-HEADS, are the *Bulk-heads* of the *Fore-Caſtle*, and the *Half-Decks*; wherein there are placed *Murdering-Pieces*, &c. to Clear the

 Decks,

Decks, *fore and aft*, upon Occasion.

CUDDY, in Great Ships, is a Place lying between the Quarter-Deck, and the Captain-Lieutenant's Cabin, under the Poop; and is divided into Partitions, for other Officers.

CULVER-TAIL'D, signifies, the fasting, or letting of one Timber into another, so that they cannot slip out; as the *Carlings*, into the *Beams* of a Ship.

CUN, or CUNNING, the same with *Cond*; which see.

CURRENTS, are the impetuous Motions of the Waters, which in certain Latitudes run, and set on particular Points of the *Compass*: And commonly, their Force is conformable to the Course of the *Moon*; so as to be more rapid, or strong, when She is in the *Change*, or at *Full*; and more weak, in her *Wain*.

CUT *a Feather*: If a Ship has too Broad a *Bow*, 'tis common to say, She will not *Cut a Feather*; that is, She will not pass thro' the Water, so swift, as to make it foam, or froth.

CUT-WATER, or *Knee of the Head*, is the Sharpness of the Head of the Ship, below the Beak: And is so called, because it *Cuts*, or Divides the Water, before it comes to the *Bow*.

D

DAILE, is the Trough wherein the Water does run over the Decks.

DAVIS's *Quadrant*. See *Back-Staff*.

DAVIT, in a *Ship*, is a short Piece of Timber, with a Notch at one end, wherein, by a Strap, hangs the *Fish-Block*. The Use of this *Block*, is to help up the Fluke of the Anchor, and to fasten it at the Ship's *Bow*, or *Looff*. The *Davit* is shiftable from one side of the Ship, to the other, as there is Occasion.

DEAD-MENS-EYES, are Blocks with many Holes, but no Shivers: Thro' them the *Lanniers* go, which make fast the *Shrowds*, below the *Chains*. And sometimes the *Main-stays* are set *taught*, by these *Dead-Mens-Eyes*, and *Lanniers*. Thro' these, also, the *Crow-Feet* do reeve.

DEAD-RECKONING, at *Sea*, is that Estimation, Judgment, or Conjecture, which the Seamen have of the Place where the Ship is, by keeping an Account of her *Way*, or *Distance Run* by the *Log*; by knowing the *Course* they have *Steer'd*, by the *Compass*; and by Rectifying all, with Allowance for *Lee-way*, *Currents*, &c. with Consideration of the Ship's *Trim*, &c. So that this *Reckoning*, is without any *Observation* of of the *Sun*, or *Stars*; and is to be

be rectified, as often as any good Observation can be had.

DEAD-RISING, is that part of a Ship which lies *aft* between the *Keel* and the *Floor-Timbers*, next adjoining to the *Stern-Post*, under the *Bread-Room* in a *Ship of War*.

DEAD-ROPES, are such as does not run in any Block.

DEAD-WATER, is the Eddy-Water just a-stern of a Ship; and is so call'd, because it does not pass away so swift, as the Water running by her Side does.

A Ship makes much Dead-Water; that is, She has a great Eddy following her Stern.

DECK, of a *Ship*, is a Plank'd-Floor *from Stem to Stern*, whereon the Guns lie; and on which the Men Walk to, and fro. Great Ships have Three *Decks*, *First*, *Second*, and *Third*, beginning to account from the Lowermost.

Half-Deck, reaches from the Main-mast, to the Stem of the Ship.

Quarter-Deck, is that aloft the Steerage, reaching to the Round-house.

Flush-Deck, is that which lies even in a Right-Line, *fore and aft*, from Stem to Stern.

Cambered-Deck, is that which lies compassing; and is by no means proper for a *Man of War*.

DECLINATION *of the Sun, or Star*, is in an Arc of a Great Circle, intercepted between the *Sun*, or *Star*, and the *Equinoctial*: And is easily found by this

PROPORTION;

As the Radius, *or Sine of* 90 *Degrees*,
To the Sine of the Sun's *greatest Declination*:
So is the Sine of the Sun's *Equinoctial Distance*,
To the Sine of the Sun's *present Declination*.

When an *Observation* of the *Sun*'s *Meridian Altitude* is made, in order to find the *Latitude of the Place*; The *Sun*'s *Declination* also must be known; as usually, and readily, by means of *Tables* carefully Calculated for that Purpose; such as those in Mr. *Jones*'s Treatise of *Navigation*, Edit. 2d.

DEEP-SEA-LINE, or *Dip-Sea-Line*, is a small Line, to Sound with, some 150 *Fathom* long; with a long hollow Plummet at the Head, and Tallow put into it, that will bring up any Gravel, Stones, Sand, Shells, and the like, from the Bottom; and to know the Differences of the Ground: Which having bin before Discover'd by other *Observations*, and Entred into their Books; they guess, by their Soundings, what Coast they are upon, tho' they cannot see Land. If it happens that no Ground come upon the Tallow, they conclude, they are upon *Ouzie-Ground*: Which they Discover again, by Sounding with a Wollen-Cloth upon

upon the Lead, whereby this Ground will be brought up. This Line is first mark'd at 20 Fathom; and afterwards encreased, by 10's, to the End; distinguished by so many small Knots upon each little String, that is fix'd at the Mark, throw the Shrowds, or midst of the Line; shewing it is so many times 10 Fathom deep, where the Plummet does rest, from drawing the Line out of One's Hand. 'Tis used only in Deep Waters, when the Seamen think they approach the Shore.

DEFLECTION, at *Sea*, is the Tendency of a Ship from the true Course, by reason of *Currents*, &c. which turn her out of her right Way.

DEGREE, the 360th Part of the Periphery of a *Circle*. See *Mercator's Chart*.

DEPARTURE, in *Navigation*, is the nearest Distance between any two *Meridians*, counted on the *Parallel of Latitude*; shewing how far one Place is to the *Eastward* or *Westward* of another, in *Degrees*, or *Miles*, proper to the *Parallel of Latitude*.

DIFFERENCE *of Latitude*, in *Navigation*, is the nearest Distance of any two *Parallels of Latitude*; shewing how far one Place is to the *Northward* or *Southward* of another: And it never exceeds 180 Degrees. See *Latitude*.

DIFFERENCE *of Longitude*, or *Meridional Distance*, in *Navigation*, is the Distance on the *Equinoctial*, which is contain'd between the *Meridians* of any two Places; shewing (in the *Equinoctial*) how far the *Meridian* of one Place is to the *Eastward* or *Westward* of the Meridian of another.

DISEMBOGUE, at *Sea*: When a Ship passes out of the Mouth of some great *Gulf*, or *Bay*, they call it *Disemboguing*. Also, 'tis said of a *River*, That at such a Place, or after it has run so many Leagues, it *Disembogues* it self into the Sea. Thus, the *Volgo Disembogues* it self into the *Caspian*; and the *Danube*, into the *Euxine-Sea*.

DISTANCE *Run*, in *Navigation*, is the Number of *Miles*, or *Leagues*, that a Ship has sailed from any given Point. 'Tis usually found by the *Log-Line*; which see.

Meridional Distance. See *Meridian*.

DIVISION, is the *Third Part* of a Fleet of *Men of War*: But sometimes 'tis the *Ninth Part*; which happens, when the Fleet is Divided into Three Squadrons; for then each Squadron is distributed into Three *Divisions*. As 'twas practis'd by the Fleets of *England* and *France*, jointly, in the Years 1672, and 1673: The *English* Fleet form'd Two Squadrons, the *Red*, and the *Blue*, each distributed into Three *Divisions*: The *French* Fleet, which form'd the *White Squadron*, was also distributed into *Three Divisions*.

In a *Sea-Engagement*, the Order of Battle, is, To place in one Line

Line, all the *Squadrons*, and all the *Divisions* of the Side: And this Order is kept, as long as Wind, Valour, and Fortune, will permit. See *Engagement.*

DOCK, is a Pit, Pond, or Creek, by the Water-side, made convenient either to Build, or Repair Ships in: And is either,

Dry-Dock, where the Water is kept out, by great *Flood-Gates*, till the Ship is Built, or Repair'd; then the Gate is open'd, and the Water let in, to Float, and Launch her.

Wet-Dock, a Place where the Ship may be haled into, out of the *Tide*'s Way; and so *Dock* her self, or *Sink* her self a Place to lie in.

DOCK-YARDS, are the Magazines of all sorts of Naval-Stores: The Principal one's, are those at *Chatham*, *Portsmouth*, *Plimouth*, *Woolwich*, *Deptford*, and *Sheerness*. In Time of Peace, *Ships of War* are laid up in these *Docks*: But the Biggest, and greatest Number of the Biggest Ships, are lodged at *Chatham*; where, and at other Yards, they receive, from time to time, such Repairs as are necessary.

These *Yards* are generally supplied, from the *Northern* Crowns, with *Hemp*, *Pitch*, *Tar*, *Rosin*, and several other Species: But as for *Masts*, particularly those of the Largest size, they are brought from *New-England.*

How much it imports the Good of the Public, to keep those Magazines constantly replenish'd, every One is able to judge: And And-it were much to be wish'd, the Improving the afore-mention'd Commodities in our *English* Plantations, might meet with all possible Encouragement; lest, one time or other, it may prove difficult to get them elsewhere. It is reasonable to think, such an Undertaking will put the Nation to some Considerable Charge, e're it be brought to Perfection; but when so, many are the Advantages that will arise from it.

DOGGAR, is a Small Ship, built after the *Dutch* Fashion, with narrow Stern; they carry commonly but one Mast.

DOMESTIC-*Navigation*, is, Coasting, or Sailing along the Shore; in which, the *Lead*, and *Compass*, are the chief *Instruments.*

DOUBLE-*Blocks*. See *Blocks.*

DOUBLE *the Cape*, or *Point*; that is, to Come up with it, Pass by it, and so leave it behind.

DRABLER, a small Sail in a Ship; being the same to a *Bonnet*, that a *Bonnet* is to a *Course*; and is only used, when the *Course* and *Bonnet* are too *Shoal* to Cloath the *Mast.*

DRAGS: *Sea-Drags*, are whatever hangs over the Ship, in the the Sea, as Shirts, Coats, or the like: And Boats, when Tow'd, or whatever else, that, after this manner, may hinder the Ship's Way, when she sails, are call'd *Drags.*

DRAUGHT

DRAUGHT of a *Ship*, is so many Feet as she *Draws*; that is, as she sinks into the Waters. So that if a Ship sink into the Water 18 Feet Perpendicular, she is said to *Draw* 18 *Feet Water*: And therefore her *Draught* is said to be more or less, as she *Draws* more or less Water.

DRIFT, at *Sea*: Any thing that floats upon the Water, is is said to *Run a Drift*.

DRIFT-SAIL. See *Sail*.

DRIVE: A Ship *Drives*, when an *Anchor* being let fall, will not hold her fast; but that she sails away with the *Tide*, or *Wind*. The best Help, in this Case, is to let fall more *Anchors*, or to *veer out more Cable*; for the more Cable she has out, the safer she *rides*.

The Ship Drives to Leeward, is usually said of a Ship that lies *a Hull*, or *a Try*.

DRIVE-BOLTS. See *Bolts*.

DRY-DOCK. See *Dock*.

DUCKING *at the Main-Yard-Arm*, is a Way of Punishing *Sea-Offenders*: And is perform'd thus; The Malefactor has a Rope fasten'd under his Arms, about his Middle, and under his Breech; and so is hoisted up to the End of the *Yard*; from whence he is violently let down into the Sea, sometimes twice, sometimes three several times, one after another.

And if the Offence be very great, he is also drawn underneath the very *Keel* of the Ship; the which they call *Keel-Haleing*, or *Keel-Raking*.

The Criminal being under-Water, a Great-Gun is fired right-over his Head; as well to astonish him the more, as to give Warning to all Others, to Look out, and Beware.

Other Punishments there are, at *Sea*; as particularly those at the *Jeer-Capstan*, and *Bilboes*.

That at the *Capstan*, is, when a *Capstan Barr* being thrust thro' the Hole of the *Barrel*, the Offender's Arms are Extended at the full length, cross-wise, and so Tied unto the *Barr*; having, sometimes, a Basket of Bullets, or some other-like Weight, hanging by his Neck: In which Posture he continues till he be either brought to Confess some Plot, or Crime, whereof he is Suspected; or that he has Suffered, what he is Censured to Undergo, at the Discretion of the Captain.

The Punishment by the *Bilboes*, is, when an Offender is *laid in Irons*, or in a kind of *Stocks*, that they use for that Purpose; and which are more or less Ponderous, as the Quality of the Offence is, which he is Guilty of.

DUCK-UP, is a Term used by the *Steer's-man*, or Man at *Helm*, when either *Main-sail*, *Fore-sail*, or *Sprit-sail*, hinders his seeing to Steer by a Land-Mark: And then his Word is, *Duck-up the Clew-Lines of those Sails!* that is, Hale the Sail out of the Way!

Also,

Also, When a Shot is made by a *Chace-Piece*; if the *Clew* of the *Sprit-sail* hinders the Sight; then they call, *Duck-up the Clew-Lines of the Sprit-Sail!*

E

EARING, Is that part of the *Bolt-Rope*, which, at the four Corners of the Sail is left open, in the shape of a *Ring*: The two Uppermost parts are put over the Ends of the *Yard-Arms*, and so the Sail is made fast to the Yard: And into the Lowermost *Earings* the *Sheats* and *Tacks* are seiz'd, or bent at the *Clew*.

EASE! That is, Make more Slack! or, Let go Slacker!

Ease the Bow-Line! or, *Ease the Sheet!* That is, Let it be be more Slack!

Ease the Helm! That is, Let the Ship go *more Large*; or *more before the Wind*; or *more Larboard*.

EAST, is One of the Four *Cardinal-Points* of the World; and is that Point of the *Horizon*, where the *Sun* is seen to Rise, when 'tis in the *Equinoctial*: So that when the *Sun* Rises *due East*, it makes Equal Days and Nights over all the World.

EASTERN-*Amplitude*, is an Arc of the *Horizon*, intercepted between the Point of the *Sun's Rising*, and the *East-Point* of the *Magnetic-Compass*.

EASTERN-*Hemisphere*. See *Hemisphere*.

EASTING. See *Departure*.

EBB; is the Reflux of the Sea, when the Water begins to Fall: And according to its several Degrees of *Ebbing*, 'tis distinguish'd into *Quarter-Ebb*, *Half-Ebb*, *Three-quarter-Ebb*, *Low* (or *Dead-Low*) *Water*.

EDDY-*Tide*, or *Water*; is, where the *Water* runs back, contrary to the *Tide*; or, which hinders the free Passage of the Stream, and so causes it to Return again.

EDDY-*Wind*; is that which returns, or is beat back, from a *Sail*, *Mountain*, or any thing that may hinder its Passage.

EDGE *in with a Ship*; is said of a *Chase*, that is *making up* to him.

END *for End*: When a Rope runs all out of the Block, that it is unreev'd; they say, *'Tis run out End for End*.

The Cable at the Hawse is run out End for End; that is, the *Cable*, or *Hawser*, is all run out at the *Hawse*.

ENGAGEMENT, *Sea-Engagement*: Whenever a Fleet of *Men of War* is *Engaged*, whether to Give, or to Take Battle, with another every way Equal unto it; then every *Squadron* of such *Fleet* does usually Order and Subdivide it self into Three Equal *Divisions*; with a *Reserve* of certain Ships out of every *Squadron*, to bring up their *Rear*. And every one of these, observing a *due Birth*, and Distance, are, in the *Engagement*, to *Second*

one

one another: And the better to avoid Confuſion, and *Falling-foul* on one another; to Charge, Diſcharge, and Fall off by Threes, or Fives, more or leſs, as the Fleet, in Groſs, is greater, or ſmaller. The Ships of *Reſerve*, being to be inſtructed either to Succour and Relieve thoſe that be any Way *Engaged*, and in Danger; or to Supply, and put themſelves in the Place of Thoſe that ſhall be made Unſerviceable. This is the Order and Courſe to be conſtantly Kept, and Obſerved, during the whole Time of Battle.

If the *Fight* ſhould continue, even within the Nihgt; it may well be maintain'd, and kept in this very Order, if ſo be that every Ship do but carefully heed the *Admiral* of his particular *Squadron*, by his *Light*; and withal, his *Leading Ship*, that is next before him; that ſo, when the *Admiral falls off*, and makes a Retreat, for the preſent, upon ſome eſpecial Occaſion; all the Ships of that *Squadron* may do the like; and Retire under their ſeveral *Diviſions*, to Amend and Repair any thing that has bin Miſcarry'd in the *Fight*, or to Speak and Adviſe with their Commanders; and ſo to be ready to renew the *Engagement*, and to Re-Charge the Enemy, according to their Inſtructions.

It is alſo carefully Obſerved by all *Admirals*, That they ſo Order and keep themſelves, in their ſeveral *Diviſions*, as they may beſt be ſeen and diſtinguiſh'd by their whole *Squadron*, and that as well by Night, as Day; that ſo Directions may beſt be given, and Notice taken from, by every particular Ship of the *Fleet*; and the Executions of Orders, perform'd with the more Facility, and Certainty.

As for a *Fleet* which conſiſts but of few Ships, and being to fight in an open Sea, it ſhould be brought up to Battle in one only Front, with the *Chief-Admiral* in the Middle of them; and on each ſide of him, the ſtrongeſt and beſt-provided Ships of the Fleet; who keeping themſelves in as convenient a Diſtance as they ſhall be able, are to have an Eye and Regard, in the Fight, to the Weakeſt and Worſt Ships of the *Party*; and to Relieve, and Encourage them, upon all Occaſions; and withal, being near the *Admiral*, may both Guard him, and aptly receive Inſtructions from Him.

ENSIGNS; are the *Colours* placed in the *Sterns*, or *Poops*, of *Ships*: And there are few Ships, whether *Men of War*, or *Merchants*, but have their *Enſigns*. Their chief Uſe, is, That when any ſtrange Ships meet at Sea, or make into any Harbour; by *heaving out the Colours* in the *Poop*, that is, putting theſe *Enſigns* abroad, of what Part, and Country they are.

Thus the *Engliſh* heave out their *Colours*, with St. *George's*-

Cross in it; the *Scotch*, with St. *Andrew*'s; and so all other Countries, with some peculiar Distinction, whereby they may be known. And these *Ensigns* serve for various other Uses, too many here to enumerate. See *Flags*.

ENTRING-LADDERS, in a *Ship*, are of two sorts; one is used by the Ship's Sides, in a Harbour, or in Fair-Weather, for Persons to go in and out of the Ship. The other is made of Ropes, with small Staves, for Steps; and is hung out of the *Gallery*, to enter into the Boat, or to come aboard the Ship from thence, when the Sea runs so high, that they dare not bring the Boat to the Ship's Side, because of the danger of *staving* her.

ENTRING-PORTS. See *Ports*.

ENTRING-ROPES. See *Ropes*.

ENTRING *a Ship*, the same with *Boarding a Ship*; which see.

EQUATOR (on the *Earth*,) or *Equinoctial* (in the *Heavens*,) is that great Circle, whose *Poles* are those of the World; and which is supposed to Divide the Globe into Two Equal Parts, call'd the *Northern* and *Southern Hemispheres*. This Circle passes thro' the *East* and *West* Points of the *Horizon*: Therefore, at the *Meridian* is raised as much above the *Horizon*, as is the *Complement of the Latitude of the Place*.

All the *Stars* which are under this Circle, that is, such as have no *Declination*, do always Rise due *East*, and Set due *West*. And the *Sun*, when 'tis said to come to this Circle, makes Equal *Days* and *Nights* all round the Globe; for then he is said to Rise due *East*, and Set due *West*; which he does at no other Time of the Year.

Those that live under this Circle, have their *Days* and *Nights* Equal: And the *Sun*, at Noon, is in their *Zenith*, and therefore casts no Shadow.

The *Declination*, or *Latitude*, is reckon'd on the *Meridian*, from this Great Circle, either *North* or *South*: And those Circles which are suppos'd to run thro' each Degree of *Declination*, or *Latitude*, are call'd *Parallels of Declination*, or *Latitude*.

The *Equinoctial*, or *Equator*, is supposed to be Divided into 360 Equal Parts, or Degrees: But a Natural Day, is Measur'd by the *Revolution* of the *Equinoctial*; and is ended, when the same Point of the *Equinoctial* comes again to the same *Meridian*, that is, in 24 Hours:

Therefore each Hour must be $\frac{360}{24}$, or 15 *Degr.*

Of the *Equator*:	Of *Time*:
$\frac{360}{24}$ or 15 *deg.* 00 *m.*	is 01 *Hour*, 00 *m.*
01 *deg.* 00 *m.*	is 00 *h.* 04 *m.*
00 *deg.* 15 *m.*	is 00 *h.* 01 *m.*
00 *deg.* 01 *m.*	is 4 *Seconds* of *Time*.

EYE *of the Anchor*; is the Hole wherein the Ring of the *Anchor* is put into the Shank.

EYE *of the Strap*; is the *Ring*, or *Round*, which is left of the *Strap*, to which any Block is feiz'd.

F

FACK; is one Round of any any Rope, or Cable, quoil'd out of the way.

FADDOM, or *Fathom*; a Meafure containing 6 *Feet*, much us'd at Sea.

FALL; is that part of the Rope of a Tackle, which is hal'd upon.

FALL *off*: When a Ship under Sail, keeps not fo near the Wind, as fhe fhou'd do; fhe is faid to *Fall off*.

FALL *not off*! A Word of Command, from him that Con's the Ship; fignifying as much as, Keep the Ship near the Wind! See *Con'd*.

LAND-FALL. See *Land*.

FALLS: A Ship that has Rifings in fome part of her Decks, more than other, is faid to have *Falls*.

FALSE-KEELS. See *Keel*.

FALSE-SHEAT. See *Sheats*.

FALSE-STEM. See *Stem*.

FARTHEL; the fame with *Furl*: Which fee.

FASHION-PIECES; are two Pieces of Timber arifing from the *Stern-Poft*, and defcribing the Breadth of the Ship, at the Stern, and are the outermoft Timbers thereof; to which are faften'd the Planks that reach to the *after-end* of the Ship.

FATHOM; the fame with *Faddom*: Which fee.

FAT; fignifies the fame with *Broad*. So a Ship is faid to have a *Fat-Quarter*, if the *Trufking in*, or *Tuck* of her *Quarter* under-Water, be deep.

FEATHER; *Cut a Feather*: Which fee.

FEAZING; fignifies, the ravelling out of any Great-Rope, or Cable, at the End.

FEND, imports the fame as *Defend*.

Fending the Boat; is, faving it from being dafh'd againft the Rocks, Shore, or Ship's-fide.

FENDERS; are Pieces of old Hawfers, Cable-Ropes, or Billets of Wood, hung over the Ship's-fides, to keep other Ships from rubbing againft her, and bruifing her.

FEND-BOLTS. See *Bolts*.

FETCH *him up*, that is, to give Chafe, or to Purfue a Ship, at Sea.

FIDD; is an Iron or Wooden-Pin, to fplice and faften Ropes together: 'Tis made Taper-wife, and fharp at one end. And that Pin in the Heel of the *Top-maft*, which bears upon the *Cheffe-trees*, is call'd alfo, a *Fidd*.

FIDD-*Hammer*; is that whofe Handle is a *Fidd*, or made Taper-wife.

FIGHT-

FIGHTING-*Sails*. See *Sails*.

FIGHTS; are the Waste-Cloths which hang round about a Ship, in a Fight, to secure the Men from being seen by the Enemy.

CLOSE-FIGHTS; are the Bulk-heads, *fore* and *aft* the Ship; put up, for the Men to stand secure behind, and Fire on the Enemy, and Scoure the Decks, in case of Boarding.

RUNNING-FIGHTS; are those *Sea-Fights* where the Enemy do not stand the Battle, but are continually Chased. Of these *Sea-Fights*, we might give several Instances; and particularly, that off *Cape-Barfleur*, between the *English* Fleet, under Admiral *Russel*, (now Earl of *Orford*; and the *French* Fleet, under Monsieur *Tourvile*: The Fight began on the 19th of *May*, 1692. about Half-an-Hour after Eleven in the Morning; And *Monsieur* (in the *Royal Sun*) stood it for about an Hour and an half; and then Tow'd off, and so made a *Running-Fight* of it: But being Chased for some Days, was forced to Run a-shore at *Cherbrooke*; where He (of 110 Guns,) and Two more, of 104 Guns each, were Burnt; as also Thirteen at *La-Hogue*, by Sir *George Rooke*, (then *Vice-Admiral* of the *Blue*,) *viz. one* of 90, *two* of 80, *four* of 76, *four of* 60, and *two* of 56 Guns.

FIRE-SHIPS; are Vessels Charg'd with Artificial Fire-works; who having the Wind of an Enemy's Ship, Grapples her, and sets her a-Fire.

FISH; is a Plank, or Piece of Timber, fasten'd to a Ship's Mast, or Yard, to strengthen it.

To Fish a Mast, or *Yard*; is, to fasten Planks, or Pieces of Timber, to it; which is done, by nailing them on with Iron-Spikes; and *Would* them, that is, winding Ropes hard about them.

FISH-BLOCK. See *Block*.

FLAGS; are the *Colours* that the *Admirals* of a Fleet carry, on their *Tops*; and are Marks of Distinction, as well of *Officers*, as *Nations*. The *Admiral* in *Chief*, carries it on his *Main-Top*; the *Vice-Admiral*, on the *Fore-Top*; and the *Rear-Admiral* his, on the *Misen-Top*. When a *Council of War* is to be held at Sea; if it be on-board the *Admiral*, they hang a *Flag* in the *Main-Shrowds*; if in the *Vice-Admiral*, in the *Fore-Shrowds*; if in the *Rear-Admiral*, in the *Misen-Shrowds*.

FLAG-OFFICERS; those who Command the several *Squadrons* of a *Fleet*, such are, the *Admirals*, *Vice-Admirals*, and *Rear-Admirals*.

To Lower, or *Strike the Flag*; that is, to Take in the *Flag*, or Pull it down upon the *Cap*; is a Respect due from all Ships, or Fleets inferior, whether in respect of Right of Sovereignty,

Place, or the like; expressing an Acknowledgment, and Submission, when they meet with others, any way justly their Superiors: And in the Case of *Sovereignty*, in our Narrow Seas, which has bin long claim'd, and made good by the Kings of *England*: So that if any Ship whatsoever, shall in any of those Parts, meet with any *Admiral* of *England*, and do not acknowledge this Sovereignty, by *Taking in her Flags*; she may, and is to be treated as an Enemy. And, in a Fight, to *Lower*, or *Strike the Flag*, is a a Token of *Yielding*.

To Heave out the Flag; that is, to Put out, or Put abroad the *Flag*.

The FLAGS *of all* States *and* Nations, *according to the best Accounts, are as follow: The* Names *of* Places, *being in an* Alphabetical *Order*.

FLAG of *Algier*, in *Barbary*; is *Red*, and Hexagonally cut, charg'd with a *Turk's-Head* and *Turban*.

FLAG of *Amsterdam*, in *Holland*; consists of Three Bands, or Cloths; the Uppermost *Red*, the Lowermost *Black*, and the Middlemost *White*; which has therein the Arms of *Amsterdam*: Bearing *Gules*, a Pale *Sable*, charg'd with three Saltiers, (or St. *Andrew*'s-*Cross*) *Argent*, with an Imperial Crown for a *Crest*, and supported by two Lions *Sable*.

FLAG of *Bergen*, in *Norway*; is *Red*, travers'd with a Cross *Argent*, charg'd in the Middle with a *Scutcheon*, bearing *Argent*, a Lyon *Gules*, holding a Sword *Azure*, with a Hilt *Sable* in the Right-Paw, and surrounded with a Garland of two *Green* Branches.

FLAG of *Bremen*, in *Lower-Saxony*; consists of Nine Bands, *viz*. Five *Red*, and Four *White*; charg'd, near the *Flag-staff*, with a Pale Checker'd *Argent* and *Gules*.

FLAG of *Brandenburg*; is *White*, charg'd with an Eagle *Gules*, holding a Sword *Azure*, with a Hilt *Sable*, in the Right-Tallon; and a Scepter *Or*, in the Left.

Another FLAG of *Brandenburg*; consists of seven Bands, four *White*, and three *Black*; charg'd with a *Scutcheon*, bearing *Argent*, an Eagle *Gules*.

FLAG of *Burgundy*; is *White*, travers'd with a Saltier, (or St. *Andrew*'s-*Cross*,) of two Ragged-Staffs *Gules*.

Another FLAG of *Burgundy*; is *Blue*, charg'd with the same Cross.

FLAG of *Calais*, in *France*; is *Blue*, charg'd, in the middle, with a Cross *Argent*.

FLAG of *China*. The *Emperor* of *China* maintains several entire Fleets, for to secure his Trade, and Navigation; And is said, by some, to bear for Ensigns Armorial, and Flags, *Argent*, charg'd with three *Black-a-moors-Heads*, placed in the Front their Bust Vested *Gules*: But,

according

according to others, two Dragons *Sable*.

FLAG of *Courland*; is *Red*, charg'd with a Crab-Fish *Sable*.

Another FLAG of *Courland*; consists of two Bands, the uppermost *Red*, and the lowermost *White*.

FLAG of *Dantzic*, in *Prussia*; is *Red*, charg'd, near the *Flag-staff*, with two Crosses *Argent*, one above the other, and over them a Crown *Argent*.

Another FLAG of *Dantzic*; is *Red*, with four Crosses *Argent*, two and two; and Crowns *Argent*.

FLAG of *Denmark*; is *Red*, but slit and travers'd with a White Cross. And the

FLAGS of *Danish* Merchant-Ships, are Square.

FLAG of *Dunkirk*, in *Flanders*; consists of six Bands intermingled, three *Blue*, and three *White* one's.

FLAG of *Elbing*, in *Prussia*; consists of two Bands, the uppermost White, with a Cross *Gules*; the lowermost Red, with a Cross *Argent*.

FLAG of *Embden*, in *Westphalia*; consists of three Bands, which are *Yellow*, *Red*, and *Blue*.

FLAG-*Royal*, or *Standard-Royal* of *England*; ought to be *Yellow* (viz. *Or*,) according to some; But others will have it, *White* (or *Argent*:) 'Tis charg'd with a Quarter'd Scutcheon of *England*, *Scotland*, *France*, and *Ireland*. 'Tis never carry'd but by the Sovereign Prince, or his High-Admiral, or Commission.

Another FLAG-*Royal* of *England*; is Quarterly: The First and Fourth Quarter Counter-quarter'd; In which the first and fourth *Azure*, Three *Flower-de-Luces Or*; The Royal Arms of *France*, Quarter'd with the *Imperial* Ensigns of *England*, which are in the Second and Third *Gules*, *Three Lions Passant Gardant in Pale*, *Or*. In the Second Place, within a *double Tressure Counter-Flower-de-Luce Or*, *a Lion Rampant Gules*, for the Royal Arms of *Scotland*. In the Third place, *Azure*, an *Irish-Harp Or*, *String'd Argent*, for the Royal-Ensigns of *Ireland*. But sometimes 'tis altered; as in setting the *English* Arms before the *French*, and the like.

UNION-FLAG of *England*; is *Gules*, charg'd with these words, FOR THE PROTESTANT RELIGION, AND FOR THE LIBERTY OF ENGLAND.

FLAG of the *Admiral* of *England*; is *Red*, charg'd with an Anchor *Argent*, set in *Pale*, entangled in, and wound about with a Cable of the same.

JACK-FLAG of *England*; is *Blue*, charg'd with a Saltier *Argent*, and a Cross *Gules*, border'd *Argent*.

FLAG of an *English Merchant-Ship*; is *Red*, and a Franc-quarter *Argent*, charg'd with a Cross *Gules*.

FLAG of *Flanders*, consists of Three Bands, the Uppermost *Red*,

Red, the Lowermost *Yellow*, and the Middlemost *White*; which is charg'd with a Saltier, (or Saint *Andrew*'s Cross) of two Ragged-Staffs *Purple*.

JACK-FLAG of *Flanders*; is *Yellow*, charg'd with a Lion in an *Orl Sable*, set in a Scutcheon fortify'd with Eight *Flower-de-Luce*; and adorn'd above with Three *Flower-de-Luce*, *Sable*.

JACK-FLAG of *Flushing*, in *Zeland*; is *Red*, charg'd with an Urn Crown'd *Argent*.

FLAG-*Royal* of *France*; is *White*, *semè* with *Flower-de-Luce Or*; and charg'd with the *Arms of France*: Which bears *Azure*, *three Flower-de-Luce Or*, two in Chief, and one in Base; The Scutcheon is environ'd with the Collars of the Order of St. *Michael* and the *Holy Ghost*: For Crest, an Helmet entirely open, whereon a Crown clos'd, after the manner of an *Imperial* Crown, with Eight inarched Rays, top'd with a Double *Flower-de-Luce*: The Supporters are, two *Angels* habited as *Levites*.

FLAG *of the Admiral of France*: Where the Admiral of *France* is a-board in Person, a *White Flag* is set upon the *Main-top-mast-head*.

FLAG-*Royal*, or *Standard-Royal* of the *French* Galleys; is *Red*, set with *Flower-de-Luce Or*.

FLAGS of the *French* Merchant-Ships; Their *Ensigns*, sometimes are *Blue*, travers'd with a *White* Cross, and the *King's-Arms* upon the Whole: Or any other Distinction as shall be thought proper; provided their *Ensigns* be not all White.

FLAG of *Genoa*; is *White*, travers'd with a Cross *Gules*.

FLAG of *Hamburg*, in *Lower-Saxony*; is *Red*, charg'd with a large Tower *Argent*, and three Turrets the same, at the Top.

Another FLAG of *Hamburg*; *Red*, charg'd with three Towers *Argent*, placed one and two, at equal distance.

FLAG of *Holland*; consists of Three Bands; The First *Orange*, the Second *White*, and the Third *Blue*.

Another FLAG of *Holland*, has twice as many Bands as the former; that is, two of each Colour.

Another FLAG of *Holland*, has Nine Bands; that is, thrice as many as the first, and three of each Colour.

JACK-FLAG of *Holland*, is like that of the *States General*, but without a *Scutcheon*.

FLAG of *Hoorn*, in *Holland*; consists of Three Bands, the Uppermost and Lowermost *Red*, the Middle *White*; whereon is placed a Horn *Gules*, garnish'd with Hoops *Or*, and hung by a String *Gules*.

FLAG of *Japan*; is said to be *Sable*, with three Trefoils *Argent*.

Another FLAG of *Japan*; bearing *Or*; six Stars *Argent*, in an Oval Shield, border'd with little Points of *Gold*.

FLAG

FLAG of *Ireland*; is *White*, charg'd with St. *Andrew*'s Croſs.

FLAG of *Legorn*, in *Tuſcany*; is *White*, charg'd with a Croſs *Gules*.

FLAG of *Lubec*, in *Lower-Saxony*; conſiſts of Two Bands, the Uppermoſt *White*, and the Lowermoſt *Red*.

FLAG of *Malta*; is *White*, charg'd with the *Croſs of Malta*, that is, a Croſs with Eight Points *Red*.

Another FLAG of *Malta*, is *Red*, travers'd with a Croſs *Argent*.

FLAG of *Middleburg*, in *Zeland*; conſiſts of Three Bands, which are, *Red*, *White*, and *Yellow*.

Another FLAG of *Middleburg*; is *Red*, charg'd with a Tower Embattl'd *Or*.

FLAG of the *Great Mogul*; is ſaid to be *Argent*, *Semè* with Beſants, *Or*.

FLAG of *Muſcovy*; conſiſts of Three Bands, the Uppermoſt *White*, the Lowermoſt *Red*, and the Middlemoſt *Blue*; which is charg'd with an Eagle diſplay'd *Or*; bearing on its Breaſt a Shield *Or*; charg'd with a Cavalier *Argent* fighting a Dragon: with a Royal Crown over the Heads.

Another FLAG of *Muſcovy*; conſiſts of Three Bands, of the ſame Colour with the former; travers'd with a Saltier (or St. *Andrew*'s Croſs) *Azure*.

Another FLAG of *Muſcovy*; is Quarter'd by a Croſs *Azure*: the *Firſt* and *Fourth* Quarter *Argent*; the *Second* and *Third Gules*.

FLAG of *Naerden*, in *Holland*; is *Blue*, charg'd with three Stars *Or*.

FLAG of *Nanquin*, in *China*: The *Junks* of *Nanquin*, carry on their *Main-top-maſt-head* a *Red* and *White Flag*; and on the *Fore-top-maſt*, a *Red Flag*: They carry alſo *Gray*, *Blue*, *Red*, and *White Enſigns*; as alſo, two *Purple Jacks*, with *Red*, *White*, and *Blue Pendants*.

FLAG of *Oſtend*, in *Flanders*; conſiſts of two Cloths, or Bands; the Uppermoſt *Red*, and the Lowermoſt *Yellow*.

FLAG of *Poland*; is *Red*, charg'd with an Arm coming out of a Cloud *Azure*, dreſs'd to the Elbow with White Cloth, and a Ruffle *Or*; holding in the Hand a Naked Sword *Argent*, and Hilt *Sable*.

FLAG of the *Pope*; is *White*, charg'd with the Image of St. *Peter*, and St. *Paul*: That of St. *Peter*, holding in his Right-Hand two Keys placed in *Saltier*, and a Book in his Left-Hand: That of St. *Paul*, holding a Book in his Right, and a Sword in his Left-Hand.

Their *Pendants* conſiſt of three Bands, one *White*, one *Yellow*, and another *Red*.

FLAG of *Port-a Port*, in *Portugal*; has Eleven Bands, *viz.* Six *Green*, and Five *White*.

FLAG of *Portugal*; is *White*, charg'd with the *Arms* of *Portugal*: Which bears *Argent*, five

Scutcheons *Azure*, placed cross-wise, each charg'd with as many Besants of the first, placed in *Saltier*, and pointed *Sable*: The Shield border'd *Gules*, charg'd with seven Towers *Or*; three in Chief, and two in each Flanch. The Crest is a Crown *Or*.

Another FLAG of *Portugal*; is *White*, charg'd with an Armillar Sphere *Or*, set on the Globe of the World *Azure*, with an Horizon *Or*, and a *Purple* Cross above.

Another FLAG of *Portugal*, is *White*, charg'd with a *Purple* Armillar Sphere, with a Cross *Gules* on each side, and one of the same above: And placed upon a Globe of the World *Azure*, with an Horizon *Or*.

Another FLAG of *Portugal*; is *White*, charg'd, towards the *Flag-staff*, with the Arms of *Portugal*; and in the middle, with a *Purple* Armillar Sphere, set on the Globe of the World *Azure*, with an Horizon *Or*; and above, a Cross *Gules*, sustain'd by a Pillar *Or*: And towards the end, is placed a Monk dress'd in Black, with a Cross *Gules* in the Right-Hand, and a Chaplet of Beads in the Left.

These *Portugueeze* Ships, that go to their *Indies*, carries one of the three last *Flags*.

FLAG of *Ragusa*, in *Dalmatia*; is *White*, charg'd with a *Scutcheon*, with this word, LIBERTAS.

FLAG of *Revel*, in *Livonia*; consists of six Bands, that is, three *White*, and three *Blue* ones.

FLAG of *Rostoc* in *Lower-Saxony*; consists of three Bands, the highest *Blue*, the middle *White*, the lowermost *Red*.

FLAG of *Savoy*; is *Red*, Quarter'd by a Cross *Argent*, within which, these four Letters, *F. E. R. T.* are placed, one in each, signifying, *Fortitudo Ejus Rhodum Tenuit*.

Another FLAG of *Savoy*, is *White*, charg'd with the Image of the *Nostre-dame*.

FLAG of *Schelling*, and *Fly*, in *West-Friezland*; consists of Ten Bands, or Cloths, which are, beginning from the uppermost, *Red*, *White*, *Blue*; *Red*, *Blue*, *Yellow*; *Green*, *Red*, *White*, and *Blue*.

FLAG of *Scotland*; is *Blue*, with a Franc-quarter *Argent*, charg'd with a Cross *Gules*.

Another FLAG of *Scotland*; is *Red*, with a Franc-quarter *Azure*, charg'd with a *Saltier*, (or St. *Andrew*'s-*Cross*.)

FLAG of *Sicily*; is *White*, charg'd with an Eagle *Sable*.

FLAG of *Spain*; is *White*, charg'd with the *Coat of Arms* of *Spain*: Which is Quarterly; The first Quarter Counter-quarter'd: In the First and Fourth *Gules*, a Castle Tripple-Tower'd *Azure*, (for *Castile*.) In the Second and Third *Argent*, a Lion Passant *Gules*, Crown'd, Langued, and Arm'd *Or*, (for *Leon*. In the Second great Quarter *Or*, four Pallets *Gules*, (for *Arragon*:) Party *Or*, four Pallets also *Gules*, betwixt two Flanches

Flanches *Argent*, charg'd with as many Eagles *Sable*, Member'd, Beak'd, and Crown'd *Azure*, (for *Sicily*.) These two great Quarters grafted in Base *Argent*, a Pomegranate *Verte*, Stalk'd, Leav'd of the same, Open'd, and Seeded *Gules*, (for *Granada*.) Over all *Argent*, five Scutcheons *Azure*, placed cross-wise, each charg'd with as many Besants in Saltier, of the first (for *Portugal*.) The Shield border'd *Gules*, with seven Towers *Or*, (for *Algarve*.) In the Third Quarter, *Gules*, a Fesse *Argent*, (for *Austria*,) Coupie and supported by Ancient *Burgundy*, which is Bandy of six Pieces *Or* and *Azure*, border'd *Gules*. In the Fourth Great Quarter, *Azure*, *Semè* of *Flower-de-Luce*, *Or*; with a Border *Company Argent* and *Gules*, (for Modern *Burgundy*;) Coupè *Or*, supported *Sable*, a Lion *Or*, (for *Brabant*.) These two great Quarters charg'd with a Scutcheon *Or*, a Lion *Sable*, and Langued *Gules*, (for *Flanders*.) Party *Or*, an Eagle *Sable*, (for *Antwerp*.) For Crest, a Crown *Or*, rais'd with Eight Diadems, or Semi circles, terminating in a *Mond Or*. The Collar of the *Golden-Fleece* encompasses the Shield; on the sides of which is placed the two Pillars of *Hercules*, one on each side, with this Motto, PLUS ULTRA.

Another FLAG of *Spain*; is *White*, charg'd with a Scutcheon Quarter'd; the first and fourth *Gules*, a Castle Tripple-Tower'd *Azure*: The second and third *Argent*, a Lion Passant, *Gules*, Crown'd, Langued, and Arm'd *Or*.

FLAG of *Stetin*, in *Upper-Saxony*; consists of two Bands, the uppermost *White*, charg'd with a Billet, (that is, the Form of a Letter folded up) *Gules*; and the lowermost *Red*, charg'd with a Billet *Argent*.

FLAG of *Stralsund*, in *Pomerania*; is *Red*, charg'd with a Sun *Or*.

FLAG of *Sweden*; is *Blue*, slit, and travers'd with a Cross *Or*.

FLAG of the *Turk*; is *Verte*, charg'd with three Crescents (or Half-Moons) *Argent*, with their Points inward.

Or, according to some, 'tis *Red*, charg'd with three Crescents *Argent*: And sometimes only one; and Crested with a Turban, charg'd with three Black Plumes of Herons Quills, with this Motto, DONEC TOTUM IMPLEAT ORBEM.

And this is never carried but by the *Grand-Signior* himself, or by his *Commission*.

Another *Turkish* FLAG; is *Blue*, charg'd with three Crescents *Argent*, with their Point outwards.

Another *Turkish* FLAG; is *Red*, charg'd with three Crescents *Argent*, in the same Order with the last.

The *Turks* have several other *Flags*, differently distinguish'd; but their Colours are always *Red*, *White*,

White, or *Green*: And are oftentimes charg'd with various Black Letters.

FLAG of the *Turkey Galleys*; is *Red*, and cut ſharp towards the End.

FLAG of *Tuſcany*; is *White*, charg'd with the Arms of the *Grand-Duke*; Who bears *Or*, five Roundles *Gules*, two, two, and one, and one in Chief *Azure*, charg'd with three Flower-de-Luce *Or*.

Another FLAG of *Tuſcany*; is *White*, charg'd with the Croſs of St. *Anthony*, *Gules*, with a Border *Or*.

FLAG of *Venice*; is *Red*, charg'd with a Lion Wing'd, Sejant *Or*, holding a Croſs *Or* in his Right-Paw; and a Book open, under his Left-Paw, with theſe words written, PAX TIBI, MARCE EVANGELISTA MEUS.

Another FLAG of *Venice*; like the former, only the Lion holds in his Right-Paw a Sword *Azure*, with a Hilt *Sable*.

Another FLAG of *Venice*; is *White*, charg'd with the ſame Lion.

FLAG of *the States General of the United Provinces*; is *Red*, charg'd with a Lion *Or*, holding with one Paw a Cutlaſs *Argent*; and in the other, a Bundle of Seven Arrows *Or* cloſely bound together, with Heads and Feathers *Azure*.

JACK-FLAG of the *States General*; conſiſts of Slips and Pieces, *Orange* and *Blue*, with a Croſs *Argent*; and charg'd in the middle with a Scutcheon bearing *Gules*, with a Lion as before.

Another FLAG of the *United Provinces*; charg'd with thrice the Letter *P.* ſignifying, *Pugno Pro Patria*.

FLAG of *Zeland*; conſiſts of three Bands, one *Orange*, one *White*, and the other *Blue*; on the middlemoſt, or *White* one, is placed the *Arms* of *Zeland*: Which is, Chief *Or*, a Lion *Gules*, riſing out of three Waves *Azure*, in a Field *Argent*.

FLAIR: When a Ship is ſomewhat *hous'd in*, near the Water, and above that the Works hangs over too much, and therefore is laid out broader than due Proportion allowes, then 'tis ſaid, *the Work do's Flair over*: And this makes the Ship more Roomy aloft, for the Men to uſe their Arms in.

FLAT *in the Fore-Sail*; that is, hale in the *Fore-Sail* by the Sheat, as near the Ship's Side as poſſible: This is done, when a Ship will not fall off from the Wind.

FLAW; ſignifies, a ſudden Guſt of wind.

FLIE; the ſame with *Fly*: Which ſee.

FLOOD; amongſt Seamen, is when the *Tide* begins to come up, or the Water to Riſe; then they call it *Young-Flood*; the next, *Quarter-Flood*, *Half-Flood*, and *Full-Sea High-* (or *Still-*) *-Water*.

FLOOK

FLOOK. See *Fluke*.

FLOOR, in a *Ship*; is so much of her Bottom, as she rests upon, when she lies upon the Ground.

FLOOR-TIMBER. See *Timber*.

FLOTA; a Name given by the *Spaniards* to the *Plate-Fleet*, which they send Yearly to some part of their *West-Indies*.

FLOTSON; signifies the Goods lost by Shipwrack, which lie floating upon the Water; and are given to the *Lord-Admiral*, by his Letters Patent.

FLOWN-SHEATS: A Ship is said to sail with *Flown-Sheats*, when her Sails are not haled Home, or close to the Blocks.

The Sheats are Flown; that is, they are let loose, or to run as far as they will.

FLUKE, or *Flook*, of an *Anchor*; is that part of it which fastens in the Ground. See *Anchor*.

FLUSH-DECK. See *Deck*.

FLUX of the *Sea*; that is, the Tide of *Flood*, See *Tide*.

FLY; is that part of the *Mariners-Compass*, on which the 32 *Winds* are drawn, and to which the *Needle* is fasten'd underneath. See *Compass*.

FLY-BOAT; is a great Vessel with a Broad *Bow*; some of which carry Seven or Eight hundred Tun weight of Goods.

Let Flye the Sheats! is a Word of Command, in case of a Gust of Wind, lest the Ship should Over-set, or spend her *Top-Sails* and *Masts*, to have the *Sheat go a-main*, and then the Sail will hold no Wind.

FOOT-HOOKS. See *Futtocks*.

FORBEAR! is a Word of Command, in a *Ship's-Boat*, to hold still any Oar, either on the Broad, or Whole Side.

FORE-CASTLE *of a Ship*; is that part where the *Fore-Mast* stands: 'Tis divided from the rest by a Bulk-head.

FORE-FOOT; signifies one Ship's lying or sailing cross another's Way: As if two Ships being under Sail in *Ken* one of another, one of them lying in her Course, with her Stem so much *a weather* the other, that holding on their several Ways, neither of them altering their Courses, the Wind-ward Ship will run a-head of the other; then 'tis said, such a Ship lies with the other's *Fore-Foot*: But as soon as she has pass'd her a-head, 'tis not said, she pass'd by her *Fore-Foot*; but, that she is gone out a-head.

FORE-JEER-BITS. See *Bits*.

FORE-LOCKS, or *Fore-lock-Keys*; are small flat Pieces of Iron, made Wedge-like, to be put into the Ends of *Bolts*, to keep down the *Cap-Squares* of of the *Carriages of Guns*.

FORE-LOCK-BOLTS. See *Bolt*.

FORE-KNIGHTS. See *Kights*.

FORE-MAST of a Ship; is a large round Piece of Timber placed in her Fore-part, or *Fore-Castle*,

Castle, and carrying the *Fore-Sail*, and *Fore-Top-Sail Yards*: Its Length is usually $\frac{8}{9}$ of the *Main-Mast*. And the *Fore top-Gallant-Mast* is $\frac{1}{2}$ the Length of the *Fore-Top-Mast*.

FORE-MAST-MEN; are those on-board that take in the *Top-Sails*, Sling the Yards, Furl the Sails, Bowse, Trice, and take their Turn at the Helm, *&c.*

FORE-RAKE. See *Rake*.

FORE-REACH: One Ship is said to *Fore-Reach* upon another, when, both sailing together, one sails better, or out-goes the other.

FORE-STAFF, or *Cross-Staff*; is an Instrument used at Sea, for taking the Height of the Sun, Moon, or Stars, with one's Face towards the Object.

FORMER; is a Piece of Wood, Turn'd round, and fitted to the Bore of a Piece of Ordnance; on which are made the Cartridges, which are the due Charge of Powder for the Gun.

FOUL-SHIP; is that which has bin long Untrimm'd, so that Grass, Weeds, Perwincles, or Barnacles, stick or grow to her Side under Water.

The Rope is Foul; that is, the Rope is entangled in it self, or hindred by another, so that it cannot run, or be haled.

The Anchor is Foul; that is, the *Cable* is got about the *Flook*.

The Ship makes Foul Water; that is, when under Sail, she comes into such Shole-Water, as to raise the Sand and Ooze with her Way; and tho' she don't touch the Ground, yet she comes so near it, that the Motion of the Water under her, raises the Mud from the Bottom, and so Puddleth or *Fouleth* the Water: Nor can the Ship then feel her Helm, as well as in Deep Waters.

To be Foul on each other; that is, when Ships come so close as to intangle their Riggings, and endamage each other.

FOUNDER: A Ship is said to *Founder*, when by an extraordinary Leak, or by a great Sea breaking in upon her, she is so fill'd with Water, that she cannot be freed of it; so that she'll neither Veer, nor Steer, but will lie like a Log; and not being able to swim long, at last will sink.

FRAIGHT of a *Ship*; that is her Burthen, or the Quantity of Goods she can carry,

FREE: The *Pump* is said to *Free the Ship*, when it throws out more Water than Leaks into her: But when it cannot throw out the Water as fast as it Leaks in, 'tis said, *The Pump can't Free her.*

Free the Boat; that is, Bailing or Lading out the Water therein.

FRESH-SHOT; signifies the Falling down of a great River into the Sea, so that there is Fresh-Water found in the Sea a good way from the Mouth of the River:

River: And sometimes this happens by a Descent of Land-Waters on a sudden; and as this is more or less, so 'tis call'd *Great* or *Small Fresh-Shot.*

FRESH *the Hawse.* See *Hawse.*

FRIGAT; is a Ship of War, light Built, and a good Sailer: They commonly have Two Decks.

Light-Frigat; is a Small Vessel of War, a good Sailer, with only One Deck.

FRIGATTON; is a *Venetian* Vessel, common in the *Adriatic-Sea*, with a Square Stern, and carrying only a Main, and Misen-Mast, and a Bow-Sprit.

FURLE, (*Bind*, or *Make up*,) as *to Furle the Sail*; that is, to wrap up, and bind it close to the Yard: Which is perform'd, by haling up the Braces, then wrapping up the Sail close together, and so binding it fast to the Yard with the *Caskets* and *Furling-Lines.*

FURLING-LINES; are small Lines made fast to the Top-sail, Top-Gallant-sail, and the Misen-Yard-Arms, to *Furl* up the Sails by.

FURRING *of a Ship*; is laying double Planks on her Sides, after she is built, as Occasion requires: Or when a Ship's Planks are ript off, other Timbers are put upon the former Timbers, and on them other Planks; and this is done, that the Ship may bear the better Sail.

FUTTOCKS in a *Ship*, are Timbers rais'd over the Keel, or the Compassing Timbers which make her Breadth.

Ground-Futtocks, are those next the Keel: The other are are call'd the *Upper-Futtocks.*

G

GAGE: *The Ship's Gage*, is so many Feet as she sinks in the Water, or so many Feet of Water as she *Draws*: Which Seamen find thus; By driving a Nail into a Pike, near the End; then put down this Pike by the Rudder, till the Nail catch hold under it: For then, as many Feet as the Pike is under Water, is the *Ship's Gage* required.

Weather-Gage: When one Ship has the *Wind*, or is to *Windward* of another, she is said to have the *Weather-Gage* of her.

GALE, at *Sea*; signifies the Blowing of the Wind.

A Brisk, *Stiff*, or *Strong Gale*; that is, when it blows very hard, or is so much Wind as the Top-sails can just endure to bear.

Fresh Gale, is a Wind that blows very Brisk; and is properly that which blows presently after a Calm, or when it begins to quicken.

Loom-Gale; is when the Wind blows not so hard, but that the Ship can carry her Topsails *a-Trip*, (that is, hois'd up to the highest.) And the best Sailing of any, is in a *Fair Loom-Gale*; for then the

the Ship can bear all her Sails, and does not run so high.

To Gale away: When two Ships are near one another at Sea, and there being but little Wind blowing, one of them finds more of it than the other; then they say, *The Ship Gales away from the other.*

GALEASS, is a heavy, low-built Vessel, using Sails, and Oars; and carrying Three *Masts*, which cannot be lower'd, as in a *Galley*, *viz.* a Main-Mast, Fore-Mast, and Misen-Mast: They have Thirty two Seats for Rowers, and Six or Seven Slaves to each: They also have Three Tire of Guns at the Head; the Lowermost has Two Pieces, of 36 Pounders each; the Second, Two Pieces, of 24 Pounders each; and the Third, Two other Pieces, of 10 Pounders each: At the Stern there are Two Tire of Guns, each of Three Pieces, and each Piece 18 Pounders.

GALEONS, or *Galions*; so the *French* formerly call'd their Great Ships of War; but now 'tis a word in Use only among the *Spaniards* and *Italians*: Tho', properly, the *Spaniards* call only those Vessels *Galeons* (whether Great or Small) that are Yearly sent to *Vera-Cruz*, in *New-Spain*; and if employ'd to any other Part, they are not call'd by that Name.

GALLERIES in a *Ship*, are, as it were, Balconies made on the Stern, without Board, with with Passages into them from within: They are more for Ornament, and the Commanders Conveniency, than for any real Use: And indeed, in Ships of War, all Open *Galleries* have the Inconveniency of facilitating the Boarding of the Ship that way.

GALLERY-LADDER. See *Ladder*.

GALLEY, is a Low-built Vessel, using both Sails and Oars: They commonly carry Two Masts, *viz.* a Main-Mast, and a Fore-Mast; which may be Struck, or Lower'd, at Pleasure: Their Length is generally about 130 Feet; and Breadth, at the middle, about 18 Feet.

GALLIOT, is a little *Galley*, or a sort of *Brigantine*, built very slight, and fit to Chase; carrying but One *Mast*, and Two or Three *Pattereroes*; it can both Sail and Row, and has Sixteen or Twenty Seats for the Rowers, with One Man to each Oar: All the Seamen on-board are Soldiers; and each has a Musket ready by him, upon quitting his Oar.

GANG, or *Crew*: The Company wherewith a Ship's-Boat is Mann'd, is call'd that *Coxswain's Gang*, or *Crew*, who has the Charge of the Boat; as the *Barge's Gang* or *Crew*, *&c.*

GANG-WAY; is the several Passages, or Ways, from one part of the Ship to the other; and whatsoever is laid in any of those Passage, is said to lie in the *Gang-Way*.

GARBOARD-*Plank*, is the Plank next the *Keel* on the Outſide.

GARBOARD-*Strake*, is the firſt Seam in a Ship, next the *Keel*.

GARLAND, in a *Ship*, is a Collar of Ropes wound about the Head of the Main-Maſt, to keep the Shrowds from galling or fretting.

GARNET, is a Tackle having a Pennant coming from the Head of the Main-Maſt, with a Block ſtrongly ſeiz'd to the Main-Stay, over the Hatch-way, wherein is reev'd a *Runner*, with a Hook at one End, in which is hitch'd the Slings; and at the other End is a Double-Block, wherein the Fall of the Runner is reev'd: So that by it, any Casks, or Goods, that are not over-heavy, may be haled and hois'd into, or out of the Ship. When this *Garnet* is not uſed, 'tis faſten'd along by the Stay, at the Bottom of it.

Clew-Garnet. See *Clew*.

GATE *of the Sea*, or *Sea-Gate*. When two Ships are aboard one another, by means of a Wave, or Billow; then 'tis uſual to ſay, That they lie aboard one another in a *Sea-Gate*.

To GATHER: At *Sea*, 'tis common to ſay, *We Gather on him*; that is, We get the Wind of him.

GEAR, as, *About your Gear!* that is, Work on all hands!

GIFT-ROPE. See *Rope*.

GIRDING-GIRT: A Ship is *Girt*, or has a *Girding-Girt*, when the *Cable* is ſo *taught*, or ſtrain'd, that upon the Turning of the Tide, the Ship cannot go over it with her Stern-Poſt, but muſt lie a-croſs the Tide's-Way.

GOARING: When a Sail is cut ſloaping, by degrees, and is broader at the *Clew*, than at the *Earing*, 'tis ſaid to be cut *Goaring*; as all *Top*, and *Top-Gallant-Sails*, are.

GOOSE-WING. Sometimes when a Ship ſails before the Wind, or with a Quarter-Wind, and in a Freſh-Gale, the Miſen-Yard is unparel'd and launch'd with the Sail over the Lee-Quarter, and *Guys* fitted at the further End, to keep the Yard ſteady; and a *Boom*, to boom out the *Miſen-Sheat*: And this is done, to give the Ship the more Way; which otherwiſe, with theſe Winds, the Miſen-Sails could not do; and the Sail thus fitted, is call'd a *Gooſe-Wing*.

GRAPLINGS, or *Grapnels*. See *Anchor*.

GRATINGS, in a *Ship*, are ſmall Ledges of ſaw'd Planks, fram'd one into the other, like the Lettice of a Priſon-Grate, lying on the Upper-Deck, between the *Main-Maſt* and *Fore-Maſt*; ſerving for a Defence in *Cloſe-Fights*, therefore are ſometimes ſo called: They alſo ſerve for Coolneſs, Light, and other Conveniency. And at the Head of the Ship, where the Neceſſary-

Houſe

Houſe ſtands, there is another *Grating*.

GRAVING *of a Ship*, or, *To Grave a Ship*; is, to bring her to lie dry a-ground, then burn off all the old Filth that ſticks to her Sides without board.

GREAT-CIRCLE-*Sailing*; is *Sailing* by, or upon a *Great-Circle* paſſing thorow the *Zeniths* of the two Places: Tho' this ſort of *Sailing* is very exact, yet 'tis very difficult, and indeed, hardly poſſible for a Ship exactly to Sail by; but it may be of good Advantage, to keep conveniently near it, eſpecially in a *Parallel* (or *Eaſt* and *Weſt*) *Courſe*.

In *Great-Circle-Sailing*, there are Three Caſes; *viz.*

Firſt, When two Places *differ* only in *Latitude*.

Secondly, When they *differ* only in *Longitude*.

Thirdly, When they *differ* both in *Latitude* and *Longitude*.

GRIPE *of a Ship*, is the Compaſs or Sharpneſs of her *Stem* under Water, chiefly towards the Bottom of it; and ſhe is ſo ſhap'd, that ſhe may *Gripe* the more, or keep good Wind: Therefore a *Falſe-Stem* is ſometimes put upon the True one.

The Ship Gripes; that is, turns her Head to the Wind more than ſhe ſhould; and this is cauſed, either by Over-loading her a-head, the Weight of which preſſes her down, ſo that it will not readily fall off from the Wind; or by *Staying* or *Setting* her Maſts too much *aft*; which always will be a fault in ſhort Ships, that Draws much Water, and will cauſe her to be continually running into the Wind: Tho' in *Floaty Ships*, if the Maſts be not Stay'd very far *aft*, they will never keep a good Wind.

GROMETS, are ſmall Rings, made faſt by *Staples* to the upper ſide of the Yard of a Ship, to tie unto it, or to faſten the *Laskets*.

GROUND-TACKLE, ſignifies a Ship's *Anchor*, *Cables*, &c. in general; or whatever is neceſſary to make her ride ſafe at Anchor.

GROUND-TIMBERS, are thoſe *Timbers* in the Ship which lie on her Keel, and are faſten'd to it with Bolts, thro' the *Keelſon*; and are ſo call'd, becauſe the Ship reſts upon them when ſhe is a-ground.

GROUNDING *of a Ship*, is bringing of her on Ground, to be Clean'd, Trimm'd, Scrubb'd, or have a Leak Stopt.

GUDGEONS, are the Eyes drove into the Stern-Poſt, into which the Pintles of the Rudder go, to hang her on.

GUEST-ROPE. See *Rope*.

GULF, is a Part of the Ocean which runs up into the Land, through Straits, or Narrow Paſſages: As, the *Gulf of Venice*, or the *Adriatic-Sea*, in *Europe*; the *Gulf of Perſia*, in *Aſia*; the *Gulf of Arabia*, or the *Red-Sea*,

in *Africa*; and the *Gulf of Florida*, in *America*.

GUNNEL, the ſame with *Gun-Waile*.

GUNNER *of a Ship*, or *Maſter-Gunner*, has the Charge of all the Ordnance the Ship carries; to ſee that they be ſerviceably Mounted, and ſufficiently ſupply'd with Spunges, Ladles, and Rammers; that, in Foul-Weather, they be travers'd within Board, (eſpecially thoſe of the Lower-Tire,) and that the Ports be ſhut, and Caulk'd up; and that at all times they may be well Laſh'd, and made faſt, leſt any one of them ſhould chance to break looſe, to the imminent Danger of Foundring the Ship: And in Time of an Engagement, he is to provide that every Piece be ſufficiently Mann'd: He is to be Cautious and Provident in the Guard of the *Powder* in the *Powder-Room*, and upon no Occaſion to ſuffer any Fire to come near it, unleſs it be a Candle in a well-glaz'd Lantern: He is to give an Account of all his Charge, upon Demand. He has a *Mate*, and *Quarter-Gunners*, for his Aſſiſtance.

GUNNERS-TACKLE. See *Tackle*.

GUN ROOM, is the Appartment under the *Great-Cabin*, which is the proper Place of Rendevous of the *Maſter-Gunner*, and his *Crew*; where they get their Carthrages ready, as alſo all other Neceſſaries belonging to them.

GUNTER's-SCALE. See *Scale*.

GUN-WAIL, or *Wun-Wale of a Ship*; is the upmoſt *Wail* of a Ship, or that Piece of Timber which reaches on either Side from the Quarter-Deck to the Fore-Caſtle, being the uppermoſt *Bend* which finiſhes the upper-Works of the Hull in that part wherein they put the *Stanchions*, which ſupport the *Waſt-Trees*.

GUY. is any Rope uſed to keep off things from bearing or falling againſt the Ship's Side, as they are hoiſted in: So that if any thing is to be haled in over the *Gun-Wale*, it's gently eaſed in by a *Guy-Rope* made faſt to the Stanchions of the Waſt-Trees.

GUY-ROPE, is that *Rope* which is made faſt to the Fore-Maſt at one end, and is reev'd thro' a ſingle Block ſeiz'd to the Pennant of the Winding-Tackle, and then again reev'd thro' another ſeiz'd to the Fore-Maſt, and whoſe Uſe is to hale forward the Pennant of the Winding-Tackle.

H

HAILE: *To Haile a Ship*; that is, to Call to her, to know whence ſhe is, and whither bound: The uſual Call is, *Hoe! the Ship?* The other Anſwers, *Halo!* &c. And to Salute another Ship with *Trumpets*, and

and the like, is also call'd *Hailing*.

HALE, signifies *Full*: As, to *Hale up*, is to Pull up; to *Hale in*, or *out*, is to Pull in, or out.

To Over-Hale a Rope, is to *Hale* it too stiff, or *Hale* it the contrary way.

Keel-Hale. See *Keel*.

HALLIARDS, are those Ropes by which all Yards are hoisted up. The *Cross-Jack*, and the *Sprit-sail-Yard* have no *Halliards*, because those are always Slung; tho' in Small Ships, they have *Halliards* to the *Sprit-sail-Yards*.

HAMMOCK; a Piece of Canvase hung up fast by the four Corners, between Decks, for the Seamen to sleep in.

HAND, or *Handing*. When a thing is to be Deliver'd away, or pass'd from one to another, or to be brought to any one; then the Word is, *Hand it this way*, or *that way!* Also, When more Men are wanted to do any Labour, as *Hoisting*, &c. they call for more *Hands*, not more Men.

HAND-SPIKE; is a Wooden-Leaver, with which we Traverse the *Ordnance* at Sea; or Heave withal in a *Windlass* to weigh up the *Anchor*.

HARPINGS; is properly the Breadth of the Ship at the *Bow*. Also the Ends of the *Bends* which are fasten'd into the Stem, are, by some, call'd *Harpings*.

HATCHES *of a Ship*, are like *Trap-Doors* in the *Midships*, or between the *Main-Mast* and *Fore-Mast*. This way, all Goods of Bulk are let down into the *Hold*.

HATCH-WAY, is that Place where the *Hatches* are: So that to lay a Thing in the *Hatch-way*, is to put it so, that the *Hatches* cannot be come at, or open'd.

Coamings of the Hatches: When the *Hatches* are rais'd up higher than the rest of the Deck, those Pieces of Timber or Planks, which raise, and bear them up, are call'd *Coamings of the Hatches*. See *Scuttles*.

HAWSER; is a great Rope, or a sort of a small Cable, serving for various Uses a-board a Ship; as, to fasten the *Main* and *Fore-Shrowds*; to *warp* a Ship by as she lies at Anchor, and wind her up to it by a Capstan, &c. And a Great Ship's *Hawser*, may serve for a Cable to the *Sheat-Anchor* of a Small Ship.

HAWSES, are great Holes under the Head of the Ship, thro' which the Cables run, when she lies at Anchor.

Bold-Hawse; that is, when the Holes are high above the Water.

Fresh the Hawse! or, *Veer out more Cable!* is when part of the *Cable* that lies in the *Hawse* is fretted or chaf'd, and 'tis requir'd that more *Cable* may be *Veer'd out*, so that another part of it may rest in the *Hawse*.

Fresh

Fresh the Hawse! that is, Lay new Pieces upon the Cable in the *Hawse*, to preserve it from fretting.

Burning in the Hawse; that is, when the Cable endures a violent Stress.

Clearing the Hawse: When two Cables that come thro' several *Hawses* are twisted and entangled one with the other, the Untwisting of them, is call'd *Clearing the Hawse*.

Thwart the Hawse, or *Rides upon the Hawse*; that is, when a Ship lies *thwart*, or cross, or with her Stern just before another Ship's *Hawse*.

HAWSE-FULL *To Ride Hawse-full*. See *Ride*.

HEAD *of a Ship*, or *Boat*; that is, the Fore-part. *The Chase stands right a-Head*; that is, right before us.

HEAD-LAND. See *Land*.

HEAD-LINES, are the Ropes of all Sails which are next to the Yards, and by which the Sails are made fast to the Yards.

HEADS, *Rung-Heads*. See *Rung*.

HEAD-SAILS, are those which belong to the *Fore-mast* and *Boltsprit*; for 'tis they that govern the *Head* of the Ship, and do make it *fall off*, and keep out of the Wind: And these, in Quarter-Winds, are the chief Drawing-Sails.

HEAD-SEA, is when a great Billow, or Wave of the Sea, meets the Ship right *a-head*, as she sails in her Course.

HEAVE! That is, to throw away: As, *Heave it Over-board!* that is, Fling it Over-board!

Heaving at the Capstan; that is, turning about the *Capstan*.

Heave and Set; that is, when a Ship being at Anchor, rises and falls, by the force of the Waves, then she is said to *Heave and Set*.

Heaving a-Peek. See *Peek*.

Heave the Lead! See *Lead*.

HEEL: If a Ship lean on one side, whether she be *a-ground*, or *a-float*; then 'tis said, *She Heels a-Starboard*, or *a-Port*; Or that she *Heels offward*, or to the *Shore*; that is, enclines more to one side than the other.

HEEL *of the Mast*; signifies that part of the Foot of any Mast which is par'd away slanting on the Aftward side thereof, in order that it may be *Stay'd aftward on*, The *Heels* of the Top-Masts are Squares.

HEIGHT *of the Sun*, or *Star*, is an Arc of a Vertical-Circle intercepted between the *Sun*, or *Star*, and the *Horizon*. The Knowlege of this, when taken at *Noon*, is of great Use in *Navigation*, in order for finding the *Latitude of the Place* the Ship is in, or Distance of the *Zenith* from the *Equator*.

Meridian-Height. See *Meridian*.

HELM, or *Tiller*, of a Ship; is that Piece of Timber which is fasten'd into the Rudder, and so comes forward into the Steerage, or Place where he at the *Helm*

 Steers

Steers the Ship, by holding the *Whipstaff* in his Hand. Some Ships have a Wheel, like those in *Cranes*, placed between the *Quarter-Deck* and *Coach*; which has several Advantages, to what the Common Methods have.

Bear up the Helm! that is, Let the Ship go more Large before the Wind.

Helm a Mid-ship! or, *Right the Helm!* that is, Keep it Even with the Middle of the Ship.

Bear up Round! that is, Let the Ship go directly before the Wind, in the middle, between her two Sheats.

Port the Helm! that is, Put the *Helm* over to the Left-side of the Ship!

Starboard the Helm! that is, Put it to the Right-side of the Ship.

Ease the Helm! See *Ease*.

HEMISPHERE; is the Half of a *Sphere*, or *Globe*, when 'tis supposed to be cut thro' the Centre, in the Plane of one of its great Circles.

Thus, the *Equator* divides the *Terrestrial* Globe into the *Northern* and *Southern Hemispheres*.

The *Meridian* divides the Globe into the *Eastern* and *Western Hemispheres*.

The *Horizon* divides the Globe into the *Upper* and *Lower Hemispheres*.

HEMISPHERES, are also, those *Maps* or *Draughts* of the *Heavens*, Constellations, *&c.* that are usually made for the Benefit of Navigators.

HITCH; is to catch hold of any thing with a Hook, or Rope, to hold it fast: As, *Hitch the Fish-Hook to the Fluke of the Anchor!* That is said, when they are about to Weigh the *Anchor*.

Hitch the Tackles into the Rings of the Boat! That is said, when the *Boat* is to be Hoised in.

HOISE; is to Hale up any thing into the Ship; or getting up a Yard, *&c.* As, *Hoise up the Yard! Hoise the Water in!* &c.

HOLD; is that part of a Ship between the *Keelson* and the *Lower-Deck*: Wherein, divided by Bulk-Heads, are the *Powder-Room*, *Bread-Room*, *Steward's-Room*, &c. And in a *Merchant-man*, all the Goods, and Lading in general, are Stow'd in the *Hold*.

To Stow the Hold; is to take Goods into the *Hold*.

Predy the Hold. See *Predy*.

To Rummidge the Hold; is to Remove, or Clear, the Goods and Things therein.

HOLD-OFF: In Heaving in the Cable at the Capstan, if it be very stiff and great, or have lain long in a Slimy or Oazy-Ground; unless that part of it which 'tis heav'd in by, be hal'd away hard from the Capstan, the Cable will Surge, or slip back: Therefore, it must be haled away as fast as it comes in, that the Cable may keep close about the *Whelps*; and this Work is call'd, *Holding-off*.

He Holds his own: That is, ſpeaking of a Ship under Sail, He keeps his Courſe Right forward.

HOOKS; are thoſe fork'd Timbers, in a Ship, which are placed directly upon her Keel, as well in her *Run*, as in her *Rake*.

Foot-HOOKS; the ſame with *Futtocks*: Which ſee.

Boat-HOOK; a *Hook* uſed to Fend-off the *Boat* from Bruiſing.

Shear-HOOKS; are *Hooks* like Sickles, fix'd in the Ends of the Yard-Arms, that if a Ship under Sail come to Board her, thoſe *Shears* may cut her Shrowds, and ſo ſpoil her Tacklings: But becauſe they are ſo ſubject to break their own Yards, and cut the Ropes that come from the Top ſails, they are now very ſeldom uſed.

Can-HOOKS, being made faſt to the End of a Rope, with a Nooſe, (like that which Brewers uſe to ſling or carry their Barrels on,) and made Uſe of for Slings.

Looff-HOOKS, is a Tackle with two *Hooks*; one, to hitch into a Cringle of the *Main* or *Fore-Sail*, in the Bolt-Rope, at the Leetch of the Sail, by the Clew; and the other is to hitch into a Strap which is ſplic'd to the Cheſtree: Their Uſe is, to pull down the Sail, and ſuccour the Tackles in a large Sail, and ſtiff Gale, that all the Streſs may not bear upon the Tack. It is alſo uſed, when the Tack is to be ſeiz'd more ſecure: And to take off, or put on a *Bonnet*, or *Drabbler*.

Cat-HOOKS. See *Cat*.

HORIZON; is that Circle of the Sphere which divides the Heavens and the Earth into Two Parts; an *Upper*, and a *Lower*.

Senſible HORIZON; is that Circle which limits our Sight, and may be conceiv'd to be made by ſome great Plain, or the Surface of the Sea.

It divides the Terreſtrial Globe into Two Parts; the one *Light*, the other *Dark*; which are ſometimes greater, or leſs, according to the Condition of the Place.

It Determines the Viſible Riſing and Setting of the *Sun*, *Moon*, or *Stars*, in any particular Latitude. For when any of theſe appear juſt at the *Eaſtern* Part of the *Horizon*, 'tis then ſaid to *Riſe*; and when it does ſo at the *Weſtern* Part, 'tis then ſaid to *Set*.

Rational or *Real Horizon*; is a great Circle which Divides the Globes into Two Equal Parts, or *Hemiſpheres*: And the Poles thereof, are the *Zenith* and *Nadir*; or the Two Points, one directly over our Heads, the other oppoſit to it under our Feet.

From this, the *Altitude* of the *Sun*, or *Stars*, is Accounted, which is their Height above the *Horizon*.

This Circle is Repreſented by the *Card* of the *Mariners-Compaſs*, which is Divided into 32 *Points*. See *Compaſs*.

Right-HORIZON; is that which cuts the *Equator* at *Right Angles*.

Oblique-HORIZON; is that which cuts the *Equator* Obliquely.

Parallel-HORIZON; is that where the *Pole* of the World is the *Zenith*: Or is that *Horizon* which either is in the *Equator*, or some Parallel to it.

HORIZONTAL; that is, Parallel to the *Horizon*.

HORSE; is a Rope made fast to the Fore-Mast-Shrowds, having a *Dead-Man's-Eye* at its End, thro' which the Pennant of the Spritsail-Sheats is reev'd: It's Use is, to keep the Spritsail-Sheats clear of the Flukes of the Anchor.

HORSE; is also, that Rope which is made fast to the Shrowds, to preserve him that heaves the Lead there, from falling into the Sea.

HORSES; the *Waps* are sometimes so called: As also, those short *Waps* that are seiz'd to the middle of the *Top-mast* and *Top-Gallant-mast-Stay*, wherein are reev'd the *Top-sail* and *Top-Gallant-sail Bow-lines*.

HOUNDS; are Holes in the Cheeks, at the Top of the Masts, thro' which the *Ties* run, to Hoise the Yards: The Top-Mast has but one Hole or *Hound*, and one *Tie*.

HOWKER, or *Houcre*; is a Vessel, common with the *Dutch*, built somewhat like a *Pink*, but Masted and Rigg'd like a *Hoy*: They'll Sail well, and that near the Wind: They Tack soon and short, and Live in any Sea. The Largest *Houkers* carry about 200 Tuns; others there are of 50 or 60 Tuns: One of these, Mann'd with five or six Sailors, is sufficient to make an *East-India* Voyage.

HOWSING-IN: After a Ship is past the Breadth of her Bearing, and that she is brought in too narrow to her Upper Works, she is said to be *Housed-in*, that is, Pinched in.

HOWLE: When the *Futtocks*, or *Foot-Hooks* of a Ship are Scarf'd into the Ground-Timbers, and Bolted; and the Plank laid on them up to the Orlop, the *Carpenters* say then, They begin to make the Ship *Howle*.

HOY, or *Huoy*; is a small Vessel, or Bark, whose Yards are not a-cross, nor the Sail square, like those of other Ships: Her Sails are cut like a Misen; so that she can sail nearer the Wind, than a Vessel with Cross-Sail can do.

HOYSE; the same with *Hoise*: Which see.

HULKS; are large Vessels, having their Gun-Decks from 113 to 150 *Feet Long*; and from 31 to 40 *Feet Broad*: They contain from 400 to 1071 *Tuns*: Their Use is chiefly for setting Masts of Ships, and the like.

HULL *of a Ship*; is the full Bulk, or Main Body of a Ship, without Masts, or any Rigging, from Stem to Stern.

To

To Strike a Hull; that is, in a Storm, to lie close and obscurely in the Sea, or tarry for some Consort, bearing no Sail, with the Helm *lash'd a-Lee*.

To Hull, or *Lie a-Hull*, or *Hulling*; is said of a Ship, when either in a dead Calm, (to prevent her beating the Sails against the Masts, by Rolling;) or in a Storm, when she cannot carry them, she takes all her Sails in, so that nothing but her Masts, Yards and Rigging are abroad; the Helm is lash'd fast to the Lee-side of the Ship: In this Condition, if she is a good Sailer, she will lie easily under the Sea, and make her Way one Point before the Beam.

HULLOCK *of a Sail*; is a Piece of the Misen, or some other Sail, part open'd or left loose, all the rest being fast made up to the Yard-Arm; its Use is, to keep the Ship's Head to the Wind. In case a Ship will not *Weather-Coil*, to lay her Head the other Way; loose a *Hullock* of her *Fore-sails*, and then change the Helm to the Weather-side; by that means, she is made to fall off, and to lay her Head where her Stern was before.

HURRICAN, is a most furious Storm, which the *West-Indies*, and especially the *Caribbe-Islands*, are subject to; and which so dreadfully afflict them, in or about the Month of *August*. Their Extent and Continuance is but small, nor do they happen Yearly; but then their Violence is unconceivably strange, and their Effects wonderfully surprising; the Sea flies in the Air in a terrible manner, and drowns all the adjacent Ground; insomuch that Ships have bin driven over the Tops of high Trees, many Leagues in the Land, and there left.

HYDROGRAPHY, is the Art which teaches how to Describe and Measure the *Sea*; giving an Account of its Tides, Counter-Tides, Soundings, Bays, Gulfs, Creeks, *&c.* As also the Rocks, Shelves, Sands, Shoals, Promontories Harbours, Distance from one Port to another; and other Things remarkable on the Coasts.

HYDROGRAPHICAL-*Charts*, are certain *Sea-Maps*, Delineated for the Use of Pilots, and other Mariners, wherein are mark'd all the *Rhumbs*, or Points of the *Compass*: As also the Rocks, Shelves, Sands, and Capes; And the Meridians drawn Parallel one to another. See *Charts*.

J

JACK; is that *Flag* that is hois'd up the *Spritsail-Top-mast-Head* of a Ship. See *Flags*.

JACOB's-STAFF; the same with the *Cross-Staff*. See *Staff*.

JEAR-CAPSTAN, or *Jeer-Capstan*. See *Capstan*.

JEER-ROPE. See *Rope*.

JEERS; As, *To be brought to the Jeers*, (speaking of one to be Punish'd at the *Jeer-Capstan.*) See under the word *Ducking*.

JETSON, is a Thing cast out of a Ship, being in Danger of Wrack, and beaten to the Shore, by the Waters, or cast on the Shore by the *Mariners*.

IMPREST. See, *To Man a Fleet*.

IRON-SICK; is said of a Ship, whose Bolts, Spikes, or Nails, are so eaten with Rust, as to stand hollow in the Planks, and to make her Leake. This is prevented, by putting Lead over all the Bolt-heads under Water.

JONKS, or *Jonques*; are Vessels very common in the *East-Indies*: They are about the Bigness of our *Fly-Boats*; but differ in Form of Building, according to the different Methods of the Nations, in those Parts, they belong to.

Their Sails, oftentimes, are only of Reeds, and of Matts; and their Anchors are made of Wood.

JOURNAL, at *Sea*, is a Book kept by the Officers of a Ship; where an Account of her Way is duly inserted, as also, the Changes of the Weather, with all Remarkable Accidents, and Occurrences.

ISLAND, or ISLE, is a Tract of Dry Land, environ'd with Water.

1. *ISLANDS belonging to* EUROPE, *are*,

The *Mediterranean* Islands, lying South of *Europe*.

The *Azores*, or *Western* Islands, lying West of *Spain*.

The *Britannic* Islands, lying North of *France*.

The *Scandinavian* Islands, lying in the *North* and *Baltic-Sea*.

The Island of *Ice-land*, lying ing West of *Scandinavia*.

2. *ISLANDS belonging to* ASIA, *are*,

Ceylon, and the *Maldives*, lying West of the Isles of *Sunda*.

The Islands of *Sunda*, lying West of the *Moluccoes*.

The *Moluccoes*, lying South of the *Philippine* Isles.

The *Ladron* Isles, lying East of the *Philippine*.

The *Philippine* Isles, lying South-West of the *Japan* Isles.

The *Japan* Islands, lying East of *China*.

3. *ISLANDS belonging to* AFRICA, *are*,

The *Madera*, lying West of *Barbary*.

The *Canary* Islands, lying West of *Bildulgerid*.

The *Cape de Verde* Isles, lying West of *Negroland*.

St. *Thomas*'s Island, lying West of *Ethiopia*.

The

The *Princess* Island, lying West of *Ethiopia*.

St. *Helena*, lying West of St. *Thomas*.

Isle of *Ascension*, lying North-East of St. *Helena*.

Madagascar, or St. *Lawrence*, lying East of *Ethiopia*.

The Isles of *Comore*, lying North-West of *Madagascar*.

4 *ISLANDS belonging to* AMERICA, *are*,

Newfoundland, lying East of *Terra Canadensis*.

California, lying West of *Nova Granada*.

Terra del Fuogo, lying South of *Terra Magellanica*.

The *Antilles*, Greater and Lesser.

The Greater *Antilles*, are *Cuba*, *Jamaica*, *Hispaniola*, *Porto-rico*, all East of *New Spain*.

The Lesser *Antilles*: As,

Caribee Isles, lying South-East of the Greater *Antilles*.

The *Sorovento*, lying North of *Terra-Firma*.

The *Lucayes*, or the *Bahama* Isles, lying South-East of *Florida*.

Bermudas, or *Summer-Isles*, lying East of *Florida*.

ISTMUS, is a narrow Neck of Land annexing a *Peninsula* to the Continent, by which People may go into the one from the other.

As in *Europe*; The *Istmus* of *Corinth*, joining *Morea* to *Greece*.

The *Istmus* of *Taurica Chersonesus*, joining *Taurica Chersonesus* to *Little Tartary*.

In *Asia*; The *Istmus* of *Malacca*, joining *Malacca* to *Peninsula Indiæ intra Gangem*.

In *Africa*; The *Istmus* of *Swez*, joining *Africa* and *Asia*.

In *America*; the *Istmus* of *Panama*, joining *Mexico* to *Peru*.

JUNKES, signifies only, Old *Hawsers*.

JURY-MAST. See *Masts*.

K

KECKLE; *To Keckle the Cable*, or *Bolt-Rope*, is to Serve it, or bind some small Ropes, or Old Clouts, about it, to prevent the *Cable* from galling ing in the *Hawse*, or *Bolt-Rope* from doing so against the Ship's Quarter.

KEDGER, or *Kedge-Anchor*. See *Anchor*.

KEDGING; is when a Ship is brought up and down in a narrow River, the Wind being contrary to the Tide, and yet she is to go with the Tide; then set the *Fore-Sail*, *Fore-Top-Sail*, and *Misen-Sail*, and let the Ship drive with the Tide, so that they may *flat her about*: And if she happen to come over near the Shore, then having a small Anchor in the Head of the Boat, with a small Hawser fasten'd to it from the Ship, which let drop in the middle of the Stream, and

this will *wend*, or turn her Head about: When ſhe is come fully about, they lift up the Anchor again; and this Working, is to *Kedge*, or *Kedging*; and this Anchor is call'd *Kedger*.

KEEL, is the firſt or loweſt Piece of Timber in a Ship; it lies in the very bottom of her Hull, one End whereof is at the Stern, the other at the Stem; and into this are all the Ground-Timbers and Hooks faſten'd and bolted fore and aft.

Falſe-KEEL; is a *Keel* put on under the firſt, in caſe the Ship be Shallow, and ſo Over-floaty, and Roll too much.

Rank KEEL; that is, a deep *Keel*; ſuch as will keep the Ship from Rolling.

KEEL-HALE, or *Keel-Rake*. See *Ducking*.

KEEL-ROPE. See *Rope*.

KEELSON, is a long Piece of Timber, like the *Keel*, laid over the *Floor-Timbers* of a Ship, lying within, as the other without, and directly over it; and faſt bound together with ſtrong Iron-Bolts thro' the Timbers and all.

KENKS, are Doublings in a Rope, or Cable, when Handed in or out, ſo that it does not run eaſie: Or when any Rope makes Turns, or Twiſts, and does not run free in the Block, then 'tis ſaid to make *Kenks*.

KETCH, is a Veſſel ſmaller than, and ſomething like a *Hoy*; but is ſo built, that 'twill endure and live in any Sea, or Weather whatſoever, and ſails very well: ſuch Veſſels are very uſeful and proper to attend upon great Ships, for carrying their Stores, and other Neceſſaries.

BOMB-KETCHES, or BOMB-*Ships*, are thoſe Veſſels, with *Mortar-Pieces* on-board, which are employ'd in *Bombarding* an Enemy's Sea-Port-Towns, *&c.*

KEVELS, or *Chevils*, the ſame with *Knevels*; which ſee.

KNAVE-LINE, is a Rope made faſt to the *Main* or *Fore-Top*, whence it comes down by Ties to the *Ram-head*, where 'tis reev'd thro' a ſmall Piece of Wood, and ſo brought to the Ship's-ſide, and hal'd *taught* to the Rails: Their Uſe is, to keep the *Ties* and *Halliards* (when New) from turning about one another.

KNECKS, the ſame with *Kenks*; which ſee.

KNEES, are Crooked Pieces of Timber, bow'd like a *Knee*, that binds the *Beams* and *Futtocks* together, being bolted faſt into them both: ſome of theſe ſtand right up and down, ſome along Ship; they are uſed about all the Decks: ſome ſaw'd or hew'd to that form, and ſome growing ſo, naturally, which are certainly the beſt for Service.

KNEE-TIMBERS, are thoſe Timbers which are fit and uſeful for making *Knees* of, in Building of Ships.

Carling-KNEES. See *Carling*.

Knee of the Head; that is, the Cut-water of the Ship.

KNETTLES.

KNETTLES; are two Pieces of Spun-Yarn, put together untwisted, with a Knot at each End, to seize a Block, Rope, or the like.

KNEVELS, or *Keve's*, are small Pieces of Wood, nail'd to the Inside of the Ship: Their Use is, to Belay the *Sheats* and *Tacks* unto.

KNIGHTS; are two thick short Pieces of Wood, commonly Carv'd like a Man's Head, with four *Shivers* in each; three for the *Halliards*, and one for the *Top-Ropes* to run in.

Fore-KNIGHTS, is that which stands fast bolted to the *Beams*, abaft the *Fore-Mast*.

Main-KNIGHT, is that which stands abaft the *Main-Mast*.

KNOTS used at Sea, are distinguish'd into these; *viz*.

Wale-KNOT, is a round *Knot* so made with the Lays of a Rope, that it cannot slip; and serves for Sheats, Tacks, and Stoppers.

Bow-Line-KNOT is so firmly made, and fasten'd to the Crengles of the Sails, that they must break, or the Sail split, before 'twill slip.

Sheep-Shank-KNOT, is to shorten a Rope without cutting it, which may be presently loosen'd, and the Rope not the worse for it.

KNOTS *on the Log-Line*, are the Divisions of it, which usually are Seven Fathom, or Forty-two Feet asunder; tho' they really should be *Fifty Feet*. And as many *Knots* as the *Log-Line* runs out in Half a Minute, so many Miles does the Ship sail in an Hour; supposing her to run at the same Rate. See *Log*.

L

LABOUR: At *Sea*, a Ship is said to *Labour*, when she Rolls and Tosses much.

LADDERS, in a *Ship*, are distinguish'd into Three sorts; *viz*.

Bolt-Sprit-LADDER, which is at the Beak-Head, made fast over the *Bolt-Sprit*, to get upon it when there is Occasion.

Entring-LADDER, made of Wood, and is placed in the *Wast* of the Ship.

Gallery-LADDER, made of Ropes, and hung over the *Gallery*, and *Stern* of Ships, for to come out, or go into a Boat by, in Foul-Weather, and High-Sea.

LADLE, is an Instrument to Load the Guns with Powder.

LAGAN, *Lagon*, or *Ligan*, is a Wrack which lies in the bottom of the Sea.

LAND-FALL; that is, to Fall in with the Land.

A Good Land-Fall, is when a Ship *makes* or sees the Land as she expected, or according to her Reckoning: And,

A Bad Land-Fall, signifies the contrary.

LAND-

LAND-*Lay'd*, or, *To Lay the Land*; that is, just to lose the Sight of it.

LAND-*Lock'd*, is when *Land* lies all round the Ship, so that no Point of the *Compass* is open to the Sea: And if she is at Anchor in such a Place, she is said to *ride Land-Lock'd*; and therefore is concluded to ride safe from the Violence of *Winds* and *Tides*.

LAND-*Mark*, is any Mountain, Rock, Steeple, Wind-mill, a Tree, or the like, near the Sea-side. As *such a Steeple is a Land-Mark to Sea-faring Men*,

LAND *is shut in*; that is another Point of *Land*, hinders the Sight of that which the Ship came from.

LAND *to*; *The Ship lies Land-to*: That is, She is so far off the Shore, that she can but just *Ken* (or discern) it.

LAND-*Turn*, is a Wind that blows from the Shore, in the Night, at certain Times, in most Hot Countries.

To Set the Land; that is, to see by the *Compass* how it bears.

A HEAD-LAND, or a *Point of Land*, is that which lies farther out than the rest. See *Point*.

LANGREL *Shot*, is a Shot sometimes used at Sea, made of two Bars of Iron, with a Joint in the middle, by means of which it may be shorten'd, and so the better put into the Gun; at each end there is a Half-Bullet either of Lead or Iron: This Shot, when Discharg'd, flies out at Length; and therefore will do the more Execution amongst the Enemy's Rigging, *&c.*

LANNIARDS, or *Lanniers*, are small Ropes reev'd into the *Dead-Men's-Eyes* of all Shrowds; either to slacken them, or to set them *taught*: The *Stays* of all *Masts* are set *taught* by *Lanniards*.

LARBOARD, is the Left-hand-side of the Ship, when you stand with your Face towards the Head.

LARGE: *A Ship goes Large*; that is, she goes neither *before* the Wind, nor *upon* the Wind, but, as it were, *Quartering* between both, with a fresh Gale, and all Sails drawing.

To Sail with a Large Wind; that is with a Fair Wind.

LASH, or *Lace*, signifies, to Bind, or Make fast; As, *To Lash the Bonnet to the Course*; or *the Drabler to the Bonnets*. Also, the *Carpenter* takes care that the *Spare-Yards* be *Lash'd* fast to the Ship's Side. And in a Rolling-Sea, the *Gunners* mind that the *Guns* be well *Lash'd*, lest they should break loose.

LASHERS, are those Ropes only, which bind fast the Tackles, and the Breechings of the Ordnance, when haled, or made-fast within Board.

LASKETS, are small Lines, like Loops, sow'd to the *Bonnets* and *Drablers* of a Ship, for to Lash or Lace the *Bonnets* to the

Courses,

Courſes, or the *Drablers* to the *Bonnets*.

LASKING; *The Ship goes Lasking*: That is, She goes neither *by* the Wind, nor directly *before* the Wind: And 'tis much the ſame with *going Large*, or *Veering*, that is, going with a Quarterly Wind.

LATCHES, or *Latchets*, the ſame with *Laskets*; which ſee.

LATITUDE *of a Place, on the Surface of the Globe*, is an Arc of the *Meridian* intercepted between the *Zenith* of the Place, and the *Equinoctial*; and therefore is always Equal to the Height of the Pole of the World from the *Horizon*. Hence, the Diſtance of a Ship, from the *Equinoctial*, either *North* or *South*, counted on the *Meridian*, is her *Latitude*: So that if a Ship ſail from the *Equinoctial*, or from a Leſſer *Latitude* to a Greater, ſhe is ſaid *to Raiſe the Pole*: But if ſhe ſail towards the *Equinoctial*, or from a Greater *Latitude* to a Leſs, ſhe is ſaid *to Depreſs the Pole*. And in order *to find the Latitude of a Place*:

Firſt, There muſt be given, the *Sun's Declination*: Which is taken from ſome Correct *Tables* of the *Sun's Declination*. See *Declination*.

Secondly, The *Sun's Meridian Altitude*: Which is taken by a *Quadrant*.

Thirdly, There muſt be given, the *Sun's Poſition*, or *Situation* upon the *Meridian*, Northwards, or Southwards, in reſpect of the *Zenith* of the *Place*: And this is taken by the *Magnetic-Compaſs*; that is, by *Setting the Sun*.

Alſo, 'tis neceſſary to Obſerve,

1. That the *Zenith* is always in the *Meridian*, and always 90 *Degrees* diſtant from the *Horizon*.

2. That if the *Meridian Altitude* of the *Sun* be Subtracted from 90 *Degrees*, the Remainder is the *Sun*'s Diſtance from the *Zenith*.

3. That if the *Meridian Altitude* of the *Sun* be 90 *Degrees*, then the *Sun* is the *Zenith*.

4. That the Diſtance of the *Zenith* from the *Equator*, is the *Latitude of the Place*; and is always Equal to the *Elevation of the Pole*.

5. That if the *Equator* Croſs the *Zenith*, then the *Place* lies under the *Equator*; and has no *Latitude*.

6. That if the *Equator* be *North* of the *Zenith*, then the *Place* is in *Southern Latitude*: Or if the *Equator* be *South* of the *Zenith*, then the *Place* is in *Northern Latitude*: And *vice-verſâ*, in each Particular.

7. That when the *Sun* is in the *Equator*, it has no *Declination*.

8. That when the *Sun* is *North* of the *Equator*, it is in *North Declination*: Or when *South* of the *Equator*, in *Southern Declination*: And *vice-verſâ*, in each Particular.

These

These being thorowly underſtood; the *Latitude* of any *Place* is readily found, by the *Rules* given for *Working an Obſervation*, under the word *Obſervation*: Which ſee.

But to find the *Latitude* of a *Place*, by a *Chart*, ſee *Mercator's Chart*.

Middle-LATITUDE, is *Half* the Sum of any Two given *Latitudes*.

Middle-LATITUDE-*Sailing*. See *Middle*.

Difference of LATITUDE, is the *Northing* or *Southing* of a Ship; or the Way gain'd, to the *Northward* or *Southward* of the Place ſhe Departed from: Or 'tis the *Difference* between the *Latitudes* of any two Places, ſhewing how far one of them is to the *Southward* or *Northward* of the other. And 'tis eaſily found, by this *Rule*, (the *Latitudes* being given:)

1. *If the Places are on the ſame ſide of the* Equinoctial, *their* Difference *is the* Difference of Latitude *ſought.*

2. *If the Places are on contrary ſides of the* Equinoctial, *that is, one in* North, *and the other in* South Latitude, *their Sum is the* Difference of Latitude *ſought.*

LAKE, is a ſmall Collection of deep Standing-Water, intirely ſurrounded with Land, and having no viſible or immediate Communication with the Sea.

LAUNCH, ſignifies, *to put out*: As, *Launch the Ship!* that is, Put her out of the Dock!

Launch out the Capſtan-Bars! that is, Put them out!

Launch aft, or *forward on*; that is, when Things are ſtow'd in the Hold, to put them more aft, or forward on.

Launch, Ho! that is, when a Yard is Hoiſted high enough, Hoiſe no more! Or, in Pumping, if the Pump ſucks, Pump no more!

LAY *the Land*. See *Land*.

LEAD, as, *Sounding-Lead*, or *Deep-Sea-Lead*, is Six or Seven Pound Weight of *Lead*, near a Foot long, and faſten'd at the End of the *Sounding-Line*, or *Deep-Sea-Line*.

Heave the Lead; that is, to *Sound*, or to find where the Ship may ſail, by the Depth of Waters. He that *Heaves the Lead*, ſtands by the *Horſe*, or in the *Chains*, and ſings the Depth he finds.

LEAGUE, is a Meaſure of Length commonly uſed at Sea, and is reckon'd to be Three *Engliſh* Miles.

LEAK, is a Hole in the Ship, thro' which the Water comes in.

Spring a Leak; that is ſaid of a Ship that begins to *Leak*.

To Stop a Leak, is to put into it a Plug wrapt in Okum, and well Tarr'd, or in a Tarpawling Clout, to keep the Water out; or nailing a Piece of Sheet-Lead upon the Place.

LEDGES, are ſmall Pieces of Timber lying a-thwartShips, from the

the *Waſt-Trees* to the *Roof-Trees*: They ſerve to bear up the *Gratings* or *Nettings* over the Half-Deck.

LEE, is a word of various Significations; tho' generally, the Part oppoſite to the Wind, is meant by it: As,

LEE-*Shore*, is that *Shore* againſt which the Wind blows.

LEE-*Latch*: *Have a Care of the Lee-Latch!* That is, Take Care that the Ship don't go to the *Lee-ward* of her Courſe, or too much to *Lee-ward*, or too near the Shore.

A-LEE *the Helm!* that is, Put the *Helm* to the *Lee-ward*-ſide of the Ship.

To Lie by the Lee; or, *To Come up by the Lee*; is to bring the Ship ſo, that all her Sails may lie flat againſt her Maſts and Shrowds, and that the Wind may come right upon her Broad-ſide.

LEE-*Fang*, is a Rope reev'd into the Cringles of the Courſes, to hale in the Bottom of the Sail, that the Bonnets may be Laced on: As alſo, to Take in the Sail.

LEE-*Seel*. See *Seel*.

Leeward-Ship, is that which ſtands not ſo near the Wind.

Leeward-Tide. See *Tide*.

Lee-Way. See *Way*.

LEETCH *of a Sail*, is the Outward Edge, or Skirt of the Sail, from the Earing to the Clew; or the Middle of the Sail, between the Earing and the Clew.

Leetch-Lines, are ſmall Ropes made faſt to the *Leetch* of the Top-ſails, (to which they only belong,) and reev'd into a Block at the Yard, cloſe by the Top-ſail-Ties: They ſerve to hale in the *Leetch* of the Sail, when the Topſails are to be taken in.

LEGS, are ſmall Ropes, of about a Foot in Length, put thro' the Bolt-Ropes of the *Main*, and *Fore-ſail*, in the *Leetch* of each; their Ends are ſpliced into themſelves: They have a ſmall Eye, into which the *Martnets* are made faſt by two Hitches.

LET-*Fall the Main-ſail, Fore-ſail*, or *Sprit-ſail*; is to put out thoſe Sails, when their Yards are hoiſed up aloft.

LEVANT: By this word is meant the *Eaſtern-part* of a Continent: But with the Seamen, it ſignifies the *Mediterranean-Sea*.

LIEUTENANT, is an *Officer* on-board a *Man of War*, who, in the *Captain*'s Abſence, Commands in Chief. See *Officers*.

LIFTS, are Ropes which belong to the Yard-Arms of all Yards: They ſerve to Top the Yards, that is, to make the Ends of the Yards hang higher, or Lower, as Occaſion requires.

Top-Sail-Lifts, ſerve as *Sheats* for the *Top-Gallant-Yards*.

Standing-Lifts, are thoſe of the *Spritſail-Yards*.

Topping the Lifts; that is, haling of the *Lifts*: As, *Top a Starboard!* or, *Top a Port!* that

that is, Hale upon the *Starboard*, or *Larboard-Lift!*

LIGAN, the ſame with *Lagan*; which ſee.

LIGHTER; is a Broad Large Boat, which goes with Sails and Oars: They are very common in the River *Thames*; and are uſed for carrying of Timber, Coals, and other Goods, up and down the River: They are alſo uſed for Carrying *Ballaſt*, &c.

LIGHT-FRIGATE. See *Frigate*.

LIMBER-HOLES, are little Holes cut thro' the Floor-Timbers of a Ship, ſerving to let the Water to the *Well* of the *Pump*: Which elſe would lie between thoſe Timbers where the Keel-Rope runs.

LINE: *Navigators* uſually call the *Equator*, or *Equinoctial-Line*, ſimply the *Line*. And at *Sea*, they have a Ridiculous Ceremony; That when *Sailors* Croſs the *Line*, or *Tropic*, that have not bin there before, they muſt Pay certain Forfeitures Demanded of them, or elſe be *Duck'd*, or *Baptiz'd*, (as they call it,) either from the Main-Yard-Arm, or otherwiſe: This Cuſtom is inviolably uſed by moſt Nations, who practiſe it indiſpenſably in *Eaſt-India* Voyages; and each practiſes it differently: Nay, thoſe of the ſame Nation puts it in Execution in different manner. It is perform'd, by ſome, thus, (by way of *Baptiſm*, as was ſaid before;) The Ship's Company Range themſelves in two Lanes, each with a Bucket in his Hand, with Tubs of Sea-Water ready by them: Then the *Boatſwain's Mate*, or ſome ſuch Officer, comes to the Foot of the Main-Maſt, with Viſage all bedawb'd, and his Body all roll'd in Ropes, ſome hanging down from his Shoulders; and after him, in Order, there follows five or ſix Sailors dreſs'd in a like manner: And holding in his Hand ſome Book of Navigation; he that is to be *Baptiz'd*, comes and Kneels before him, and puts his Hand upon the Book, and is obliged to Swear ſolemnly, and ſincerely, That as often as there is Occaſion for *Baptizing* or *Ducking* Others, he will exerciſe upon them, the ſame Ceremony as they are about to exerciſe upon him, without ever thinking to exempt them from it. After this, he ariſes, and goes towards the Head of the Veſſel, thorow the Lane of Tubs, and Seamen who Attend with Buckets full of Water, ſo that as he goes along, they throwing it upon him, he receives his *Baptiſm* ſufficiently.

LINES, are ſmall Ropes belonging to the Ship; of which there be ſeveral, as *Bow-Lines*, *Bunt-Lines*, *Knave-Lines*, *Log-Lines*, *Smiting-Lines*, *Leetch-Lines*, *Sounding-Lines*, &c. which ſee.

LINE of *Battle*; is the Diſpoſition of the Stations of a Fleet, at the Time of an *Engagement*: For the Order of *Battle*, is to draw as much as poſſible all the

Ships into a Right Line, as well to gain and keep the Advantage of the Wind, as to run the same Board. Those of the Van, Centre, and Rear, place themselves in the same *Line*, when the *Squadron*, or *Division*, are United. See *Engagement*.

LIVE; *To Live*; that is, To Endure the Sea: As they say of a Boat, That it will *Live* in any tolerable Sea; that is, It will Endure, or bear with, any Sea.

LOCKERS, are those little Boxes, or Seats, in a Ship, contrived to put or Stow any Thing in; as in little Cupboards, or Chests.

LOG, is a little Piece of Wood of a Triangular Form, with as much Lead in one end thereof, as will serve to make it swim upright in the Water; at the other End is made fast to the *Log-Line*.

LOG-LINE, is a small *Line*, having the *Log* tied to one End: 'Tis kept wound about a Reel for that Purpose. The *Log-Line* for about 10 Fathom, in Small Ships, but more in Great ones, from the *Log*, has, or ought to have, no *Knots*, or *Divisions*; because so much should be allowed, that the *Log* may be out of the Eddy of the Ship's *Wake*, before the Glass be turn'd up: But then the *Knots* begin, and ought to be at least 50 Feet from one another; tho' the common Practice at Sea, is to have them but 7 *Fathoms*, or 42 *Feet* distance. The Use of the *Log* and *Log-Line*, is to make an Estimate of the *Ship's Way*, or *Distance run*, by *Heaving the Log* every Hour, or every two Hours.

To Heave the Log, is first to throw it into the Water, and let it run away so far, as to be out of the Eddy of the Ship's *Wake*; then One having a Half-Minute-Glass ready in his Hand, turns it up, just when the first Knot runs off the Reel; and then the *Line* running easily off as the Ship move, when the Glass is out, he cries, *Stop!* The other stops the Reel: Then they count the Knots run out, as also the odd *Fathoms*, or *Feet*, (if there be any.) And they reckon; *For as many Knots as run out in Half a Minute of Time, so many Miles the Ship sails in an Hour; and for every 5 odd Feet, a 10th Part of a Mile more.* Thus 4 *Knots* in *Half a Minute*, is 4 Miles an *Hour*: And 3 *Knots*, 45 *Feet* in *Half a Minute*, is $3\frac{9}{10}$ Miles an *Hour*.

The Common Division of the *Log-Line*, is grounded upon this *Supposition*; That a *Mile* contains 5000 *Feet*, and 60 of such *Miles* a *Degree*; whence a *Degree* would contain 300000 *Feet*. Now Half a *Minute* being the 120th Part of an *Hour*, and the 120 Part of 5000 *Feet* is 41,6, &c. or near 42 *Feet*. So that as many times 42 *Feet* as the Ship runs in Half a *Minute*, so many *Miles* she runs in an *Hour*: Therefore, according to this

Suppo-

Supposition, 42 *Feet* must be the Distance between *Knot* and *Knot* upon the *Log-Line*.

But this *Supposition*, and consequently the *Practice* from it, is altogether Erroneous: For a *Degree* is now actually found to contain 360000 *Feet* at least; wherefore, a *Minute*, or a *Mile*, must contain 6000 *Feet*, (which is the True *Sea-Mile*.) And since Half a *Minute* is the 120th Part of an *Hour*, the 120th Part of 6000 *Feet* is 50 *Feet*: Therefore as many times 50 *Feet* as the Ship runs in Half a *Minute*, so many *Miles* must she go in an *Hour*, (supposing she goes at the same rate;) and for every 5 odd *Feet*, a 10th Part of a *Mile* more; therefore, the Distance between *Knot* and *Knot* must be 50 *Feet*.

LOG-BOARD, is a *Board*, or Table, divided usually into Five Columns; The First *Column* contains the *Hours of the Day*, from *Noon* to *Noon*: In the Second *Column* is placed the *Ship's Course*: In the Third and Fourth *Column* is placed the *Distance run in Knots*, *Fathoms*, and *Half-Fathoms*, or sometimes *Feet*: In the Fifth *Column* is placed the *Winds*, *Weather*, *Accidents*, &c.

LOG-BOOK, is a Book Column'd and Ruled, by some, like the *Log-Board*, wherein the *Log-Board*'s Account is every *Noon* Entred, with the *Observations* then made; And from hence 'tis Corrected, and Transcrib'd into the *Journals*.

LONG-BOAT. See *Boat*.

LONGITUDE *of a Place*, is its Distance, (measured upon the *Equator*, or some Parallel to it,) from the *First Meridian*, which may be taken, (at pleasure, either *Eastward*, or *Westward*; whence 'tis call'd *Eastern*, or *Western Longitude*. And because the *Meridian* of any Place Divides the Globe (whose Whole Circumference is reckon'd 360 *Degrees*) into Two Equal Parts, call'd *Hemispheres*, the one *Eastern*, and the other *Western*: Therefore the Greatest *Longitude* a Place can have, is that of $\frac{360}{2}$, or 180 *Degrees*.

And since the *Longitude* of Places is Measured either upon the *Equator*, or some one of its *Parallels*, which continually decrease towards the *Poles* of the *Equator*, or of the *World*, and yet are each of them divided into 360 Equal Parts, or *Degrees*; therefore the *Degrees of Longitude* (according to this *Definition*) are not every where Equal one with the other, and consequently do vary in their Proportion to the same sort of Miles, as the *Parallels*, on which they are Measured, are more or less distant from the *Equator*, according to this Proportion:

As the Co-Sine of the Latitude,
To the Longitude in Degrees of the Parallel;
So is the Radius,
To the Longitude in Degrees of the Equator.

Whence the following

TABLE,

TABLE, *Shewing to every* Degree *of* Latitude, *the Exact Number of* Miles, Seconds, *and* Thirds, *that are Answerable to* One Degree *in the* Equator.

Latitude.	*Min. Sec. & Th.* of the *Equator*, Equivalent to 1 *Deg.* of the *Parallel.*			*Latitude.*	*Min. Sec. & Th.* of the *Equator*, Equivalent to 1 *Deg.* of the *Parallel.*			*Latitude.*	*Min. Sec. & Th.* of the *Equator*, Equivalent to 1 *Deg.* of the *Parallel.*		
D.	M.	S.	T.	D.	M.	S.	T.	D.	M.	S.	T.
1	59	59	28	31	51	25	48	61	29	5	19
2	59	57	48	32	50	52	59	62	28	10	5
3	59	55	4	33	50	19	13	63	27	14	22
4	59	51	13	34	49	44	33	64	26	18	8
5	59	46	17	35	49	8	56	65	25	21	26
6	59	40	16	36	48	32	28	66	24	24	15
7	59	33	11	37	47	55	6	67	23	26	37
8	59	24	58	38	47	16	50	68	22	28	36
9	59	15	41	39	46	37	44	69	21	30	8
10	59	5	19	40	45	57	35	70	20	31	16
11	58	53	52	41	45	16	57	71	19	32	21
12	58	41	20	42	44	35	20	72	18	32	28
13	58	27	44	43	43	52	51	73	17	32	32
14	58	13	3	44	43	10	1	74	16	32	18
15	57	57	21	45	42	25	36	75	15	31	45
16	57	40	32	46	41	40	46	76	14	30	55
17	57	22	43	47	40	55	12	77	13	29	49
18	57	3	49	48	40	8	52	78	12	28	29
19	56	43	52	49	39	21	49	79	11	26	55
20	56	22	53	50	38	34	3	80	10	25	8
21	56	0	53	51	37	45	33	81	9	23	9
22	55	37	51	52	36	56	23	82	8	21	1
23	55	13	56	53	36	6	33	83	7	18	44
24	54	48	59	54	35	16	3	84	6	16	18
25	54	22	43	55	34	24	53	85	5	13	47
26	53	55	39	56	33	33	5	86	4	11	8
27	53	27	38	57	32	40	43	87	3	8	3
28	52	58	37	58	31	47	43	88	2	5	39
29	52	28	38	59	30	54	9	89	1	2	49
30	51	57	42	60	30	0	0	90	0	0	0

'Tis plain, That by the preceding *TABLE*, having the *Longitude* of a *Place*, (whose *Latitude* is Known ;) the *Meridional Distance*, or its *Easting*, and *Westing* in *Miles*, from the *Meridian* of another *Place*, counted in the Proper *Parallel* of *Latitude*, is easily found : That is, The *Longitude* may be turn'd into *Miles*, by Multiplying the *Degrees* in the *Longitude* given, by the Number of *Miles* Proportional to *One Degree* in the *Latitude* given.

Thus the *Longitude* of *Lisbon* from *London*, is 10 Deg. *West*, its *Latitude* 39 Deg. *North* : At which Distance from the *Equator* (by the *Table*) *One Degree* of *Longitude* is esteemed Equal to 46', 37'', 44'''. Therefore the Distance of *Lisbon* from the *Meridian* of *London*, is,

$$(\overline{46' \times 10} + \overline{37'' \times 10} + \overline{44''' \times 10} =)\ 466\tfrac{7}{25}\ \textit{Miles West.}$$

In Order for finding the Longitude of any Place, *'tis necessary to Observe*,

1. That the Whole *Circumference* of the *Heavens*, which is reckon'd to be 360 *Degrees*, passes from *East to West*, thro' the *Meridian* of any *Place*, in 24 *Hours* ; and consequently, by Proportion, 15 *Degrees* in One *Hour*, One *Degree* in 4 *Minutes* of an *Hour*, &c.

2. That the *Longitude of any Place*, (being the Distance of its *Meridian* from the *First Meridian*, or some Fixed one,) is found, by finding the *Difference of Time* between the coming of any *Point* of the *Heavens*, or any *Cœlestial Body*, to one *Meridian*, and to the other : For the *Difference of Time* turn'd into *Degrees*, is the *Longitude* sought.

3. That if the *Cœlestial Body* comes sooner, or earlier, to the *First Meridian*, than it does to *Meridian of the Place* whose *Longitude* is sought ; then that *Place* lies in *Western Longitude* ; if later, then in *Eastern Longitude*.

And the *Difference of Time* between the coming of any *Cœlestial Body* to the *First Meridian*, and to the *Meridian* of any other Place, and consequently, the *Longitude* of that other *Place*, is found by these following Methods :

I. *To find the* Longitude *of a* Place, *by an* Eclipse *of the* Moon.

By a *Clock*, or *Watch*, duly Rectified, Observe at what Time the *Moon*, or any remarkable Spots thereof, enters into, or comes out of the Shade of the Earth ; and compare the Time when any of these Circumstances happen at the *Place* where you make the *Observation*, with the Time of their happening at the

First

First Meridian; The *Difference* of these two *Times*, being turn'd into *Degrees* and *Minutes*, is the *Longitude* sought.

II. *To find the* Longitude *of a* Place, *by the* Satellits *of* Jupiter.

By a *Clock*, or *Watch*, duly Rectified, Observe the Time of the Immersion, or Emersion, of any of the said *Satellits*: Which being compar'd with the Time of Immersion, or Emersion, of the same *Satelles* at the *First Meridian*; the *Difference of Time* reduced into *Degrees*, gives the *Longitude* sought.

EXAMPLE.

The Beginning or End of an *Eclipse of the Moon*; The *Immersion*, or *Emersion*, of a *Satelles* of *Jupiter*, is Observed at a certain *Place*, to be at *Eleven* at *Night*; But at *London*, it happens to be at 52 *min.* past *Nine*. *Requir'd*, *The* Longitude *of that* Place *from* London?

Answ. 17 Deg.

For 11 *h.* 00 *m.* — 9 *h.* 52 *m.* = 1 *h.* 8 *m.* = 17 *Deg.*

III. *To find the* Longitude *of any Place, by a* Clock, *or other* Automaton, *so Contriv'd, and Made, as to keep the same Uniform, Just, or Regular Motion, in all Parts of the* Earth.

The *Clock* being Rectified to the *Time* at the *First Meridian*, or any *Place* from whence you *Depart*,) shall, if duly attended afterwards, shew, in any Part of the World, the True Time at the *First Meridian*, (or the *Place* from whence you *Departed*.) Wherefore, having found likewise (either by the *Sun*'s *Altitude* in the *Day*, or by some *Star*'s *Altitude* in the *Night*.) the True *Time* at that *Place* to which you are come; The *Difference* between the *Time* thus found, and the *Time* of the *Clock*, being converted into *Degrees* and *Minutes*, shew the *Longitude* of the *Place* where you are.

Thus, in the former *Example*; The *Index* of a *Regular Clock*, or *Automaton*, Rectified to the *Meridian* of *London*, and carried to the *Place* where the *Observation* was made, would point to 52 *min.* past *Nine*; when the *Hour* of the *Night* being found at that *Place*, by taking the *Altitude* of a *Star*, would be *Eleven*.

For finding the *Longitude*, at Sea, by means of the *Course* Steer'd, and *Distance* Run, *&c.* either by the *Meridional Parts*, or by the *Middle-Latitude-Sailing*: (See *Mercator*'s *Sailing*, or *Middle-Latitude Sailing*.) And the same is found very readily, and sufficient for our Daily Practice at *Sea*, by the help of the *Table* of *Difference of Latitude and Departure*; as may be seen in the *Use* of that *Table*, in Mr. *Jones*'s *Treatise of Navigation*, Edit. 2[d].

LOOF, or *Louf* of a *Ship*, is that Part of her aloft, which lies just before the *Chest-Trees*.

LOOF-*Pieces*, are those Guns which lie at the *Loof* of the Ship.

LOOF-*Hook*. See *Hook*.

LOOF-*Tackle*, is a small *Tackle*, serving to lift all small Weights in or out of a Ship.

LOOF, or *Luff*, is also a Word used in *Conding* of a Ship: As,

Luff! Keep your Loof! That is, Keep the Ship Near the Wind!

Loof-up! That is, Keep Nearer the Wind!

To Loof into a Harbour; is to Sail into it, close by the Wind.

To Spring the Loof; is when a Ship, that before was going *Large* before the Wind, is brought close by the Wind.

LOOM: The *Looming* of a Ship, is her Perspective, as she appears at a Distance, Great or Little.

LOOM-*Gale*; is a gentle, easie Gale of Wind, in which a Ship can carry her *Top-sails*.

LOOP-*Holes*, are Holes made in the *Coamings* of the *Hatches*, for *Close-Fights*, and other Conveniencies.

LOUF, the same with *Loof*; which see.

LOW-*Water*. See *Water*.

LOXODROMIC-*Line*, is the *Line* of the Ship's Way, when she sails upon a *Rhumb* Oblique to the *Meridian*.

LOXODROMIC-*Tables*, are the *Tables* of *Rhumbs*, or the Traverse *Table* of *Miles*, with the *Difference of Latitude* and *Longitude*; by which the *Sailor* may easily find his *Course*, *Distance*, *Latitude*, or *Longitude*; and Practically Resolve all the *Cases of Sailing*.

LUFF, the same with *Loof*; which see.

LUST: *The Ship has a Lust to the Starboard*, or *a-Port*; that is, She is inclin'd to *Heel* that way.

LYE *under the Sea*; is said of a Ship, when her Helm is lash'd fast *a-Lee*, and she lies so *a-Hull*, that the Sea breaks upon her *Bow*, or *Broadside*.

LYE *a-Hull*. See *Hull*.

LYE *a-Try*. See *Try*.

M

MAGNETIC-*Needle*, is that *Needle*, or Wire, in the *Card*, or *Flye* of the *Sea-Compass*, which is Touch'd by the *Magnet*, or *Loadstone*; and hence has the wonderful Property of Pointing towards the *Poles* of the World. This Property of the *Magnet* was Discover'd by *Roger Bacon*, an *Englishman*, who lived in the Time of K. *Edward* the Third.

MAGNETICAL-AMPLITUDE. See *Amplitude*.

MAGNETICAL-AZIMUTH. See *Azimuth*.

MAGNETICAL-*Meridian*. See *Meridian*.

MAIN-*Beam*. See *Beam*.

MAIN-

MAIN-*Capſtan.* See *Capſtan.*

MAIN-*Maſt*, MAIN-*Top-Maſt*, MAIN-*Top-Gallant-Maſt.* See *Maſt.*

MAIN-*Miſen.* See *Maſts.*

MAN *of War*; that is, a *Ship* of *War.* See *Rate.*

MAN *a Ship*, or *Fleet*; is to Provide them with a ſufficient Number of *Men*, for an Expedition.

In *Manning* the *Navy*, 'tis uſual to Promiſe, by *Proclamation*, a Bounty to all Seamen, and Able-bodied Landmen, who come into the Service by a certain Time; which is frequently Two Months Pay, and ſeldom more. This does indeed prevail upon Many; but yet great Nnmbers do conceal themſelves until the Fleet is at Sea, and Others lurk about even till the Time limited, for ſuch Bounty, is near expired; which does in no little meaſure prevent the *Fleet*'s being in a Readineſs for an Early Campaign.

And as Seamen are thus Encourag'd to Enter themſelves Voluntarily, ſo is there another Method uſed, to Compel them to it; and that is *Preſſing*, by *Warrants* from the Lord *High-Admiral* to the *Captains*, and by them Aſſign'd to their *Lieutenants*: And to render this the more Effectual, Veſſels are purpoſely Hired into the Service, to proceed from Place to Place, with thoſe *Officers*, and their *Preſs-Gangs*, not only to Receive *Voluntiers*, but to Impreſs what Men they can light on: But their Succeſs has bin very uncertain, and always very Expenſive; Therefore it were much to be wiſh'd, in a Matter of ſo great a Conſequence to the Nation, that more Speedy and Effectual Methods could be taken for *Manning the Fleet.*

MAN *the Capſtan!* See *Capſtan.*

MAN *the Top!* or *Yard!* That is, when the Men are Commanded to go up to the *Top*, or *Yard*, for ſome particular Service.

MAN *the Side!* or *Ladder!* That is, when an *Officer*, or any Perſon of Faſhion, is at the Ship's Side, ready to come aboard, the Men are Commanded to Wait, and help him up the Side.

MANGER, is a Circular Place, made with Planks faſten'd on the Deck, right under the *Hawſes*; for to receive the Sea-Water, beating in at the *Hawſes*, in a Streſs of Weather.

MARINE, of or belonging to the Sea; from *Mare*, the *Sea.*

MARINER, a *Seaman.* See *Sailor.*

MARINERS-COMPASS. See *Compaſs.*

MARINES, are *Sea-Soldiers.*

MARK, *Land-Mark.* See *Land.*

MAR-LINE, a ſmall Line of untwiſted Hemp, very plyable, and well Tarr'd; ſerving to ſeize the Ends of Ropes, and keep them from Ravelling out; or the Straps at the *Arſe* (or Lower-end) of the *Blocks.*

MAR-LINE *a Sail*; that is, when the Sail is rent out of the *Bolt-Rope*, to make it fast with *Mar-Line*, put thro' the Eye-let-Holes made in it for that purpose, unto the *Bolt-Rope*, till it can be Mended.

MAR-LINE-*Spike*, is a little Piece of Iron, to splice small Ropes together; as also to open the Bolt Rope, when the Sail is sow'd unto it.

MARTNETS, are small Lines made fast to the Leetch of the Sail, and reev'd thro' a Block at the Topmast-Head, and so they come down by the Mast to the Deck. They serve, in Furling the Sail, to bring that part of the Leetch, which is next the Yard-Arm, close up to the Yard, so that the Sail may the better be Furl'd up.

Top-MARTNETS, are those which belong to the *Top sails*; they are made fast to the Head of the *Top-Gallant-Mast*, and their *Fall* comes down only to the *Top*.

Top the Martnets! That is, Hale them up!

MASTER-ATTENDANT. See *Officers*.

MASTER *of a Ship*, is the next Officer to a *Lieutenant*; he has the general Conduction of the Way, and Sailing of the Ship, in his Charge and Care: He Directs, and Shapes the Course that she is to Sail; and Commands all the Sailors, for Steering, Trimming, and Sailing the Ship: He is to keep the most accurate Account of the Ship's Way, and be at all times able to give Estimate, or Prick off her Place on the True *Sea-Chart*. He has *Mates*, to assist him in his Office.

In *Merchant-men*, the Master is the Chief-Officer.

MASTS *of a Ship*; of which, the Principal ones, are, the *Main-Mast*, *Fore-Mast*, *Misen-Mast*, and *Bowsprit*.

Main-MAST of a *Ship*, is a long Piece of Round Timber, standing upright in the Middle, or Wast of the Ship; it carries the *Main-Yard*, and *Main-Sail*. Its Length, according to some should be $2\frac{1}{2}$ of the Length of the *Midship-Beam*. Others give this *Rule*, for finding what Length it should be; *viz*.

Multiply the Breadth of the Ship (in Feet) by 24: From the Product, cut off the last Figure towards the Right-Hand; the rest shall be the Length of the Main-Mast *in Feet*.

As for *EXAMPLE*: Suppose the Length of the *Midship-Beam* was 30 *Feet*; 24 times 30 is 720; then cutting off the last *Figure*, there rests 72 *Feet*, the Length of the *Main-Mast* required.

As for the *Thickness* of the *Main-Mast*, 'tis usual to allow an *Inch* to every *Yard* in Length.

Fore-MAST of a *Ship*, is a Round large Piece of Timber, standing in the *Fore-part*, or *Fore-Castle*, of the Ship; it carries

carries the *Fore-Sail*, and *Fore-Yard*. It's Length, by ſome, is to be $\frac{8}{9}$ of the *Main-Maſt*; and by others, $\frac{4}{5}$ thereof.

Miſen-MAST of a Ship, ſtands aft, in the Sternmoſt-part of the Ship: In ſome Great Ships there are two of theſe; that next the *Main-Maſt*, is call'd, the *Main-Miſen*; and that next the *Poop*, the *Bonaventure-Miſen*. The Length of the *Miſen-Maſt*, is, by ſome, accounted the ſame with the Height of the *Main-Top-Maſt* from the *Quarter-Deck*; or Half the Length of the *Main-Maſt*, and Half as Thick.

Top-MASTS of a *Ship*, are thoſe made faſt, and ſecure unto the Heads of the *Main-Maſt*, *Fore-Maſt*, *Miſen-Maſt*, and *Bowſprit* reſpectively. See *Top*.

Top-Gallant-MASTS of a Ship, are thoſe ſet faſt on the Head of the *Main*, and *Fore-Top-Maſts*; whence are call'd, the *Main-Top-Gallant-Maſt*, and *Fore-Top-Gallant-Maſt*: They carry *Flag-Staffs* on their *Tops*, whereon the *Flags*, *Pendants*, &c. hang.

Jury-MAST: When a *Maſt* is *born by the Board*, in a *Storm*, or *Fight*, the Seamen ſet up, in the room of it, another made of Yards, or other Pieces of Timber, ſuch as they can get, ſpliced, or *fiſh'd* together, *woulding* them with Ropes: With this they make a ſhift, till better provided; and they call it, a *Jury-Maſt*.

Armed-MAST; that is, made of more than one Tree.

Maſt a Ship, is to ſet up her *Maſts*; wherein great Care ought to be taken.

Maſted: A Ship is ſaid to be *Maſted*, when ſhe has all her *Maſts* compleat.

Over-Maſted, or *Taut-Maſted*, when her *Maſts* are either too-Long, or too-Big; which makes her lie too-much down by the Wind, and Labour too-much *a-Hull*.

Under-Maſted, or *Low-Maſted*; when her *Maſts* are either too-Small, or too-Short: in which caſe, ſhe cannot bear ſo great a Sail as ſhould give her true Way.

To Spend a Maſt; is when it is broke by Foul-Weather.

To Spring a Maſt; is when 'tis Crack'd in any Place.

MAST-*Steps*. See *Steps*.

MATES, are Aſſiſtants to the ſeveral *Officers* a-board a Ship; as, *Maſter*'s *Mates*, *Surgeon*'s *Mates*, *Gunner*'s *Mate*, *Carpenter*'s *Mate*, *Boatſwain*'s *Mate*, *Cook*'s *Mate*, *Corporal*'s *Mate*.

MATS, are a kind of thick Clouts, wove out of Spun-Yarn, Sinnet, or Thrums; they are uſed, to preſerve the *Main* and *Fore-Yards* from galling againſt the *Maſts*, at the *Tyes*, and at the *Gunnel* of the *Loof*: They alſo ſerve to keep the Clew of the Sail from galling there; and to ſave the Clews of the *Fore-Sail* at the *Beak-head* and *Bowſprit*.

MERCATOR's or WRIGHT's *Chart*, or *Projection*, is the True Projection of the Globe *in Plano*: It differs from the Common *Plain Chart*, in this; That on the *Plain Chart*, the *Meridians* are Right Lines, all Parallel one to the other, and consequently do never meet; yet they cut the *Equator*, and all *Circles of Latitude*, at Right Angles, in the *Globe*: And all *Parallels* to the *Equator* (being *Lesser Circles*) are here made Equal to the *Equator* it self (being a *Great Circle*,) and therefore the *Degrees* of those *Parallel* (or *Lesser Circles*) are Equal to the *Degrees* of the *Equator*, or any other *Great Circle*; which is manifestly false, and contrary to the Nature of the *Globe*. For the *Meridians* on the Globe do all meet in the *Poles* of the *World*, cutting the *Equator*, (and therefore all its *Parallels*,) at *Right Angles*: And therefore all such *Parallels* do grow lesser towards either *Pole*, decreasing from the *Equator*: As for Instance; 360 *Degrees*, or the whole Circle in the *Latitude* of 60 *Degrees*, is but 180 *Degrees* of the *Equator*: Whereas in the *Plain Chart*, that *Parallel*, and all others, are made Equal one to the other, and to the *Equator*.

The *Meridians* in *Mercator*'s *Chart* are also Right Lines, all Parallel one to the other, and cross the *Equator*, and its *Parallels*, at Right-Angles, as in the *Plain Chart*. But here, tho' the *Circles of Latitude* are Equal to the *Equinoctial*, yet they keep the same Proportion with the *Meridian*; because the *Degrees* thereon are Lengthen'd, as the same *Parallels* on the *Globe* do Lessen: In the *Plain Chart*, the *Degrees* of the Greater and Lesser Circles are Equal; and in this, tho' the *Degrees* of the *Circles of Latitude* are Equal, yet the *Degrees* of the *Meridian* are Unequal, being enlarged from the *Equinoctial* towards either *Pole*, to retain the same Proportion as they do on the *Globe* it self: For, as *Two Degrees* of the *Parallel* of 60 *Deg.* is but *One Degree* of the *Equinoctial*, (or any *Great Circle*;) so here, *Two Degrees* of the *Equinoctial* is Equal but to *One Degree* of the *Meridian* betwixt the *Latitude* of 59 $\frac{1}{2}$ *deg.* and 60 $\frac{1}{2}$ *deg.* and so of the rest. See *Chart*.

The *USE* of this *Chart*.

I. *To find the* Latitude *of any Place in the* Chart.

RULE. Take the nearest Distance of the Place to any Parallel, or East and West Line.

Then lay that Distance on the Graduated *Meridian*, setting one Foot of the Compasses in the said Parallel, and turning the other Foot the same Way the proposed Place lies from it; the last Foot shews the *Latitude* required.

II. To

II. *To find the* Longitude *of any Place in the* Chart.

RULE. Take the nearest Distance from the proposed Place to the *Meridian.*

Move the Compasses (being kept at that distance) with one Foot on the said *Meridian*, till both Feet come to the *Equator*; and the Foot which stood on the Proposed Place, shews its *Longitude* sought.

III. *To find the Course or Bearing of one Place from another.*

1. If the *Chart* is made with *Rhumb-Lines*, or *Points of the Compass*, drawn in it.

RULE. Lay a Ruler on the two Places given.

Take the nearest Distance from the Centre of any Compass to the Ruler's Edge.

Move the Compasses, (being at that Distance,) with one Foot close to the Ruler, and the other Perpendicular to it: In so moving the Perpendicular Foot, among the *Rhumb-Lines*, will shew the Course, or Bearing, of the Proposed Places.

2. If the *Chart* have only *Meridians* and *Parallels* drawn thereon.

In such *Charts*, one *Quarter of the Compass* is drawn in one of the Squares, and sometimes Half Points, and Quarter Points: Whereby the *Course* may be found more accurately than by the *Rhumb-Lines* only, by this

RULE. Lay a Ruler on the two Proposed Places, keeping it steady there.

Observe (by the Ruler's Edge) whether two *Meridians*, or two *Parallels*, are nearest to one another: If two *Parallels* be nearest, the Course is Less than *Four Points* from the *Meridian*; but if *Two Meridians*, 'tis More.

Then take the Distance of those which be nearest, and apply it to the Square, having Part of the Compass in it; setting one Foot in the Centre of the Compass, the other being turn'd to the side of the Square, will, according to the foresaid Conditions, shew the *Course*, or *Bearing*, required.

IV. *To find the Distance of any two Places on the* Chart.

CASE 1.

Two Places under one Meridian, i.e. *such as differ only in* Latitude; *to find their Distance.*

RULE. Find the *Difference of Latitude* between the given Places; and that will be the Distance required.

CASE 2.

Two Places under the Equinoctial; *to find their Distance.*

RULE.

RULE. Find the Difference of *Longitude* between them; and that will be the Distance sought.

CASE 3.

Two Places in one Parallel, i. e. *such as* Differ *only in* Longitude, *being given*; *to find their Distance.*

RULE. Take the Distance between the given Places in the Compasses.

Lay that Distance on the *Graduated Meridian*, so that one Foot may be as many Degrees above the *Latitude* of the given Places, as the other below it: There stay the Compasses.

Count the Degrees between the Feet of the Compasses; and that will be the *Distance* required.

CASE 4.

Two Places Differing *in* Latitude *and* Longitude *being given*; *to find their* Distance.

RULE. Take their Difference of *Latitude* from the *Equator*.

Laying a Ruler on both given Places, apply that Distance so to the Edge thereof, that when one Foot is placed close to the Ruler, and the other turn'd round about, it may just touch some *East* and *West Line* cross'd by the said Ruler's Edge: There stay the Compasses.

Then the *Distance* (by the Ruler's Edge) from the Place where the Compasses rested, to that Place where the Ruler crosses the aforesaid *East* and *West Line*, measured on the *Equinoctial*, gives the *Distance* required.

MERCATOR's *Sailing*. See *Sailing*.

MERIDIAN, is a great Circle passing thro' the *Poles of the World*, and both the *Zenith* and *Nadir*; crosses the *Equinoctial* at Right Angles, and divides the Sphere equally into a *Western* and *Eastern Hemisphere*; Its *Poles* are the *West* and *East* Points of the *Horizon*: 'Tis call'd *Meridian*; because when the Sun comes to that part of this Circle which is above the *Horizon*, 'tis then *Meridies*, *Mid-Day*, or *High-Noon*.

First MERIDIAN, is that from whence the *Longitude* is Reckon'd: For the *Meridians* are Various, and Change according to the *Longitudes* of Places; and therefore may be said to be infinite in Number, since all Places from *East* to *West* have their several *Meridians*.

The *French Geographers* and *Hydrographers*, begin to Reckon their *Longitude* from the *West*-part of *Fero*, one of the *Canary Isles*, making the *Meridian* of that Place the *First Meridian*. The *Dutch*, from the *Teneriff*.

Ptolemy

Ptolemy placed the *First Meridian* One *Degree* beyond the *Fortunate*, or *Canary Isles*. After the Discovery of *America*, it was fix'd in St. *Nicolas*, one of the *Cape Virde Isles*, by the *Portugueeze*. *Hondius* placed it at *St-Jago*. *Mercator*, at *Corvo*, one of the *Western Isles*.

So that the *First Meridian* is altogether Arbitrary: And indeed, 'tis enough for all Purposes, if we know the *Difference of Meridians*; which will (or should) be found the same in all Authors. And therefore, every *Astronomer* and *Geographer*, generally makes his own *Meridian* the *First*. But at Sea, we commonly reckon our *Longitude* from the *Meridian* of the Place departed from, or last seen, making that our *First Meridian* till we see another known Land, and no longer.

MERIDIAN-*Altitude*, or *Height of the Sun*, or *Star*; is the *Altitude* of the *Sun*, or *Star*, when they are on the *Meridian* of the Place: Or an Arc of the *Meridian* intercepted between the *Horizon* and the *Sun* or *Star*.

Magnetical MERIDIAN, is a great Circle which the *Magnetic-Needle*, or the *Needle of the Mariners-Compass*, or the *Meridian of the Compass*, only respects.

Meridional-Distance, is the *Difference of Longitude* between the *Meridian* under which the Ship is at present, and any other *Meridian* she was under before.

Meridional-Parts, *Minutes*, or *Miles*, are the *Parts* by which the *Meridians* in *Wright*'s or *Mercator*'s *Chart* do Encrease, as the *Parallels of Latitude* Decrease.

The *Co-Sine* of the *Latitude* of any Place being Equal to the *Semi-diameter*, or *Radius* of that *Parallel*; Therefore, in *Mercator*'s, or the *True Sea-Chart*, this *Radius* being the *Radius* of the *Equinoctial*, or *Sine* of 90 *Degrees*; the *Meridional Parts* at each *Degree* of *Latitude* must Encrease, as the *Secants* of the Arc contain'd between the *Latitude* and the *Equinoctial* do Decrease. Therefore these *Meridional Parts* shew how many *Parts* every *Degree* and *Minute* of *Latitude* is from the *Equator*, that is, of such *Parts* as a *Degree* of the *Equator* contains 60 of them. And *Tables* of these *Meridional Parts* you have ready Calculated in mosts Books of the *Piloting Part of Navigation*; they are not only Useful in Working the several Cases in *Mercator*'s *Sailing*, but also, in Making or Graduating *Mercator*'s *Charts*.

MESS: The Ship's Company is generally divided into several *Messes*, of *Three* or *Six* in a *Mess*; that is, so as their Victuals might be the more easily distributed and given out to so many at once, who jointly Diet together; and their Share, or Proportion of Victuals, is call'd, a *Mess*.

MIDDLE-*Latitude*. See *Latitude*.

MIDDLE-LATITUDE-SAILING, is a Method of Working the several Cases in *Sailing*, nearly agreeing with *Mercator's* Way, but without the help of *Meridional Parts*.

CASE 1.

Given, *Latitudes and Longitudes, of any Two Places*;
Required, *Their Bearing and Distance*?

1. To find the *Bearing*, or *Course*, say,

As the Difference *of* Latitude,
*Is to the Co-Sine of the Middle-*Latitude;
So is the Difference *of* Longitude,
To the Tangent of the Course.

2. To find the *Distance*, say,

As the Co-Sine of the Course,
Is to the Radius;
So is the Difference *of* Latitude,
To the Distance.

CASE 2.

Given, *Latitudes, and Bearings*;
Requir'd, *Difference of Longitude*?

The Proportion is,

As Co Sine of the Middle Latitude,
Is to the Difference *of* Latitude;
So is the Tangent of the Course,
To the Difference *of* Longitude.

CASE 3.

Given, *Latitudes, and* Departure;
Requir'd, *Difference of Longitude*?

Find the *Course*, by *Case* 6. of *Plain Sailing*:
Then find the *Difference of Longitude*, by *Case* 2. of this.

CASE 4.

Given, *Latitudes*, and *Distance*;
Requir'd, *Difference of Longitude*?

1. Find the *Course*, by *Case* 5. of *Plain-Sailing*:
2. Then find the *Difference of Longitude*, by *Case* 2. of this.

And after the same manner, all the Common *Cases* in *Mercator's Sailing* are wrought.

MID-SHIP-*Beam*. See *Beam*.

MID-SHIP-*Men*, are *Officers* on-board a Ship; their Station, when on their Watch, is some on the *Quarter-Deck*, others on the *Poop*, &c. They mind the *Braces*, Look-out, and give the Word of Command from the *Captain*, and their Superior Officers: They assist on all Occasions; both in Stowing and Rummidging the Hold, and Sailing the Ship: They are generally Gentlemen upon their Preferment, having Serv'd the limited Time in the *Navy* as *Volunteers*.

MINUTE,

MINUTE, or a *Mile*, is the 60 Part of a Degree of a *Great Circle*.

Half-Minute-Glasses, are such as the Sand *Hour-Glasses*, only these run out in *Half a Minute* of *Time*; and are used in Heaving the Log. See Log.

MISEN, *Misson*, or *Mizen*, is either *Mast*, or *Sail*. See *Mast* and *Sail*.

Note, That at *Sea*, when we use the word *Misen*, we always mean the *Sail*.

The Use of the *Misen*, is to keep the Ship close to a Wind: Wherefore, if a Ship be apt to gripe too-much, they use no *Misen*. But 'tis often used, when a Ship rides at Anchor, to back her a-stern, so that she may not foul her Anchor, on the Turning of the Tide. And sometimes a Ship lies *a-Try* with her *Misson* only.

Set the Misen! That is, Fit the *Misen-Sail* right as it should stand!

Change the Misen! That is, Bring the *Misen-Yard* over to the other side of the Mast!

Peek the Misen! That is, Put the *Misen-Yard* right up and down by the Mast!

Spell the Misen! That is, Let go the Shear, and Peek it up!

Main-Misen, } See *Mast*.
Bonaventure-Misen, } See *Mast*.
Misen-Top-Mast. } See *Mast*.

Misen-Sail. See *Sail*.
Misen-Stay. See *Stay*.
Misen-Yard. See *Yard*.

MOAR, See *Moor*.

MONSOONS, are Periodical Winds in the *Indian Sea*; that is, Winds that blow for Half the Year one way, and the other Half upon the opposite Points: And those Points, and Times of Shifting, are different, in different Parts of the Ocean. And in some Places, 'tis Constant for Three Months one way, then Three Months more the contrary way, and so all the Year.

MONKS-*Seam*, is that made by sewing the Edges or Selvedges of the Sails together, one over the other; sewing it both Sides, to make it the stronger.

MOOR; *To Moor a Ship*, is to lay out her Anchors, so as is most convenient for her safe and secure Riding. A Ship is not said to be *Moor'd*, unless she has at least Two *Anchors* out; except when she is *Moor'd a-Proviso*.

To Moor a Fair Birth; that is, in a Place free from any Annoyance.

To Moor a-Cross, or *a-Thwart*; is to lay one Anchor on one side of the Stream, and the other right against it on the other side; so that they *bear* equally, as well at *Tide of Ebb*, as at *Flood*.

To Moor a-Longst; is to lay an Anchor in the middle of the Stream a-Head, another a-Stern; when 'tis fear'd the Ship may drive a-shore.

To

To Moor *Water-Shot*; is to Moor neither *a-Longst* nor *a-Thwart* the Tide, but *Quartering* between both.

To Moor *a Proviso*; is to have one Anchor out, and a Hawser a-shore; then the Ship is *Moor'd* with her Head a Shore. And Two *Cables* is the least, and Four the best to Moor by.

Mooring *for North*, *West*, &c. In an open Road, the *Master* and *Pilots* will Moor, or lay out an Anchor on that Point of the *Compass* they think the Wind is likeliest to endanger the Ship.

MUNITION-*Ships*, are those which have Stores on-board, for to Supply the Necessaries required by a Fleet of Men of War at Sea,

In Time of an *Engagement*, all the *Munition-Ships* and *Victuallers* attending the Fleet are to take their Places and proper Stations in the Rear of all the rest, and not Engage themselves at all in the Fight; but to attend such Directions as shall be sent unto them, at all times, by the *Admiral*.

MURDERERS; are small Pieces of Ordnance, either of Brass or Iron, having Chambers, or Charges, made of Brass or Iron, put in at the Breeches: They are used at the Bulk-heads of the Fore-Castle, Half-Deck, or Steerage, in order to Clear the Deck, when an Enemy Boards the Ship; they are fasten'd, and travers'd, by a Pintle, which is put into a Stock.

N

NADIR, is that Point in the Heavens, which is Diametrically opposite to the *Zenith* Point directly over our Head; or is the Lower Pole of the *Horizon*, because distant from it, every way, 90 *Degrees*.

NAILING *of a Gun*; is driving of a Nail, Iron-Spike, or the like, by Force, into the Touch-hole of a Piece of Ordnance, so as to render it useless to the Enemy.

NAILS; as *Tree-Nails*. See *Tree*.

Skupper-NAILS. See *Skuppers*.

NAVIGATION, is the Art of Conducting a Ship from one Place to another: And may be distinguish'd into Two Parts, *viz*. the *Piloting*, and the *Working-part*.

Piloting-part of Navigation, may be also, either *Common*, or *Proper*.

Common Piloting, is Sailing in sight of Land, or Coasting nigh Shore: The *Lead*, and *Compass*, with a competent Knowlege of the *Lands*, are chiefly required for the Performance thereof.

Proper Piloting, gives Directions upon what *Course* to Steer to any Place desired: As also, when at *Sea*, shews the Place the Ship is in; how far she has sailed; how far she has to sail; and how all Places bears from her,

her, at any time: And this by means of the *Latitude*, *Log*, and *Compass*; and by the several Methods of *Sailing*, as *Plain*, *Wright*'s, &c. See *Sailing*.

Working-part of Navigation, shews the Method of giving that Noble Fabrick, a *Ship*, all the *Motions* and *Directions* its capable of, so as to perform the *Navigator*'s Pleasure, and be govern'd at his Will, in all Cases.

The *Theory*, (which is the True Foundation of all Practice) of this Part of *Navigation*, as it is more Useful, so it is more Intricate, and has more of Art, than the other Part; yet 'tis never Taught by any, nor perhaps thought of by many, who Profess the Art. Therefore the *Practice* thereof is usually gain'd at the Expence of the greatest and Choisest part of Man's Time; and that only by a constant habit of a necessitated Form of Working, without being much beholden to Reason, or its Laws, or any prescribed Rules deduced from them.

Thus, they who would be sufficiently acquainted with the *Practices* at Sea, and Management of a Ship; must receive their Instructions from our Common Mistress, *Experience*, who teaches Truth, by Tryal, to all alike. So that it would be particularly Advantageous for those Gentlemen that design for the *Sea*, to be perfectly acquainted with the *Theoric-part*; since it will lead them directly to, and abundantly shorten the Time of attaining the *Practice*.

NAVIGATORS: By this word is usually meant, Persons capable of Carrying or Guiding a Ship to any Place desired.

NAUTICAL-*Chart*, or *Planisphere*. See *Chart*.

NAUTICAL-*Compass*. See *Compass*.

NEALED: If the *Sounding* be *Nealed to*; that is, if it be Deep-Water, close the Shore; or if the Lee-shore be Sandy, Clayie, Oasie, or Foul and Rocky-Ground.

NEAP-*Tides*. See *Tides*.

NEAR! *No Near!* a Word of Command from him that Con's the Ship, to the Man at Helm, requiring him to let her fall to the Lee-ward.

NEEDLE. See *Magnetic-Needle*.

NETTINGS, are small Ropes seiz'd together Gratingwise with Rope-Yarn, and sometimes made to stretch upon the Ledges from the Wast-Trees to the Rouf-Trees, from the Top of the Fore-Castle to the Poop: And sometimes are laid in the Wast of a Ship, to serve instead of *Gratings*.

NETTING-*Sail*. See *Sail*.

NIPPERS, are small short Ropes, with a little Truck at one end, and sometimes only a Wall-Knot: They serve, to help *hold off* the Cable from the Main, or Jeer-Capstan, when 'tis so slimy, so wet, or so great, that

that they cannot ſtrain it, to hold it off, with their bare Hands.

NOCTURNAL, is an Inſtrument made of Wood, Ivory, or Braſs, divided on both ſides: It's Uſe is, to take the *Altitude* or *Depreſſion* of the *Pole-Star*, in reſpect to the *Pole* it ſelf; in order to find the *Latitude* of the Place; or making an Eſtimate, or near Gueſs at the *Hour of the Night.*

NORTH, one of the Four *Cardinal Points* of the *Mariners-Compaſs*; 'tis generally mark'd with a *Flower-de-Luce*, and is directly oppoſite to the *South Point.*

NORTH-*Declination.* See *Declination.*

NORTHERN-*Hemiſphere.* See *Hemiſphere.*

NORTHERN-*Tropic.* See *Tropic.*

NORTHING, is the *Difference of Latitude* a Ship makes in Sailing to the *Northward.* See *Difference of Latitude.*

NORTH-*Pole*; a Point in the *Northern-Hemiſphere* of the Heavens, 90 *Degrees* every way Diſtant from the *Equinoctial.*

NORTH-*Star*, or *North-Pole-Star*, is in the Tail of the *Urſa-Minor*; and is ſo called, as being not above Two *Degrees* and a Half diſtant from the *Pole*, and ſeems, to the naked Eye, as if in the ſame Place.

NORTHWARD; that is, towards the *North.*

O

OAKUM, as *Okum*, or *Ockham*, are old Ropes Untwiſted, and torn in Pieces, and pull'd again out into looſe Hemp, like Hurds of Flax; that it may be driven into the *Seams*, *Trennels*, and *Rends* of a Ship, to ſtop, or prevent a Leak.

OAR, is a long Piece of Wood, whoſe End which is in the Water is made Thin and Broad, for the eaſie cutting and reſiſting the Water, and conſequently, for moving the Veſſel.

OAZY, *Oaſie-Ground*; that is, Soft, Slimy, or Muddy-Ground: This is no good Anchoring-Ground; becauſe the Anchor here cannot hold firm, but will come home, or give way in ſtreſs of Weather; and will alſo rot the Cables, if a Ship ride long over ſuch Ground: But ſuch Ground is good to bring a Ship a-ground upon, becauſe ſhe can there Dock her ſelf, and lie ſoft; but yet if ſhe lie long, ſhe will rot her Plank, and ſpoil the Oakum in her Seams.

OBLIQUE-*Horizon.* See *Horizon.*

OBLIQUE-*Sphere.* See *Sphere.*

OBSERVATION *of the Sun*, or *Star*, is uſually made when the *Sun* or *Star* is in the *Meridian*, by a *Quadrant*, *Croſs-Staff*, or the like

like Inſtrument, in order to find the *Latitude of the Place*.

To Work an Obſervation; is, To Find the *Latitude of the Place*, by having the *Sun* or *Star's Meridian Altitude*, by *Obſervation*, and the *Declination* given. It has Two *Caſes*.

CASE I.

When the Obſerv'd Object does both *Riſe* and *Set*, and conſequently has but One *Meridian Altitude* in the ſpace of 24 *Hours*.

RULE

The *Meridian Altitude* and *Declination* of { the ſame / contrary } Kind; the { *Difference* / *Sum* } of the *Zenith Diſtance*, and *Declination*, is the *Latitude of Place* ſought.

NOTE.

1. Where the *Meridian Altitude* and *Declination* is of the ſame Kind:

If *Decl.* be { Greater / Leſs } than the *Zenith Diſt.* The *Latitude* is of { the ſame / contrary } Name to the *Declination*.

2. But where the *Meridian Altitude* and *Declination* is of Different Kind, the *Latitude* is of the ſame Name with the *Declination*.

CASE 2.

When the Obſerv'd Object does not *Riſe*, or *Set*, and therefore has Two *Meridian Altitudes* in the ſpace of 24 *Hours*.

RULE.

The *Obſerv'd Object* being on the *Merid.* { below / above } the *Pole*; The { *Sum* / *Difference* } of the *Meridian Altitude*, and *Complement* of the *Declination*, is the required *Latitude*, of the ſame Name with the *Declination*.

OCEAN, [Gr. 'Ωκεανὸς, from ὠκέως, *citò*; and νάω, *Fluo*,] is that Vaſt Collection of Salt and Navigable Waters, which Compaſſes the whole Earth: 'Tis join'd to the *Mediterranean-Sea* by the *Strait of Gibraltar*; And its ſeveral Parts, have Names of Diſtinction from thoſe of the adjoining *Continents*, or their Bearing from them. As,

1. *With respect to* EUROPE, *there are*,

The *Hyperborean Ocean*, enclosing *Europe* on the North.

The *Western Ocean*, enclosing *Europe* on the West.

2. *With respect to* ASIA, *there are*,

The *Tartarean* and *Chinean Oceans*, enclosing *Asia* on the North, and East.

The *Indian*, *Persian*, and *Arabic Oceans*, enclosing it on the South.

3. *With respect to* AFRICA, *there are*,

The *Atlantic*, and *Ethiopic Oceans*, enclosing *Africa* on the West, and South.

The *Oriental Ocean*, on the East.

4. *With respect to* AMERICA *there are*,

The *Pacific Ocean*, enclosing *America* on the West.

The Vast *Eastern Ocean*, enclosing it on the East.

But the Universal *Ocean* may more properly be Divided into Three Parts, *viz.*

1. The *Atlantic* and *Ethiopic Ocean*; lying between *Africa*, and *America*.

2. The *Indian Ocean*; lying between *Africa*, and the *Indian Islands*, and *Hollandia Nova*.

3. The Great *South Sea*, or the *Pacific Sea*; lying between the *Philippine Isles*, *China*, *Japan*, and *Hollandia Nova* on the *West*; and the Coast of *America* on the *East*.

OFFICERS and *Offices* belonging to *Naval Affairs*, are those that have the Care of Giving Orders and Directions, also, of Providing and Fitting out of Ships, and Naval Stores; as those of the *Admiralty*, *Navy-Office*, *Victualling-Office*, *Dock-Yards*, &c. The *Sea-Officers*, are the *Flag-Officers*; as, *Admiral*, *Vice-Admiral*, and *Rear-Admiral*: And other *Commission'd-Officers*, as *Captains*, and *Lieutenants*; with *Warrant-Officers*, as *Master*, *Gunner*, *Surgeon*, *Purser*, *Boatswain*, *Carpenter*, *Cook*, &c. And other Petty *Officers*, as *Mates*, *Midship-men*, *Corporal*, *Coxswain*, *Quarteers*, &c. Who have each their several Duties assign'd them, as may be seen, under the Words of their respective Titles.

At each *Dock-Yard* there is a particular *Store-Keeper*, for the Receipt and Issue of all Naval Stores: And as other *Officers*, namely, the *Master-Attendant*, *Master Shipwright*, *Clerks of the Cheque*, and *Survey*, are more or less a Cheque on the said *Store-Keeper*; so is there One of the *Principal Officers & Commissioners of the Navy* particularly Appointed to Reside at, and Inspect into Affairs of the Principal *Yards*; and Another, who Assists

at

at the *Board* in the *Navy-Office*, is charged with the Examining and Adjusting the *Store Keeper's Accounts*, as a Second is those of the *Treasurer of the Navy*, and a Third the *Accounts of the Victualling*. And as each of the aforesaid *Warrant-Officers* in the *Yards* have their particular Duties assign'd them; *viz.* The *Master-Shipwright*, in Building of Ships; the *Master-Attendant*, in the Proportioning Rigging for them, and putting it over Head, laying out Moorings, and removing Ships from one Mooring to another, with several other Services; so are these *Clerks of the Cheque* Employ'd in the Mustering the Workmen in the Yards, and the Ships Companies within their Reach, and Chequing them out of Wages when Absent: And both They, and the *Clerks of the Survey* Inspect into all Stores Deliver'd, to see that they Answer to Contract, both in Quantity, and Quality.

Besides these *Principal Officers* in the Yards, there are Others inferior to them, but All are under the immediate Command of the *Commissioners* residing on the Place. Those Inferior *Officers* are Assistants to the *Master-Shipwright*, *Master-Caulker*, *Master-House-Carpenter*, *Master-Joiner*, *Master-Boat-builder*, *Master-Mast-maker*, *Fore-man* both on Shore, and a-Float, *Boatswain of the Yard*, *Master-Sail-maker*, and the like, who have also each of them their particular Duties assign'd.

OFFING, or *Offin*, is an open Sea, a good Distance from the Shore, where there is Deep-Water, and no need of a *Pilot* to Conduct the Ship into the Port, or Harbour. Also, the Middle-part of any great Stream is call'd *Offing*.

The Ship stands for the Offing; that is said of a Ship, seen from Shore, sailing out to Sea-ward.

The Ship is in the Offing; that is, She has the Shore near her, and having another a good way without her, or towards the Sea.

OFF-WARD; that is, contrary to the Shore.

The Ship Heels Off-ward; that is said of a Ship, that being a-Ground, does *Heel* towards the Water-side.

The Ship lies with her Stern to the Off-ward, and Head to the Shore-ward; that is, Her Stern lies towards the Sea, and Head to the Shore.

ORLOPE, in a *Three-Deck'd-Ship*; the *Second*, and *Lowest Decks*, are sometimes call'd *Orlope*. Also, the Spare-Deck in a Great Ship reaching from the *Main-mast* to the *Misen*.

OVER-*Blow*; that is, when the Wind blows so very hard, that the Ship can bear no Top-sails.

OVER *Grown*: When the Waves of the Sea grow high, the Mariners call it *Rough-Sea*; but when the Surges and Billows grow higher, 'then 'tis an *Over-grown-Sea*.

OVER-*Hale the Runner.* See *Runner.*

OVER-*Hale the Sheat.* See *Sheat.*

OVER-*Rake.* See *Rake.*

OVER-*Set*, or *Over-Throw* a Ship, is to bring her Keel upwards, either by bearing too-much Sail; or by Grounding her, so that she falls upon one Side.

OUTLICKER, is a small Piece of Timber, made fast to the top of the Poop, and so stands out right a-Stern: At the outmost end thereof there is a Hole, into which the standing part of the Sheat is reev'd thro' the Block of the Sheat; and then again reev'd thro' another Block which is seiz'd close by the end of this *Outlicker*: 'Tis seldom used in great Ship, except the *Misen-Mast* is placed so far aft, that there is not room enough within board to hale the Sheat flat.

OX-EYE, a Name given by Seamen, to those dreadful Storms, that are sometimes met with on the Coast of *Guinea*; for at first it appears of the Form of an *Ox*'s-*Eye*, and not much bigger; But it descends with such Celerity, that in a very little space, and many times before they can prepare themselves for it, it seems to them to overspread the whole *Hemisphere*; and at the same time, forces the Air with so much Violence, that the Ships are sometimes scatter'd several ways, sometimes directly contrary, and sometimes are sunk downright.

P

PANCH, or *Pantch.* See *Paunch.*

PANTER, *Shank-Panter.* See *Shank.*

PARALLELS *of Latitude*, are Circles imagin'd to be drawn *Parallel* to the *Equator*, thro' every Point of *Latitude*; and therefore are Lesser Circles, consequently a *Degree*, or the 360th Part of any *Parallel of Latitude*, must be less than an *Equinoctial Degree*, or that of any Great Circle, and that in the same Proportion as the *Radij* of these Circles, that is, as the *Co-Sine* of the *Latitude*, to the *Radius* or *Sine* of 90 *Degrees.*

PARALLEL-*Horizon.* See *Horizon.*

PARALLEL-*Sailing*: By *Parallel-Sailing*, is meant, *Sailing* under a *Parallel* of *Latitude.* It has these *Cases.*

CASE I.

Given, *Difference of Longitude, and Distance of two Places in the same Latitude*;
Requir'd, *the Latitude*?

Proportion:

As the Difference *of Longitude,*
Is to the Radius;
So is the Distance,
To the Co-Sine of the Latitude.

CASE

CASE 2.

Given, *The Difference of Longitude between two Places in the same Latitude* ;

Requir'd, *Their Distance?*

Proportion.

As the Radius,
Is to the Difference of Longit.
So is the Co-Sine of the Lat.
To the Distance.

CASE 3.

Given, *The Distance between two Places in the same Parallel* ;

Requir'd, *The Difference of Longitude?*

Proportion.

As the Co-Sine of Latitude,
Is to the Distance ;
So is the Radius,
To the Difference of Longitude.

PARALLEL-*Sphere.* See *Sphere.*

PARBUNCLE, is a Rope somewhat like a Pair of Slings; 'tis seiz'd both Ends together, then put double about any heavy thing that is to be hois'd in, or out of the Ship: By means of this, a full Vessel may be hois'd in, without spilling; by putting it upon these Ropes set cross-wise, and bringing the Loops over the upper-end of the Cask, then fixing the Tackle to them, the Vessel will stand upright.

PARCEL: *To Parcel a Seam*; that is, after a *Seam* is Caulk'd, to lay over it a narrow Piece of Canvase, and pour on it hot Pitch and Tar.

PARRELS, are Frames made of Trucks, Ribs, and Ropes, which go round the Masts, and made fast so to the Yards, that they may slip up and down the Masts the easier; and with the help of the Breast-Ropes, keep the Yards close to the Masts.

PARTNERS, are strong Pieces of Timber, bolted to the Beams, incircling the Masts, to keep them steady in their Steps. These *Partners* are also at the Second Deck, for the same Purpose: But the Misen-Mast has only one Pair of *Partners*; and yet that Mast is so firmly wedg'd in them, that it cannot stir, by any means. Tho' some Ships do not sail well, except their Masts be loose, and have leave to play in the *Partners*: Yet, in a *Storm*, this is very Dangerous; lest the *Partners* should be *wrong'd*, and forced out of their Places; for then there is hardly any other Remedy, than Cutting the Mast by the Board.

PASSARADO, is a Rope whereby all the Sheat-Blocks of the Main and Fore-Sails are haled down Aft; The Clew of the *Main-Sail* to the *Main-Mast*, and the Clew of the *Fore-Sail* to the *Cat-Head.* This is to be done, when the Ship goes large; and they are also kept down firm, and hindred from flying up, by this *Passarado-Rope.*

PAUNCH, *Pantch*, or *Panch*, are

are broad Clouts woven of Thrums and Sinnets together, to ſave things from galling and fretting; therefore are made faſt to the Main, and Fore-Yards, for that Purpoſe.

PAWL, is a ſmall Piece of Iron bolted to one end of the Beams of the Deck, cloſe by the Capſtan, but yet ſo eaſily, as that it can turn about. Its Uſe is, to ſtop the Capſtan from turning back, by being made to catch hold of the *Whelps*.

Pawling the Capſtan. See *Capſtan.*

PAY: *To Pay a Seam*; that is, after Caulking, to lay hot Pitch and Tar on, without Canvaſe. So that,

Paying a Ship, is laying a Coat of hot Pitch, or the like, over the Seams of a Ship: And when a Ship is well Caulk'd, and *Pay'd over* with hot Pitch, ſhe is much the lighter for it.

The Ship is Pay'd; that is, when in Tacking about, a Ship's Sails being Back-ſtay'd, fall all flat againſt the Maſts and Shrowds.

Pay more Cable! That is, Let out more Cable!

Pay Cheap! is a Command, in turning an Anchor out of the Boat, to turn it over-board faſter.

PEEK, is a Word uſed at Sea, in theſe various Sences: As,

The Anchor is a-Peek; that is, when a Ship, being about to weigh, comes ſo over her Anchor, that the Cable is Perpendicular between the Hawſe and the Anchor.

Heaving a-Peek; is to bring the Ship ſo as the Anchor may hang *a-Peek*.

To Ride a-Peek; that is, when a Ship lies with her Main and Fore-Yards hois'd up, and then having one End of the Yards brought down to the Shrowds, the other is raiſed up an end: This is done to contrary Sides, that is, the Star-board Yard-Arm of the Main-Yard comes down to the Star-board-Side, and ſo does the Lar-board-End of the Fore-Yard; ſo that the Yards appear a-croſs each other like St. *Andrew*'s *Croſs*. It is perform'd thus; Let go the Starboard Topſail-Sheats from the Main-Yard, and Top-up the Larboard Lifts: And ſo quite contrary, for the Fore-Yard.

The Uſe of *Peeking up the Yards*, is, leſt lying in a River with the Yards a-croſs, another Ship ſhould be foul upon them, and break the Yards.

To Ride a-broad Peek, is much after the ſame manner, only the Yards are rais'd up but half ſo high.

Peek the Miſen! See *Miſen*.

PEEK, alſo, is a Room in the Hold of a Ship, from the Bitts forwards to the Stem.

PENDANTS, or *Streamers*, are thoſe long *Colours* which are hung at the Heads of the Maſts, or at the Yard-Arm-ends: Their Uſe is chiefly for Ornament.

PENINSULA, is a Part of the Dry-Land, which is almoſt an Iſland, being join'd to the

Con-

Continent by a narrow Neck of Land; as *Juitland* in *Germany*, *Morea* in *Greece*, *Taurica Chersonesus* in *Little-Tartary*, *Peninsula Indiæ intra & extra Gangem* in *Asia*.

PENNANT, is a ſhort Rope made faſt at one end to the Head of the Maſt, or to the Yard-Arm, with a Block at the other end, and a Shiver to reeve ſome running Rope into: And all the Yard-Arms, except the *Miſen*, have *Pennants*.

PILLOW, is that Piece of Timber whereon the Bowſprit reſts, at its coming out of the Hull aloft, cloſe by the Stem.

PILOTS, are thoſe, who, upon Coaſts and Shores unknown to the Maſters, are Employ'd to Conduct Ships into Roads or Harbours, or over Bars and Sands, and thro' Intricate Chanels, &c. And this they do, by their Knowlege of the true Depth of Water, and Flowings of the Tides, and how they Set from Point to Point, with the Difference of thoſe a-Board from thoſe in the Chanel, or a-Shore; and by the Blowings of the Winds, where the Sands Moveable; and by Land-Marks which they are acquainted withal, when they are to paſs thorow any Chanel. And theſe *Pilots* are ſeldom Entertain'd whilſt Ships are abroad at Sea, or for a whole Voyage; but having done their Parts, they return a-Shore, where they have their Reſidence, and get their Living in this manner.

PILOTING *Part of Navigation*. See *Navigation*.

PINK, is a Veſſel Maſted and Rigg'd like others; but built with a round Stern, the Bends and Ribs compaſſing ſo as the Sides bulge out very much; which renders theſe ſort of Ships very difficult to be Boarded, and to carry greater Burdens than others. They are often uſed for Store-Ships, Hoſpital-Ships, or the like.

PINNACE, is a Smal Veſſel with a Square Stern, going with Sails and Oars, and carrying Three Maſts: They are uſed, as Scouts, for Intelligence, for Landing of Forces, and the like.

PINNACE, alſo, is one of the Boats that belong to Great Ships; they ſerve for Officers to go a-ſhore, and the like.

PINTLES, are the Hooks by which the Rudder hangs to the Stern-Poſt.

PIRATS, are Ships affected to no Party, and reſpecting Friends and Enemies alike; and therefore Takes either, as they come in their Way: They ſhew Commiſſions from contrary Parties, and the Colours of any Nation.

PITCH: *The Maſt is Pitch'd*; that is, The *Maſt* is put and let down into the *Step*.

The Maſt is Pitch'd too-far aft; that is, The *Maſt* is placed too-near the Stern.

The Ship will Pitch her Maſts by the Board; that is ſaid, when a Ship falls with her Head into the Sea, or beat violently againſt the

the Sea, so as to endanger her Top-Masts.

PLAIN-*Charts.* See *Charts.*

PLAIN-*Sailing*, shews the various Motions of a Ship upon a *Plain*, where the *Meridians* are all made *Parallel* one to the other; the *Parallels* at Right-Angles to the *Meridians*, and the *Degrees* of each *Parallel* Equal to those of the *Equinoctial.* In which, tho' according to this way of Working, each *Parallel* is falsly supposed Equal to the *Equinoctial*, and the *Terraqueous* Globe as falsly to be Plain and Flat; yet if a Ship sail under, or near the *Equinoctial*, or *Meridian*, a short Voyage, or a long one cut into many short ones, may indifferently be perform'd by it.

The several Cases in *Plain-Sailing*, with their *Solutions*, are as follow:

CASE 1.

Given, *Course and Distance run*;

Requir'd, *Departure, and Difference of Latitude?*

Proportion.

1. As the Radius,
   To the Distance run;
   So is the Sine of the Course,
   To the Departure.

2. As the Radius,
   To the Distance run;
   So is Co-Sine of the Course,
   To the Difference of Lat.

CASE 2.

Given, *Course and Departure*;

Requir'd, *Distance and Difference of Latitude.*

Proportion.

1. As the Sine of the Course,
   To the Departure;
   So is the Radius,
   To the Distance.

2. As the Sine of the Course,
   To the Departure;
   So is Co-Sine of the Course,
   To the Difference of Lat.

CASE 3.

Given, *Course and Difference of Latitude?*

Requir'd, *Distance, and Departure?*

Proportion.

1. As Co-Sine of the Course,
   To the Difference of Lat.
   So is the Radius
   To the Distance.

2. As Co-Sine of the Course,
   To the Difference of Lat.
   So is the Sine of the Course,
   To the Departure.

CASE 4.

Given, *Distance, and Departure*;

Requir'd, *Course, and Difference of Latitude?*

Proportion.

1. As the Distance run,
   To the Radius;
   So is the Departure,
   To the Sine of the Course.

2. As

2. As the Radius,
To the Diſtance run;
So Co-Sine of the Courſe,
To Difference of Latitude.

CASE 5.

Given, *Diſtance, and Difference of Latitude.*
Requir'd, *Courſe, and Departure?*

Proportion.

1. As the Diſtance run,
To the Radius;
So is Difference of Latitude,
To Co-Sine of the Courſe.

2. As the Radius,
To the Diſtance run;
So is the Sine of the Courſe,
To the Departure.

CASE 6.

Given, *Difference of Latitude, and Departure,*
Required, *Courſe, and Diſtance?*

1. As the Difference of Lat.
To the Radius;
So is the Departure,
To the Tangent of the Courſe.

2. As Sine of the Courſe,
To the Departure;
So is the Radius,
To the Diſtance run.

PLAIN-SCALE. See *Scale.*

PLANKS, are thoſe Timbers which go Fore and Aft on each Side of the Ship, whereon lie the Beams of the Firſt *Orlop.*

Garboard-PLANK. See *Garboard.*

PLATFORM, is a Place on the Lower-Deck of a Man of War, abaft the Mainmaſt, between it and the Cockpit.

PLATS, are flat Ropes made of Rope-Yarn, and weav'd one over the other: They ſerve to ſave the Cable from galling in the Hawſe, or to wind about the Flooks of the Anchors, to ſave the Pennant of the Fore-Sheat from galling againſt them.

PLOW, is an Inſtrument made commonly of Box, or Pear-Tree; for the Taking of the *Sun*, or *Star*'s *Height*, in order to find the *Latitude of the Place*: It admits of large Graduations; and is eſteem'd of, by ſome Artiſts.

PLUMMET. See *Lead.*

POINT of the *Compaſs*, is the 32d Part of the Circumference of the Circle on the *Card*, or 11 *Degrees*, 15 *Minutes*: Therefore,

Half a Point, is 5 *Degrees*, 37 *Minutes.*

Quarter-Point, is 2 *Degrees*, 48 *Minutes.*

POINT *of Land.* See *Promontory.*

POINT *of Traverſe*; *To Caſt a Point of Traverſe.* See *Caſt.*

POINTING *the Cable*; is untwiſting it at the End, and leſſening the Yarn, and twiſting them again, and making all faſt with a Piece of Marline: This is done, in order to keep the Cable from raveling out; as alſo,

also, that none of it may be cut off.

POLES *of the World* are two *Points* 90 Degrees each distant from the *Equinoctial*: They are the two Extremities of the *Axis*, (or that Imaginary Line which is supposed to pass thro' the Centre of the *Globe*, and upon which the whole is conceiv'd to turn round;) one whereof is call'd the *North*, or *Arctic*; and the other, the *South*, or *Antarctic Pole*. They are call'd *Poles*, from Πολέω, (*Verto*;) for upon them, the whole Frame of the *Globe* turns about.

Arctic, or *North Pole*, is so term'd, from ἄρκτος, signifying a *Bear*; because so very near a noted *Star* in the Constellation call'd the *Little Bear*.

Antarctic, or *South Pole*, is the *South Point*; and is so call'd, from ἀντὶ, (*contra*,) and ἄρκτος, (*Ursa*;) because of its being Diametrically opposite to the other.

POOP *of a Ship*, is the Highest, or Uppermost Part of her Hull a-stern.

PORT! or, *Port the Helm!* is a Word of Command to him at the *Helm*, to put it to *Larboard*, and the Ship will go to *Starboard*. See *Helm*.

The Ship Heels a-Port. See *Heel*.

PORT, is also a *Haven*, or *Harbour*. And,

PORTS, are those Holes in a Ship's-Side, thorow which the Great Guns are put out.

PORT-*Last*; the same with the *Gunwail* of a Ship.

The Yard is down a-Port-Last, when it lies down on the Deck.

PORT-*Ropes*. See *Ropes*.

PORTOISE; *To Ride a-Portoise*. See *Ride*.

PORT; *Sally-Port*. See *Sally*.

POWCHES, are small Bulk-heads, made in the Hold of the Ship, to Stow Goods, that they may not shoot from one side to the other.

POWDER-ROOM, is that part of the Hold, wherein the *Powder* is Stow'd.

PREDY *the Ship!* That is, Make Ready to set Sail! or to Fight!

Predy the Hold! That is, Lay, or Stow every thing there in its Due Order, and Proper Place.

PRESS, or *Pressing*. See, *To Man a Fleet*.

PREST-SAILS; as, *To Carry a Prest-Sail*; that is, The Ship Carries all the *Sail* she can croud: This is sometimes practis'd by Ships, upon Sight of an Enemy, whom they think they are not able to Encounter. But carrying too much *Sail*, if it Blow, may prove of ill Consequence, as the hazard of Oversetting the Ship; or endangering the Mast, and Yards: For if any of them *come by the Board*, the Enemy can lie upon the Bow, Counter, or Thwart one's Hawse, and Rake one Fore and Aft at his Pleasure. So that Meeting an Enemy, in such a Case,

Case, 'tis most rational to go away with an *easie Sail*, and make a Running Fight, till Night favour an Escape.

PREVENTER-ROPE. See *Rope*.

PROMONTORY, is a High Part of a Land, stretching it self into the Sea; its Extremity is usually call'd a *Cape*, or *Head-Land*.

In EUROPE, *the Chief* Promontories, *or* Head-Lands, *are*,

Cape St. Vincent, *Cape Finister*, *Cape Roxant*, Extending from the West of *Spain*.

Cape La-Hgue, Extending from the North of *France*.

Cape North, Extending from the Northmost part of *Norway*.

The *Lands-End*, Extending from the South-west of *England*.

The *Lizard*, and *Start-Point*, Extending from the South of *England*.

In ASIA, *the Chief* Promonmontories, *or* Head-Lands, *are*,

Cape Sumber, Extending from the *East* of *China*.

Cape Cmorin, Extending from *Penin. Indiæ inter Gangem*.

Cape Razalgate, Extending from the South-East-part of *Arabia*.

In AFRICA, *the Chief* Head-Lands, *are*,

Cape Gardefuir, Extending from the North-East-part of *Ethiopia Exterior*.

Cape de Bonne Esperance, Extending from the South of *Ethiopia Exterior*.

Cape de Verde, Extending from the West of *Negroland*.

Cape Spartel, Extending from the West of *Barbary*.

In AMERICA, *the Chief* Head-Lands, *are*,

Cape Horn, Extending from the South of *Terra del Fuego*.

Cape Frio, *Cape St. Augustine*, Extending from the East of *Brazil*.

Cape de Coriente, Extending from the West of *New-Spain*.

Cape Florida, Extending from the South of *Florida*.

PROPER PILOTTING. See *Navigation*.

PROVISO; *To Moor a-Proviso*. See *Moor*.

PROW, is properly the Head of the Ship.

PUDDINGS, are Ropes nail'd to the Arms of the *Main* and *Fore-Yards*, near the Ends, at some Distances from one another. Their Use is, to save the *Robbins* from galling, or wearing asunder upon the Yards, when the Top-sails are haled Home.

Pudding of the Anchor, are those Ropes which are wound about the Rings of the Anchors, to save the Clinch of the Cable from being galled by the Iron.

PULLET,

PULLET, is a Room within the Hold, in which Pigs of Lead, or the like weighty things, are put, that the Ship may be sufficiently Ballasted, with the loss of little Room.

PUMPS, used at Sea, are of several sorts; as some have such as those a-Shore, standing by, or near the Main-Mast.

Chain-PUMPS, are those used in great Ships; they go with more Ease, and yield more Water than the Ordinary *Pumps*; and are sooner mended.

BARE-PUMP, is made of a Cane, or a Piece of Hollow Wood, or the like: 'Tis used, for to Pump Beer, or Water, out of the Casks.

BUR-PUMPS, are those used commonly by the *Dutch*, who have them by their Ships-sides; in these there is a long Staff, with a *Bur* at the End like a Gunner's Spunge, to Pump up the Bilge-Water.

In *Pumping*, 'tis usual to make *Spells*, that is, to relieve the Men with fresh ones, and count how many Strokes they *Pump* each Watch, and by that means know if the Ship be Stanch, or how her Leaks encrease.

BILGE-PUMP. See *Bilge*.

PUMP's-CAN, is the *Bucket* whereby they pour Water into the *Pump*, to fetch it, and make it work, when 'tis to be used.

PUMP's-VALE, is the *Trough* by which the Water runs from the *Pump*, along the Ship's-sides, that it may go out at the *Scupper-holes*.

PUMP's-BRAKE, is the *Handle* of the *Pump*.

The Pump Sucks; that is, when the Water being out, draws up nothing but Froth, and Wind.

PURCHASE, signifies only to *draw in*: As,

The Capstan Purchases a-pace; that is, draws the Cable in apace.

The Tackle will not Purchase; that is, when they cannot draw, or hale in any thing, with the Tackle.

PURSER of a *Man of War*, is an Officer, who has the Charge of the Victuals, receives it, takes care that it be in good Condition, and well lay'd, and stow'd up: He keeps a *List* of the Ship's Company, and sets down exactly the Days of each Man's Admittance into Pay, *&c.*

PUTTOCKS, or *Puttock-Shrouds*, are small *Shrouds* which go from the those of the Main-Mast, Fore-Mast, and Misen-Mast, to the Round-Top of those Masts; for where the Shrouds come near the Top, they fall in so much, that One could not get into the Top, without the help of the *Puttocks*. And if there be any Top-Gallant-masts to the Top-masts, there are *Puttocks* to go from the Top-mast-Shrouds to those: These *Puttocks* at the Bottom are seiz'd to a Staff, or to some Rope which is seiz'd

to

to a Plate of Iron, or to a *Dead-man's-Eye*, to which the *Lanniards* of the Fore-mast-Shrouds do come.

Q

QUADRANT, *Sea-Quadrant*, or *Davis's-Quadrant*, is an Instrument contriv'd very commodiously; and at this time, is the most approv'd One, for Observing the *Sun's Meridian Altitude*, at Sea.

These *Quadrants* are made of any Length between 18 Inches and 3 Feet; but now they are generally made of 26 Inches *Radius*, with one Arch 65 Degrees, the other 25 Degrees; and a Glass to the Shadow-Vane.

The Principal Parts are Three *Vanes*, the *Horizon*, *Shadow*, and *Sight-Vane*: And Two *Arches*; on which Arches are the Degrees, both together making 90 Degrees; from whence 'tis call'd a *Quadrant*.

The *Horizon-Vane*, respects the *Horizon*, in time of Observing.

The *Shadow-Vane*, is that which gives the Shadow, and is placed on the *Sixty Arch*.

The *Sight-Vane*, is that thro' which you are to look for both *Shadow* and *Horizon*: 'Tis placed on the *Thirty Arch*.

The *Sixty Arch*, is that with the Shortest *Radius*; so call'd, because formerly 'twas divided into 60 Degrees: but now it often contains 65, or 70 Degrees. This Arch is of a Small *Radius*; and is so design'd, that the *Shade-Vane* used on it, being at so small a Distance from the *Horizon Vane*, on which its Shadow (in time of Observing) is to fall, might be more Visible to the Eye of the Observer.

The *Thirty Arch*, is of a Large *Radius*, that it might contain the Lesser Sub-divisions of a Degree; and being of a competent Breadth, thereon usually are describ'd 10 *Concentric Circle*, intersected with 3 Diagonal Lines in each *Degree*, making each Intersection 2 *Minutes*. The Limb of this Arch is divided into *Degrees* by long Strokes; each again subdivided into Six Equal Parts: Therefore each small Division is 10 *Minutes*.

The Use of the QUADRANT.

This Instrument is seldom used, otherwise than to Observe the *Sun's Meridian Altitude*; which is perform'd thus:

1. Put the *Horizon-Vane* in its proper Place, close home to the Centre of the *Quadrant*: Put the *Sight Vane* on the *Thirty Arch*, close to the Back of it: And the *Shade-Vane* on the *Sixty Arch*, close to the Back of it; set the upper Edge of this *Vane* to a Number of Degrees less than the Complement of the *Altitude* by 15 or 20 Degrees.

2. The *Vanes* being thus fix'd upon the *Quadrant*, turning your Back towards the Sun, and holding the *Quadrant* as upright

as you can, look thro' the *Sight-Vane*, and bring the upper Edge of the Shadow of the *Shade Vane* to lie upon the upper Edge of the Slit in the *Horizon-Vane*; at the same time, if the *Horizon* appear thro' the said Slit in the *Horizon-Vane*, the *Vanes* on the *Quadrant* stands at the *Sun's present-Altitude*.

3. But if the *Sky* appear, instead of the *Horizon*, slide the *Sight-Vane* a little higher: On the contrary, if the *Sea* appear instead of the *Horizon*, then slide the *Sight Vane* lower; continue so to do, till the *Horizon* appear thro' the *Horizon-Vane*.

4. In order to obtain the *Meridian Altitude*, continue Observing; as the Sun rises, the Sea will appear thro' the *Horizon-Vane*, therefore slide the *Sight-Vane* down lower accordingly: Thus do, Observing as often as may be convenient, till the Sun is at the Highest.

5. When the Sun begins to fall, the *Sky* will appear thro' the *Horizon-Vane*, instead of the *Horizon*, then desist Observing for that Day.

6. Having thus done, add the Degrees the Upper-Edge of the *Shade Vane* stands at, to the Degrees and Minutes cut by the Inside of the *Sight Vane*; their Sum is the *Complement of the Sun's Altitude*, or the *Distance* of the Upper Edge of the *Sun* from the *Zenith*: To which add 16 *Min.* the Sun's Semi-Diameter; and you'll have the *Distance* of the *Sun's Centre* from the *Zenith*, being the true Complement of the *Sun's Meridian Altitude*.

For the Upper-Edge of the Shadow of the *Shade-Vane* respects the Upper-Limb of the *Sun*; and the Lower-Edge of the said Shadow, answers to the Lower-Limb of the *Sun*: So that Observing by the first, you are to Add 16 *Min.* to what is on the *Quadrant*; but Observing by the latter, Substract 16 *Min.* from it; the *Sum*, or *Difference*, is the *Zenith Distance*; that is, the *Distance* of the *Sun's Centre* from the *Zenith*, or the *Complement* of his *Meridian Height*.

But by having a *Convex-Glass* put in the Middle of the Breadth of the *Shade-Vane*, which, in time of Observing, casts an illuminated Spot on a round black Spot made in the *Horizon-Vane*, there will be no occasion of making that Allowance in Adding or Subtracting 16 *Min.* Only setting that Part of the *Shade-Vane* which is right-against the Middle, or Centre of the Glass, to the Degrees in the *Sixty Arch*; and then cause the Enlighten'd Spot of the Glass to lie on the Black Spot, or Circle on the *Horizon-Vane*; at the same time look for the *Horizon* thro' it, as before Directed, so will you have (Observing by this Glass in the *Vane*) the true Complement of the *Sun's Altitude*, or the Distance of his Centre from the *Zenith*.

Besides,

Besides, this *Glass-Vane* is Useful, when the Edge of the Shadow of the *Shade Vane* is not Conspicuous, as in Hazy-Weather; yet he Glass may collect so much of the *Sun's* Beams, as will cast an apparent Brightness upon the *Horizon-Vane*.

Sinical-QUADRANT. See *Sinical*.

QUARTER-*Deck*. See *Deck*.

QUARTER-*Masters*, or *Quarteers*, are Ship-Officers, whose Business is Rummidging, Stowing, and Trimming the Ship, in the Hold; to Over-look the *Steward*, in his Delivery of Victuals to the *Cook*; and in Pumping, and drawing out Beer, or the like: They also are to keep their Watch duly, in Conding the Ship, or any other Duty.

QUARTER *of a Ship*, is that Part of the Ship's Hull which lies from the Steerage-Room to the Transom.

Fat-QUARTER. See *Fat*.

Close-QUARTERS, are where the Seamen *quarter* themselves, in case of Boarding, for their own Defence, and for Clearing the Decks, *&c.*

QUARTER-POINT. See *Point*.

QUARTER, signifies also, a good Treatment given to an Enemy Taken: As, *The Enemy asked Quarter. We gave no Quarter*; and the like.

QUARTERING; *The Ship goes Quartering*; that is, She goes upon a Quarter-Wind; or, She goes neither by a Wind, nor before the Wind, but betwixt both.

QUARTER-*Wind*; is when the *Wind* comes in, abaft the Main-Mast-Shrouds, even with the *Quarter*.

QUARTERING *the Men*; to Dispose of the Ship's Company so at the Time of an Engagement, as each may readily know where his Station is, and what he's to do: As, some to the *Master*, for the Management of the Sails; some to assist the *Gunners* to Traverse the Ordnance; some for Plying of the Small-Shot; some to fill Powder in the Powder-Room; others to carry it from thence to the Gunners, in Cartherages, *&c.*

QUOIL, or *Coile*, is a Rope, or Cable, laid up round, one Fack (or Turn) over another, so that they may the more easily be Stow'd out of the way, and also run out free and smooth, without *Kencks*, that is, without Twistings or Doublings; then the Cable is said to be *Quoil'd* up.

Weather-Quoil, or *Coil*. See *Weather*.

QUOINS; the same with *Coins*: Which see.

Canting and *Standing-Quoins*. See *Coins*.

RAB-

R

RABBETING, is the Letting-in of the Planks of the Ship into the Keel, which, in the Rake and Run of a Ship is hollow'd away; that the Planks may Join the better, and closer.

RABBET *of the Keel*, that is, the Hollowing there.

RAG-BOLTS. See *Bolts*.

RAINS; that Tract of Sea, to the Northwards of the *Equator*, between Four and Ten Degrees of *Latitude*, lying between the Meridians of *Cape Verde*, and of the Easternmost Island of that Name, is usually call'd, by our *Navigators*, the *Rains*: For it seems to be a Place condemn'd to Perpetual *Calms*, and, in a manner, constant *Rains*, with *Thunder* and *Lightning*, after an extraordinary manner; the little Winds that are, be only some sudden uncertain Gusts, of very small Continuance, and less Extent; so that sometimes each Hour you shall have a different Gale, which vanishes away into a Calm, before another succeeds: And here, a Fleet of Ships in sight of one another, shall have each a Wind from different Points of the *Compass*; with these weak Breezes, Ships are obliged to make the best of their Way to the *Southward*, thorow these Six Degrees.

And the *Atlantic Ocean* near the *Equator*, is very much subject to, nay, is always attended with these *Calms*; into which Ships frequently fall, and there detain'd (as 'tis reported) whole Months, for want of Wind.

RAKE *of a Ship*, is so much of her Hull as hangs over both ends of her Keel.

Fore-Rake, or *Rake-forward*, is that Part of it which is Before; and is usually more than a Third, but less than Half the Length of the Keel; a *Long Rake-forward* gives the Ship a good Way; But then, if she has not a Full Bow, 'twill make her apt to pitch much a-head into the Sea; besides, is a great Charge and Stress to a Ship, because she hangs over so a-head. But when a Ship has a *Small Rake forward*, she is so Bluff-headed, that the Sea meets her so often upon the Bow, that she cannot cut the Water so readily, and consequently, will make but small Way.

Rake-aft, or *Rake-aftward*, is that Part which is at the setting on of the Stern-Post; and is generally about a Fourth or a Fifth of her *Fore-Rake*.

The Ship Rakes so much Forward, or *Aftward*; that is, her *Rake* is so much Forward, or Hindward.

RAKE of the *Rudder*. See *Rudder*.

RAM'S-HEAD, in a *Ship*, is a great Block, belonging to the Fore and Main-Halliards; it has

in

in it three Shivers, into which the Halliards are put; and at the End of it, in a Hole, are reev'd the *Ties*.

RANGES, in a *Ship*, are two Pieces of Timber going a-crofs from Side to Side: One aloft on the Fore-Caftle, a little abaft the Fore-Maft; 'tis faften'd into the Timbers of the Ship's-Sides; and has two Knees about the Middle, on either fide of the Fore-Maft, faften'd to the Deck and the Timber, in which the Topfail-Shears run in a Shiver: There are in it alfo, feveral Wooden Pins, to belay the Fore-Bow-Line, the Fore-Tack, and the Fore-Loof-Hook.

The other *Range* is in the Beak-head, before the Wouldings of the Bowfprit: It lies in the Form of the other, and has the Spritfail, and Spritfail-Topfail-Shears and Ropes belay'd about its Pins.

RANK-*Keel*. See *Keel*.

RAT; is a Place in the Sea, where there are Rapid Streams, and Dangerous Currents, or Counter-Currents.

RATE *of Ships*, is the Diftinction of the Bignefs and Capacity of Ships of War, which is into *Six*: The Difference is ufually reckon'd by the *Length* and *Breadth* of the *Gun-Deck*; *Number of Tuns* they contain; *Number of Men*, and *Guns*, they carry.

RATE *Firft*, have their *Gun-Decks* from 159, to 174 *Feet* Long; and from 44, to 50 *Feet* Broad.

They Contain from 1313, to 1882 *Tuns*.

They Carry from 706, to 800 *Men*:

And from 96, to 110 *Guns*.

RATE *Second*, have their *Gun-Decks* from 153, to 165 *Feet* Long; and from 41, to 46 *Feet* Broad.

They contain from 1086, to 1482 *Tuns*.

They Carry from 524, to 640 *Men*:

And from 84, to 90 *Guns*.

RATE *Third*, have their *Gun-Decks* from 142, to 158 *Feet* Long; and from 37, to 42 *Feet* Broad.

They Contain from 871, to 1262 *Tuns*.

They Carry from 389, to 476 *Men*:

And from 64, to 80 *Guns*.

RATE *Fourth*, have their *Gun-Decks* from 118, to 146 *Feet* Long; and from 29, to 38 *Feet* Broad.

They Contain from 448, to 915 *Tuns*.

They Carry from 226, to 346 *Men*:

And from 48, to 60 *Guns*.

RATE *Fifth*, have their *Gun-Decks* from 100, to 120 *Feet* Long; and from 24, to 31 *Feet* Broad.

They Contain from 259, to 542 *Tuns.*

They Carry from 45, to 190 *Men:*

And from 26, to 44 *Guns.*

RATE *Sixth*, have their *Gun-Decks* from 87, to 95 *Feet* Long; and from 22, to 25 *Feet* Broad.

They Contain from 152, to 256 *Tuns.*

They Carry from 50, to 110 *Men:*

And from 16, to 24 *Guns.*

RATIONAL-*Horizon.* See *Horizon.*

RAT-LINES, or *Ratlings*, are those *Lines* which make the Ladder-Steps, to get up the Shrowds, and Puttocks.

REACH, is the Distance between any two Points of Land, that lie in a Right-Line one from the other.

Fore-REACH. See *Fore.*

REAL-*Horizon.* See *Horizon.*

REAR-ADMIRAL. See *Officers*, and *Flag-Officers.*

REARED, *Wale-Reared.* See *Wale.*

RECTIFIER, is an Instrument consisting of two Parts, which are two Circles, either laid upon, or let into the other, and fasten'd together in their Centres; they represent two Compasses, one *Fix'd*, the other *Moveable*, each Divided into the 32 Points of the *Compass*, and 360 *Degrees*; and Numbred both Ways, from the *North* and the *South*, ending at the *East* and the *West*, in 90 Degrees.

The *Fixed Compass*, represents the *Horizon*; in which, all the Points of the *Compass* are Fix'd, and Immoveable.

The *Moveable Compass*, represents the *Mariners Compass*, as liable to *Variation.* In the Centre of the *Moveable Compass*, is fasten'd a Third, or sometimes an *Index*, long enough to reach the Outside of the *Fix'd Compass.* This Instrument serves to find the *Variation* of the *Magnetic-Compass*, and to Rectifie the *Course* at Sea; having given, the *Sun*'s *Amplitude*, or *Azimuth.*

REEF: When there is a great Gale of Wind, they commonly roll up part of the Sail below; so by that means it may become the narrower, and therefore not draw too much Wind. This Contracting, or Taking up of the Sail, is call'd *Reef*, or *Reefing.*

REEFT *Top-Mast*: When a *Top-Mast* is *sprung*, that is, crack'd, or almost broken in the Cap, they cut off the lower Piece that was near broken off; and setting the other Part, now much shorter, in the Step again, they call it a *Reeft Top-Mast.*

REEVE; is to draw a Rope thorow a Block, to run up and down.

To Unreeve a Rope, is to pull a Rope out of a Block.

REFLUX

REFLUX *of the Sea*, is the Ebbing of the Water off from the Shore: As its coming on upon it, or *Tide of Flood*, is call'd, the *Flux of the Sea*. See *Tide*.

RENDS in a *Ship*; are the ſame as *Seams* between the Planks. See *Seams*.

RESERVE. See *Engagement*.

RESOLVE *a Traverſe*. See *Traverſe*.

RIBS *of a Ship*, are the Timbers of the Futtocks, when the Planks are off; ſo call'd, becauſe they are Bending like the *Ribs* of a Carkaſs.

Alſo, thoſe little long Wooden Pieces which belong to the Parrels of the Yards, and have Holes in them, like the Comb under the Beak-head, are call'd, the *Ribs* of the *Parrels*.

RIDE: A Ship is ſaid to *Ride*, when her Anchors hold her faſt, ſo that ſhe drives not, by the force of the Wind, or Tide. And a Ship is ſaid to

RIDE-*Well*, when ſhe is built ſo, that ſhe does not over-beat her ſelf into a-Head-Sea, as that the Waves Over-Rake her, that is, Over-Waſh her) from Stem to Stern.

RIDE-*a-croſs*, when ſhe *Rides* with her Main-Yards and Fore-Yards hoiſted up to the Hounds; and both Yards and Arms Top'd alike.

RIDE-*a-Peek*. See *Peek*.

RIDE-*a-thwart*, is to *Ride* with her Side to the Tide; and then ſhe never Strains her Cable.

RIDE *betwixt Wind and Tide*, is to *Ride* ſo as the Wind has equal force over her one way, and the Tide the contrary way; which makes a Ship roll and tumble very much, yet not ſtrain her Cables much.

RIDE-*Wind-Road*, or, *To Ride a great Road*; that is, to *Ride* ſo as the *Wind* has more Power over the Ship, than the *Tide*.

RIDE *Hawſſul*, or, *To Ride a-Streſs*; that is, when in a Streſs of Weather, the Ship falls deep into the Sea with her Head, that the Water runs in at her Hawſes.

RIDE *Land-Lock'd*. See *Land-Lock'd*.

RIDE *a-Portoiſe*, is when the Yards of a Ship are ſtruck down upon the Deck; or when they are down *a-Port-laſt*.

RIDE *by the Stoppers*. See *Stoppers*.

RIDERS, are great Timbers, both in the Hold, and aloft, which are bolted on to the other Timbers, to ſtrengthen them, when 'tis diſcover'd a Ship is too weak, or ſlightly Built.

RIGGING of a *Ship*, is all the Cordage or Ropes whatſoever belonging to her Maſts, or Yards, or any Part about her.

The Ship is Well-Rigg'd; that is, when ſhe has all her Ropes of fit Size and Proportion to her Burden.

The Ship is Over-Rigg'd; that is, when her Ropes are too-many, or too-great for her; which wrongs her much in her Sailing, and is apt to make her Heel: For a small Weight Aloft, beyond due Proportion, is more prejudicial than a much greater Below; and the more Upright any Ship goes, the better she Sails.

RIGHT the *Helm!* That is, Keep the *Helm* even with the Middle of the Ship! See *Helm*.

RIGHT-*Horizon*. See *Horizon*.

RIGHT-*Sphere*. See *Sphere*.

RIGHT-*Sailing*, is when a Voyage is perform'd on some one of the Four *Cardinal Points*.

If a Ship sails on the *North*, or *South* Points, that is, under the *Meridian*, she Varies not in Longitude at all; but only changes the Latitude, and that just so much as the Number of Degrees she has run.

If a Ship sails directly *East*, or *West*, she alters not her Latitude, but only *Longitude*; and that as the Co-Sine of the *Latitude* is to the Sine of 90 Degrees.

RING-*Bolts*. See *Bolts*.

RISING-*Timbers*. See *Timbers*.

RISINGS in a *Ship*, are those thick Planks which go *fore* and *aft*, on both Sides, under the Ends of the Beams and Timbers of the Second Deck, unto the Third Deck, Half-Deck, and Quarter-Deck; and on them the Beams and Timbers of the Deck do bear, at both Ends, by the Ship's Side.

ROAD, is any Place near the Land, where Ships may Ride at Anchor.

ROADER, is a Ship that Rides in a *Road*.

ROARING *of the Sea*, is the Noise that attends its Agitation: And this *Roaring* is frequently observ'd a-Shore, both before and after a Storm.

ROBBINS, in a *Ship*, are those small Lines which make the Sail fast to the Yards, being reev'd into Eyelet-holes in the Sail, under the Head-Ropes, for that Purpose.

ROOF-*Trees*, or *Ruff-Trees*. See *Trees*.

ROOM, in a *Ship*; are Places distinguish'd by Partitions, or Bulk-heads: As the

Gun-Room, is an Appartment *abaft*, under the Great-Cabbin, where the Gunner and his Crew prepare all their Necessaries, and is their proper Place of Rendevous.

Bread-Room, lies under the *Gun-Room*: Here the Bread is kept, and preserv'd Safe, and Dry.

Cook-Room, is in the Fore-Castle: Here the Victuals are Dress'd.

Powder-Room, lies near the *Bread-Room*, and serves as a Magazine for the Powder.

ROOMY, a *Roomy Ship*; that is, so Spacious, that Men may pass with Ease to and fro.

ROPES of a *Ship*, are in general all her Cordage; but at Sea they call by this Name only, the

BOAT-ROPE, that which the Ship Tows her Boat by, at her Stern.

BOLT-ROPE, wherein the Sails are Sew'd.

BREAST-ROPE, is that which ſerves to keep the Yards cloſe to the Maſts, or to laſh the Parrels to the Maſts.

BUCKET-ROPE: That which is tyed to the Bucket for drawing of Water in.

BUOY-ROPE: That which is tyed to the *Buoy* by one End, and the Anchor's Flook by the other End.

CAT-ROPE, is that uſed to hale up the *Cat*.

CHEST ROPE, *Gueſt-Rope*, or *Gift-Rope*, is a Rope added to the *Boat-Rope* when ſhe is tow'd to the Ship's Stern, to keep her from ſhearing, that is, from ſwinging to and again.

ENTRING-ROPE, is tyed by the Ship's Side, for one to hold by that goes up the *Entring-Ladder*.

JEER-ROPE, is a Piece of a Hawſer made faſt to the Main-Yard; another to the Fore-Yard cloſe to the Ties, reev'd thro' the Block which is ſeiz'd cloſe to the Top, and ſo comes down by the Maſt, and is reev'd thro' another Block at the Bottom of the Maſt cloſe by the Deck. Its Uſe is to ſuccour the Ties, by helping to hoiſe up the Yards; ſo that tho' the Ties ſhould break, yet they would hold up the Maſt.

KEEL-ROPE, is a *Hair-Rope*, running between the Keelſon and the Keel of a Ship, to clear the Limber-Holes, when they are choak'd up with Ballaſt, and the like.

PORT-ROPES, are thoſe made uſe of to hale up the Ports of the Ordnance.

PREVENTER-ROPE, is a *Small Rope*, ſeiz'd croſs over the Ties, cloſe to the Ram head; ſo that if one Part of the Tie ſhould break, yet the remaining Part ſhould not run thro' the Ram-head, and endanger the Yard.

RUDDER-ROPE, is reev'd into a Hole of the *Rudder* near the Head, and alſo thro' the Stern-Poſt; both the Ends thereof are ſplic'd together: Its Uſe is to ſave the *Rudder*, if by any Accident it ſhould be ſtruck off the Irons.

Running-ROPES, are thoſe which *Run* on Blocks and Shivers.

Standing-ROPES, the *Shrouds* and *Stays* are ſo called, becauſe they are not removed, unleſs to be Eas'd or ſet Taught.

TOPE ROPES, thoſe wherewith the Sea-men Set or Strike the *Main* or *Fore Top-Maſts*. See *Top*.

ROPE YARN. See *Yarn*.

ROUFE-*Trees*. See *Trees*.

ROUGH-*Sea*. See *Over grown-Sea*.

ROUND-*House*, is one of the Apartments Abaft on board a Ship.

ROUND-*in*, or *Round-aft the Sail*, properly belongs to the *Main-Sail* and *Fore-Sail*; for when the Wind largeth, then they say,

Let rise the Main-Tacks and Fore-Tacks!
Hale aft the Fore-Sheat to the Cat-head!
Hale the Main-Sheat to the Cubbridge-head!

And when these Sheats are thus haled down, they are kept from flying with the *Passarado-Rope*. This Work is called *Rounding-in*, or *Rounding-aft* the Sails.

ROUND-*Splice*. See *Splice*.

ROUND-*Seam*. See *Seam*.

ROUND-*Top*. See *Top*.

ROWING: Low-built Vessels are made to go from one Place to another by means of Oars, managed usually by Men sitting in Rows in the Vessel; whence this Work is called *Rowing*: 'Tis practis'd in all *Boats* whatever; As also in *Galleys* and the like Vessels, and would be of great Use for Ships in Calms, if it could be conveniently practis'd.

ROWING-*Engine*: Several Contrivances have bin given by Persons concern'd in these Affairs, for to facilitate the Labour of *Rowing* Boats at Sea or in Rivers, and of applying the same to Ships, to be used in Calms, or when there is but little Wind; Among the rest *De Chales* mentions a pretty Invention in his *Treatise of Navigation*; The same was since somewhat improved by another Ingenious Person; And is after this manner:

A Wheel is fitted to the Drum-head of the Cap-stan, whose Teeth turn a Trundle-head, thro' which an Iron-bar is run that reaches a-cross, and goes thro' the Ship's-Side; and on its Ends without Board, at a convenient Distance from the Side of the Ship, are fasten'd two Drum-heads, like that on the Capstan; in which are fitted six or eight Paddles, so as to be taken out at Pleasure. And at the outermost Ends of the Paddles is fasten'd an Iron-Pin, with a Head on it: by which means, and by the Help of a Cord, taking an half Turn round about all these Pins, both the Paddles may be twisted, or strain'd and strengthen'd so, that they shall all work proportionably; and also the Paddles may, with a Luff-Tackle, be the more handily and easily lifted in and out, in order to be fitted into, or taken out of the Drum-heads of the Bar. Now if the Paddles be made proportionably large, according to the Number of Men that can be brought to the Capstan, who if they but work, the Vessel may make very good Way in smooth Water.

ROWLE, is a round Piece of Wood or Iron, wherein the

Whip goes, being made to turn about, that it may carry over the Whip the Easier from Side to Side.

ROWSE *in*, signifies to hale or pull in; but it properly belongs to the Cable or Hawser.

ROWSE *in the Cable!* or, *Rowse in the Hawser!* that is, when a Cable or Hawser lies too slack in the Water, and it be requir'd to have them made more Taught, lest on the Turning of the Tide, the Cable should happen to be foul of the Anchor.

RUDDER of a Ship, is a Piece of Timber hung at the Stern-Post upon Hooks and Hinges: In a Ship under Sail, the *Rudder* is, as it were, the Bridle of her, for by means of it, she is turn'd about at the Pleasure of him that stands at the Helm.

A Narrow *Rudder* is best for a Ship's Sailing, provided she can feel it; that is, be guided and turn'd by such a *Rudder*; For a broad *Rudder* will hold much Water, when the Helm is put over to any side: But yet if a Ship have a Fat Quarter, so that the Water cannot come quick and strong to her *Rudder*, she will require a *Broad Rudder*.

Rake of the Rudder; that is; the Aft-most Part of the *Rudder*.

RUDDER-*Irons*, are the Cheeks of that Iron whereof the Pintle is Part, which is fasten'd and nail'd down about the *Rake of the Rudder*

RUDDER-*Rope*. See *Rope*.

RUMB, *Rhumb*, or *Course of a Ship*, is the Angle which she makes in her Sailing with the Meridian of the Place where she is. See *Course*.

Complement of the Rumb, is the Angle made by the Line of the Ship's Way, with any Parallel to the *Equator*.

RUMB signifies also, one *Point* of the *Mariners Compass*, or the two and thirtieth Part of the *Card* or Horizon, that is, 11 *deg.* and 15 *minutes*, 11 $\frac{1}{4}$ *degrees*.

RUMB-*Line*, is the Line that is described by the Ship's Motion, on the Surface of the Sea, Steer'd by the Compass, making the same or equal Angles with every Meridian. These *Rumb-Lines* are *Helispherical* or *Spiral Lines*, proceeding from the Point where we stand, winding about the Globe of the Earth, and after infinite Revolutions come to the Pole, where they end. See *Loxodromic-Line*.

In *Plain* and *Mercator*'s Sailing, these *Rumb-Lines* are represented by strait Lines. Their Use is to shew the Bearing of any two Places one from another; that is, upon what Point of the Compass any Place lies from another.

RUMMIDGE, signifies to remove Things from one Place to another. This is one of those Sea Terms which are brought into common Use.

RUMMIDGE *the Hold.* See *Hold.*

RUN *of a Ship*, is that Part of her Hull under Water, which comes narrower by Degrees from the Floor-Timbers to the Stern-Poſt. This is alſo called *the Ship's Way aft-ward*; And according to the Ship's *Run*, ſhe Steers well or ill, by reaſon of the Eaſineſs or Difficulty of the Water's Paſſage to the Rudder. If too ſhort and too full below, the Water comes but ſlowly to the Rudder, becauſe the Force of it is broken by her Breadth: This is ſometimes Remedied by putting on a falſe Stern-Poſt, tho' 'tis better to lengthen her.

Good-Run, a Ship is ſaid to have a *Good-Run*, when ſhe comes off handſomly by Degrees, and her Tuck not lying too Low.

Bad-Run, when the Tuck of a Ship lies too low, it hinders the Paſſage of the Water to the Rudder, and therefore ſhe cannot Steer well, nor keep a good Wind, nor make any good Way thro' the Sea, but will be ſtill falling to Leeward.

RUN, *Diſtance Run.* See *Diſtance.*

RUNG-*Heads*, a peculiar Name given to thoſe Heads of the Ground-Timbers of a Ship, which are made a little bending, or where they begin to compaſs, and do direct the Sweep or Mould of the Futtocks and Navel-Timbers; for here the Lines, which make the Compaſs and Bearing of a Ship, begins.

RUNGS, are the Floor-Timbers, or Ground-Timbers of a Ship, thoſe that thwart the Keel, and are Bolted to it, and conſtitute her Floor, their Ends are the *Rung-Heads*.

RUNNER, is a Rope belonging to the Garnet, and to the two Bolt-Tackles, *viz.* that before, which comes to the Aftermoſt Shrouds of the Fore-Maſts, and that Abaft which comes to the Fore-moſt Shrouds of the Main-Maſt. This *Runner* is reev'd in a ſingle Block which is ſeiz'd to the End of a Pennant, and has at one End a Hook to hitch into any Thing, and at the other End a double Block, into which is reev'd the Fall of the Tackle or the Garnet, by which means it Purchaſes more than the Tackle or Garnet can do alone.

Overhale the Runner! That is, bring down that End which has the Hook to it, that it may be Hitch'd into the Sling, *&c.*

RUNNING-*Ropes.* See *Ropes.*

RUNNING-*Fights.* See *Fights.*

RUT *of the Sea*, is where it daſhes againſt any Thing.

SAIL,

S

SAIL, is made of ſeveral Pieces of ſtrong Cloth, and faſten'd to the Yards and Stays, for receiving the Wind that ought to Impel the Ship along. There are ſeveral ſorts of Sails belonging to Great Ships, moſt of which take their Names from their ſeveral *Yards*.

Main-SAIL, *Main-Top-Sail*, and *Main-Top-Gallant Sail*; are thoſe that belong to the *Main-Yard*, *Main-Top-ſail-Yard*, and *Main-Top-Gallant-Yard*.

Fore-SAIL, *Fore-Top-Sail*, and *Fore-Top-Gallant-Sail*; belonging to the *Fore-Yard*, *Fore-Top-ſail-Yard*, and *Fore-Top-Gallant-Yard*.

Miſen Sail, and *Miſen-Top-Sail*; thoſe that belong to the *Miſen-Yard*, and *Miſen-Top-ſail-Yard*.

Sprit-SAIL, and *Sprit-Top-Sail*; belonging to the *Spritſail-Yard*, and *Spritſail-Topſail-Yard*.

All theſe *Sails* are uſually Cut in Proportion as the *Maſts* and *Yards* are in Length.

Drift SAIL, is a *Sail* uſed Under Water, veered out right-a-Head by Sheats, as other *Sails*: It ſerves to keep the Ship's Head right upon the Sea in a Storm, and to hinder her Driving too-faſt in a Current.

Head-SAILS, are thoſe which belong to the *Fore-Maſt* and *Bowſprit*: They keep the Ship from the Wind.

After-SAILS, are thoſe that belong to the *Main-Maſt* and *Miſen*: They keep the Ship to Windward.

Therefore 'tis common, that Ships Sailing on a Quarter-Wind, require a *Head-Sail*, and an *After-Sail*; one to Countermand the other.

Netting-SAIL, is only a *Sail* laid over the *Nettings*.

Studding-SAILS, are Bolts of Canvaſs, or any Cloth that will hold Wind, extended in a Fair-Gale along the ſide of the *Main-Sail*, and Boom'd out with a Boom.

They are ſometimes uſed to the Clew of the *Main-Sail*, *Fore-Sail*, and *Spritſail*, when the Ship goes before the Wind, or Quartering.

Preſt-SAILS. See *Preſt*.

Short-SAILS, or *Fighting-Sails*, are the *Fore-Sail*, *Main-Sail*, and *Fore-Top Sail*: Theſe alone are uſed in a *Fight*, leſt the reſt ſhould be fired, or ſpoil'd; beſides, they would be troubleſome to handle, hinder the Sight, and the Uſe of Arms.

The *Sails* which are made of *Engliſh* Canvas, comes far ſhort of that which we have from *Holland*, as to Strength, and Goodneſs: And therefore, ſince the Uſe of this Manufacture was introduced into the Navy, for the Benefit of our Country; it were much to be wiſh'd, that it might ſoon be brought to

Perfection,

Perfection, as of late, indeed, 'tis much improv'd; for otherwise, the Use thereof, may prove greatly Prejudicial, not only with respect to the Danger our Great Ships may be expos'd to thereby, but by obliging the Smaller to come into Port, for New Supplies, when the Safety of the Trade in our *Channel*, and up and down the *Soundings*, does, in Time of War, absolutely require their keeping out at Sea. Besides, the Opportunity of Service may be lost, not only with whole *Squadrons*, but single *Ships*, upon Meeting an Enemy: For if the *Sails* prove bad, it may, on the one hand, impede their getting up with them; and, on the other, subject them to a Surprize, when Outnumber'd: And the Consequence may be Fatal to a *Fleet*, or any Number of Ships, on a *Lee-Shore*.

SAIL; besides its proper Signification, is often us'd for a Ship: As,

To Spy (or *See*) *a Sail*; that is, to Descry a Ship; therefore, he that Discovers it, usually calls, *A Sail! A Sail!* So when we speak of a *Fleet*, or Number of Ships together, we say, It consist of Forty or Fifty *Sail*, more or less; and not, Forty or Fifty Ships.

SAILING: The *Art of Sailing*, shews how to Conduct a Ship at Sea, or to Guide and Direct her from one Place to another. See *Navigation*.

GREAT-CIRCLE-SAILING. See *Great-Circle*.

MERCATOR's-SAILING. See *Wright's*.

MIDDLE-LATITUDE-SAILING. See *Middle-Latitude*.

PARALLEL-SAILING. See *Parallel*

PLAIN-SAILING. See *Plain*.

At *Setting* SAIL the wind being Fair, and Steering right Before it, the Working *Phrases*, or *Terms* then used, are generally such as these:

Let fall your Fore-sail!
Heave out Main-top-sail!
Hoise up Fore-top-sail!
Hoise up Main-top-soil!
Lose Spritsail!
Heave out Misen-top-sail!
Square your Spritsail!

It continuing a Fair *Loom-Gale*; and it be required to have the Ship in all her Canvase, then,

Heave out Spritsail!
—— *Top-sail!*
—— *Fore-top-gallant-sail!*
—— *Main-top gallant-sail!*
Hoise up your Small Sails!
Hale aft your Fore-sheats!

Note, That the *Main-sail* is now kept Furl'd: For if the *Fore-sail* and *Fore-Top-sail* be good *Sails*, the Ship will now make better Way, than if the *Main-sail* was down; which would Becalm the *Fore-sail*, and *Fore-Top-sail*; and the Ship steers best

beſt with her *Head-ſails*. Now, at *Setting Sail*, 'tis ſuppos'd that the *Sheats* are all Haled home, and the *Yards* Hois'd up; and then you have a *Ship* Compleat, Under-Sail right-afore.

If it prove a Freſh Gale, or that the Wind blow Freſh, then,

Hale down your Fore-top-gallant-ſail!
Hale down Main-top-gallant-ſail!
In Spritſail-top ſail!
Let go Spritſail-top-ſail-Sheats!
Hale home his Clew-Lines!
In Fore-top-gallant-ſail!
In Main top gallant-ſail!
In Miſen-top-ſail!
Let go Top gallant-Sheats!
Caſt off Top-gallant Bow-Lines!
Hale home Top-gallant Clew-Lines!
Veer out ſome of the Weather-ſheat of the Fore-ſail!
Let go your Weather-Braces!
Top your Spritſail!
Loſe Main-ſail!

If the Wind Veers forward, then,

Get too your Fore-Tack!
Caſt off your Weather-ſheat!
Let go your Weather-Brace!
Veer out ſome of your Lee ſheat!
Let fall Main ſail!
Get too Main-Tack!
Caſt off Main-Brace, and Main-Top ſail!
Hale aft Main-ſheat!

If the Wind is Sharp, then,

In Spritſail!
Square Spritſail-Yard!
Let go Spritſail-Sheats!
Hale up Spritſail Clew-Lines!
Get Main-Bow-line in Block!
Hale forward Main-Bow-line!
Hale Main-top-ſail Bow-line!
Hale tau't Fore-Bow-line, and Fore-top ſail Bow-line!
Hale aft Main-ſheat!
Hale a-board Miſen!
Set in your Lee-Braces!
And keep her as Near as ſhe will Lie!

Then you have all your Sails *Trimm'd Sharp*, or *by a Wind!*

If it prove a Stiff Gale, then,

Settle down your Fore-top-ſail!

If much Wind, then,

Hale down Fore-top-ſail!
Hale down Main-top-ſail!

If it Blows ſtill Harder, then,

Take in Top-ſails!
Let go your Lee-Braces!
Caſt off your Bow-lines!
Brace your Weather-Braces!
Let go Top ſail-ſheats.
Hale home Top ſail Clew-lines.

The *Sails* being Furl'd, then,

Square your Top-ſail-Yards!

Here the Ship is brought into her *Low-Courſes*, or *Low Sails*.

If it prove to be Stormy-Weather, then,

See that your Main-Halliards be Clear!
Make all your Geer Clear, to Lower the Main-Yard.
Hale down Misen!
Cast off Top sail-sheats!
—— Clew-Garnets!
—— Bunt-Lines!
—— Leetch Lines!
—— Lifts, and all your other Geer.
Hale too the Capstan, to (Lower) or bring down the Yard!
Furl the Sail Sure!
Make fast the Yard for Traversing!

If it be a *Grown-Sea*, and very Foul-Weather, then it's better *Spooming*, or putting her right Before the Wind.

See the Guns be fast Lash'd!
Put the Helm a-Weather!
Right your Helm!
Let rise Fore-Tack!
Settle the Fore-Yard!

If the *Fore Sail* give-way, then,

Lower Amain!
Hale the Sail into the Ship, and Loose it from the Yard!
Get too the Fore-Bonnet!
Make all Clear, and Hoise the Fore-Yard!

Thus you have the Ship brought from her *Canvas*, to a *Fore-Bonnet*, Spooming before the Wind. And this is the usual Manner of Handling a Ship *By*, and *Large*, in *Fair*, and in *Foul-Weather*.

SAILORS, are those Men that perform the Work of a Ship, as Hoising the Sails, getting the Tacks a-board, Steering the Ship, Furling the Sails, Slinging the Yards, and the like, as required of them on all Occasions.

And as these Men are absolutely Necessary, for the Good of our Nation; so several Things are (or should be) very carefully Perform'd with respect to them; As the Looking well after them, when Wounded, or Sick, at Sea; and when they are put on-Shore, under those Circumstances: To this End, there are *Physicians* in the Fleet, with suitable Salaries. When they are put Sick a-Shore, they are Provided for, by Officers at the several Ports, by *Commissioners of the Sick and Wounded*, whose Business is to take Care, not only for their Lodgment, and Nurses, but also for all other Things Necessary for the Cure of their Wounds, and Diseases.

Besides, Care is taken for the Convenient and Comfortable Reception of Poor, Maim'd, and Disabled Seamen, and the Widows and Children of such who have lost their Lives in the Public Service, into the Hospital at *Greenwich*. Nor is the Government less wanting in making them

Easie,

Easie, as to their Pay and all other Particulars: Therefore they do expect from the Sailors a strict Compliance with their Duty; and so to behave themselves, that not only their Country-men, but the rest of the World may be convinc'd, an *English* Sea-man does still retain his wonted Zeal and Bravery.

SALLY-*Port*, is a great Hole in the Side of a *Fire Ship*, made on purpose for the Men to escape when they have Grappled an Enemy, and fired their Train.

SALUTE, is a Deference, and Respect, which at Sea, is render'd not only between Ships of different Nations, but also between those of the same Nation, according to the Degree and Quality of the Commanders.

And this Respect is shew'd by passing under the Lee, *Saluting* them with so many Guns, Trumpets, or Hale them thrice with a Joint Shout of the whole Ship's Company, and the like, according to the different Occasion requiring it.

And if any Ship or Fleet, either of our own or Strangers, whether Merchant-Men or Men of War, shall come up any thing near, or within reach of a Cannon shot, of any of Her Majesty's Ships, either at Anchor or under Sail, it becomes them to pass under her Lee, after the Custom of the Sea, and in their going by, to *Salute* her, with some odd Number of Guns, the which are to be answer'd with fit Correspondency.

And the Number of Odd Guns is, at Sea, so punctually and strictly observ'd, that whensoever they are given *Even*, 'tis receiv'd for an infallible Sign, that either the Captain, or some noted Officer, is dead in the Voyage.

These *Salutes* are of Use, not only in regard of an Acknowledgment of Superiority to Her Majesty's Ships, especially in our Channels, but that by an Expectance, and looking out after this, all treacherous Attempts, that may be Plotted, by stealing upon them to the Windwards, of laying of them a-board, either with Fire-Ship, Mine-Ship, or the like, may seasonably be prevented and avoided.

And as Her Majesty's Admiral Ships are always to be *Saluted* with Guns by all Ships whatsoever; so also, when any Ship or Ships comes to an Anchor under the Command of any Fort or Castle, they are to give some Guns; the which are to be respectively answered by the Castle or Fort.

And when Ships that have been long in Consortship at Sea, are to part several Ways upon their Occasions, they usually *Salute* one the other with some Guns.

Also, 'tis a general Custom, upon the Death of any Commission'd

mission'd Officer, at Sea, at his throwing over Board, to Ring his Knell with some even Number of Guns.

SCALE, signifies any *Measure*, or *Numbers* which are commonly used; or the *Degrees* of any Arc of a Circle, or of such Right Lines as are divided from thence; as the *Sines*, *Tangents*, *Chords*, *Secants*, &c. drawn or plotted down upon a Ruler, for ready Use and Practice in Geometrical Operations.

SCALE *of Equal Parts*, is a Line equally divided, whose *Parts* may represent any *Measure* or *Numbers* whatsoever, as *Leagues*, *Miles*, &c. It serves to lay down any Measure taken; or a Line being laid down, to find how much of the Measure that Line contains.

DIAGONAL-*Scale*, is only a *Scale of Equal Parts*, divided with the utmost Accuracy, by means of a *Diagonal Line*.

PLAIN-*Scale*, is a *Scale* having on one Side thereof, the *Scale of Chords*, *Natural Sines*, *Tangents*, *Semi-Tangents*, *Secants*, *Rumbs*, *Hours*, *Leagues*, *Longitudes*, &c. and on the other Side, the *Diagonal Scale*. And by means of this, the several *Cases* in *Trignometry*, and consequently in *Sailing*, are *Projected* and laid down.

GUNTER's-*Scale*, is a large Scale, having on one Side the Common *Plain-Scale*; that is to say, the *Diagonal Scale* of Equal Parts, and those of *Natural Sines*, *Tangents*, &c. On the other Side are drawn the *Lines of Artificial Sines*, and *Tangents*, fitted so to a *Line of Numbers*, or *Gunter*'s *Line*, (so called from the Inventor Mr. *Gunter*) that all Questions in *Plain-Sailing*, &c. as also all *Proportions*, or where there are *Three Terms* given and a *Fourth* required, are easily wrought with Compasses; and that by this

RULE.

The Extent of the Compass from the First Term to the Second, will reach from the Third Term to the Fourth Term. Or

The Extent from the First Term to the Third, will reach from the Second to the Fourth Term.

Note, That the *Meridian Line*, and *Line of Equal Parts* adjoyning, on the *Gunter*; the First is or may be the *Meridian*, and *Degrees of Latitude* on *Mercator*'s *Chart*; the latter, the *Equinoctial* and *Degrees of Longitude*.

SCARFED, signifies *Pieced*, *Fasten'd*, or *Joyn'd* in; and is said of one Piece of Wood let into another, by cutting away as much from the one as the other: And when any of the Floor or Ground-Timbers are not long enough of themselves, they are *Scarf'd* into one another, to make two or three as one.

The Stem of a Ship is Scarfed into her Keel; that is, the two Pieces

Pieces are ſhaped away ſlanting, ſo as to joyn with one another cloſe and even; and this the Carpenters call *Wood and Wood.*

SCOPER or *Scuper-Holes*, are little Holes cloſe to the Decks, thorow the Ship's Sides: They ſerve to carry the Water off, that comes from the *Pump*, or any other Way.

SCOPER-*Leather*, or *Scuper-Leathers*, ſuch as are nail'd over the *Scopers* upon the Lower Deck, ſerving to keep out the Sea from coming in, and yet to give way for it to run out.

SCOPER-*Nails*, or *Scuper-Nails*, are little ſhort *Nails*, with broad Heads, made purpoſely to *Nail* the *Scuper-Leather*, and the Coats of *Maſts*, and *Pumps*.

SCUTTLES, are little ſquare Holes, cut in the Deck big enough to let one Man thorow; they ſerve to let People down below on Occaſion: Or from Deck to Deck. They are generally before the Main-Maſt, before the Knight in the Fore-Caſtle: In the Gun-Room to go down to the Stern-Sheats: In the Round-Houſe, to go down into the *Captain*'s Cabbin, when forc'd by the Enemy, in a Fight A-loft. Some ſmall *Scuttles* have Grattings over them, to give Light to them betwixt Decks, and for Out-lets to the Smoak of the Guns: But all *Scuttles* have Covers to them, that Men may not tumble in, when 'tis Dark.

SCUTTLES, alſo are little Windows, and Long Holes which are cut out in Cabbins, to let in Light.

SCUTTLE-*Hatch*, is the little *Hatch* that covers the *Scuttle*. See *Hatch*.

SEA, all the vaſt Collection or Body of Salt Water in General, is called *Sea*: But this Name is promiſcuouſly apply'd to any of the Parts, whether *Ocean*, *Gulfs*, *Straits*, *Bays*, *Roads*, &c.

The Seas, *with reſpect to* EUROPE, *are*

The *Euxine Sea*, enclos'd with Part of *Europe* on the North and Weſt, and Part of *Aſia* on the South and Eaſt.

The *Mediterranean-Sea*, encloſed with *Europe* on the North, and *Barbary* on the South.

The *Iriſh-Sea*, encloſed with *Ireland* on the Weſt, and *Britain*, on the Eaſt.

The *German-Sea*, encloſed with *Britain* on the Weſt, and *Scandinavia* on the Eaſt.

The *Baltic-Sea*, encloſed with Part of *Germany* on the South, Part of *Poland* on the Eaſt, and *Swedeland* on the Weſt.

The *Seas* in the other Parts of the World, are different Parts of the *Ocean*, variouſly named, according as they lie adjacent to different Countries, except the *Caſpian-Sea* in *Aſia*.

SEA-

SEA-BOARD. See *Board.*

SEA-CHART. See *Chart.*

SEA COMPASS. See *Compass.*

SEA-DRAGS. See *Drags.*

SEA-GATE. See *Gate.*

SEA-MEN. See *Mariners.*

SEA-QUADRANT. See *Quadrant.*

SEA-YOKE. See *Yoke.*

HEAD SEA See *Head.*

Overgrown-SEA. } See *Over-*
Rough-SEA. } *grown.*

SEAMS of Ships, are the Places where the Planks meet, and are joined together.

To Parcel a Seam. See *Parcel.*

To Pay a Seam. See *Pay.*

SEAMS: This Word is also used, according to common Acceptation, when speaking of *Sails*; and of these *Seams*, there are two sorts, *viz.*

Monk-SEAM, is a *Flat Seam.* See *Monk.*

Round-SEAM, is so called, because Round like the common *Seam.*

SEASE, *Seaze*, or *Seize*, is to bind fast any Rope together, with small Rope Yarn, or the like: And the fastening of a Block at the End of a Pennant, Tackle, or Garnet, *&c.* is called *Seizing*, or *Seazing.*

SEASEN, or *Seasing*, is the Name of a Rope by which the Boat Rides by the Ship's-side, when in Harbour, *&c.*

SEEL, signifies much the same as *Heel*; for as 'tis called *Heeling*, when a Ship lies down constantly or steadily on one Side; so 'tis called *Seeling*, when she tumbles on one Side violently and suddenly, by reason of the Sea's forsaking her, that is, the Waves leaving of her for a time in a Rolling Sea.

Lee SEEL; that is, when the Ship tumbles to *Lee-ward:* In this, even in a Storm, there is not much Danger, for the Sea will presently right her up again.

But in Case a Ship *Seels*, or Rolls to *Wind-ward*, then there is Fear of her coming over too short, or suddenly, and so be foundred, by having the Sea break right into her, or else have some of her upper Works carried away.

SEND, when a Ship falls deep into the *Trough*, or *Hollow of the Sea*, then 'tis said she *Sends* much that Way, whether *a-Head*, or *a-Stern.*

SENSIBLE-*Horizon.* See *Horizon.*

SEPARATION; the same with *Departure:* Which see.

SERVE: *To Serve a Rope*, is to wind something about it, to keep it from fretting out.

To Serve Ropes with Plats, or Sinnets, is to lay *Sinnet*, *Spun* or *Rope-Yarn*, or a Piece of *Canvase*, upon *Ropes*; then roll it fast about the *Rope*, to keep it from galling, or fretting in any Place.

SET: To Observe on what Point of the *Compass*, the Sun, Land,

Land, *&c.* bears, is call'd, *Setting* the *Sun*, *Land*, &c.

SET-BOLTS. See *Bolts*.

SETTEE, is a Veſſel, very common in the *Mediterranean*, with One Deck, and a very Long and Sharp Prow; they carry, ſome Two Maſts, ſome Three, without Top-Maſts; their Yards and Sails are all like the Miſen: The leaſt of them are of 60 Tuns Burthen: They ſerve to Tranſport Cannon, and Proviſion for Ships of War, and the like Service.

SETTLE a *Deck*; is to take it down lower than it was at firſt.

SEW: *The Ship is Sewed*; that is, when a Ship, at Low-Water, comes to lie Dry on the Ground. If ſhe be not left quite Dry, then they ſay, She *Sews* to ſuch a Part.

And where the Ship cannot all lie Dry, then 'tis ſaid, *She cannot Sew there*.

SHACKLES, are the Oblong Iron-Rings, bigger at one End, End than at the other, with which the Ports are ſhut faſt, by thruſtiug the Wooden-Bar of the Port thorow them.

SHACKLES, alſo, are Rings like the former, faſten'd at the Corners of the Hatches, to lift them up with.

SHALLOP, is a ſmall Light Veſſel, with only Main-Maſt and Fore-Maſt, and Sails to be haled up and down on Occaſion: They are generally good Sailers; and conſequently, very fit to ſerve as Tenders on *Men of War*, as they are often uſed.

SHANK *of an Anchor*. See *Anchor*.

SHANK-*Panter*, is a ſhort Chain, faſten'd under the Fore-Maſt-Shrouds, by a Bolt, to the Ship's-Side; having at the other End a Rope faſten'd to it: It ſerves to make faſt the Anchor at the Bow; and the whole Weight of the aft-part of the Anchor reſts thereon, when it lies by the Ship's Side: And the Rope by which 'tis haled up, is made faſt about a Timber-head.

Sheep-SHANK-*Knot*. See *Knot*.

SHARP; As, *Sharp the Main-Bow-Line!* That is, Hale it tau't.

SHEAR; that is, to Swing too and again.

The Ship Shears; that is, Goes in and out, and not right forward; either by not Steering ſteadily, or by means of the ſwift running of the Tide, *&c.* Then they ſay, *She Shears*; or *goes Shearing*.

SHEAR-OFF; that is, to get Away.

SHEATH: *To Sheath a Ship*, is to Caſe, as it were, that Part of her Hull which is under Water, with ſomething to keep the Worms from Eating into her Planks; 'tis uſually done, by laying Tar and Hair mix'd together, all over the Old Planks, and then nailing on thin New Boards: But this hinders a

Ship's Sailing; therefore, of late, some have bin *Sheath'd* with Mill'd Lead.

SHEATS, are Ropes bent to the Clew of the Sails; serving in the *Lower Sails*, to *Hale aft*, or *Round off* the Clew of the *Sails*: But in *Top-sails*, they serve to *Hale Home*, that is, to Hale the Clew of the Sail close to the Yard-Arm.

Standing Part of the SHEAT, is that Part of it which is made fast to the Ring of the Ship's Quarter.

Over-Hale the SHEAT! that is, Hale upon the Standing-part of the *Sheat*.

To Hale aft the Main-Sheats, is in order to make the Ship keep by a Wind.

To Hale aft the Fore-Sheats, is, that the Ship may fall off from the Wind.

Ease the Sheat! That is, Veer it, or let it go out gently; See *Ease*.

Tally the Sheats! That is, Hale aft the *Sheats* of the *Main*, and *Fore-Sail*.

Let Fly the Sheat! that is, Let it go all at once, and run out as fast as it can; Then the Sail will hang loose, and hold no Wind. See *Fly*.

False-SHEAT, is a Rope bent to the Clews of the *Main*, and *Fore-Sail*, above the *Sheat-Block*, to Succour and Ease the *Sheat* in a violent Gust of Wind.

SHEATS, also, are those Planks under-Water, which come along the Ship's *Run*, and are closed into the *Stern-Post*.

Stern-SHEATS, is that Port within Board in the *Run* of the Ship.

Flown-SHEATS. See *Flown*.

SHEAT-*Anchor*. See *Anchor*.

SHEEP-*Shank-Knot*, or *Sheer-Shank-Knot*. See *Knot*.

SHEER-*Hooks*. See *Hooks*.

SHEER, or *Sheering*; the same with *Shear*: Which see.

SHEERS, are two Mast-Yards, or Poles set up an end, and at a pretty distance off at the Bottom, and seiz'd a cross one another near the Top; this they call *a Pair of Sheers*: They are placed Below, upon the Chain-Wails of the Shrouds, and lash'd fast to the Ship's-side, to keep them steady Aloft: They serve either to take in, or let out a Mast; or else for to hoise in or out into Boats, that have no Masts, such Goods as are to be taken in; and for that End, there is fasten'd, at the place where they cross one another, a strong Double-Block, with a Strap.

SHIFTERS, are those Men on board a *Man of War*, who are Employ'd by the *Cooks*, to *Shift* or Change the Water in which the Fish or Flesh is put and laid for some time, in order to fit it for the *Kettle*.

SHIVERS, are those little round Wheels in which the Rope of a Pulley or Block runs: They turn with the Rope, and have Pieces of Brass in their

Centres

Centres (call'd *Cocks*) with Holes in them, into which the Pin of the Block goes, and on which they turn. These *Shivers* are usually of Wood; but some are of Brass, as those in the *Heels* of the *Top-Masts*.

SHOALE, signifies the same as *Shallow*.

Good-SHOALING; that is, when as a Ship sails towards the Shore, she finds by her Sounding, it grows shallow by degrees, and not too suddenly; for then a Ship may go in safely.

SHOOT: *The Ballast Shoots*; that is, runs over from one Side to another.

SHORT-SAILS. See *Sails*.

SHOT *of a Cable*, is the splicing of two Cables together, that a Ship may Ride safe in deep Waters, and great Roads: For a Ship will Ride easier by one *Shot of a Cable*, than by three Short *Cables* out a-Head.

SHOT for *Ordnance*, are of several sorts: As,

Case-SHOT, is any thing of Small Bullets, Nails, Old-Iron, and the like, to be put into a *Case*, to Shoot out of *Ordnance*.

Chain-SHOT, are made of two Bullets, with a *Chain* betwixt them, so contriv'd, that they will spread their full Length in Flying.

Cross-Bar-SHOT, is a Round *Shot*, with a long Spike of Iron cast into it, as if it went thro' the Middle of it.

Langrel-SHOT. See *Langrel*.

Round-SHOT, are Round Bullets fitted in Proportion to the Bore of the Piece.

Trundle-SHOT, is only a Bolt of Iron, 16 or 18 Inches long, sharp-pointed at both Ends; and about a Hand's-breadth from each End, has a Round broad Bowl of Lead cast upon it, according to the Bore of the Piece.

SHOT *between Wind and Water*; that is, close by the Surface of the Water.

SHROUDS, are great Ropes in a *Ship*, that go up on both Sides of all the *Masts*, except the *Bowsprit*: They are made fast below by Chains to the Ship's-Sides; and aloft, over the Head of the Mast, their Pennants, Fore-Tackle, and Swifters being first put under them: And they are *served* there, to prevent their galling and fretting the Mast. The Top-Mast *Shrouds* are fasten'd to the Puttocks by Plates of Iron, and by Dead-Men's-Eyes and Lanniers also, as the others are.

Ease the Shrouds! That is, Get them Slacken'd.

Set Tau't the Shrouds! That is, Set them Stiffer.

Puttock-SHROUDS. See *Puttocks*.

SIDES of a *Ship*, are distinguish'd into the *Starboard*, or *Larboard-Side*; that is to say, into the Right-Hand, and Left-Hand-Side. See *Starboard* and *Larboard*.

To give a Broad-ſide; that is, to Fire all the Guns on one Side of the Ship.

SIDE-*Wind*. See *Wind*.

SIGNALS, are the ſeveral *Signs*, made either by Firing off certain number of Guns, putting out ſuch Colours, or ſuch a Light; whereby it may be known, when the *Admiral* Tacks about, when He intends to lie a-Try, when a-Hull, when to Fight, when to Chaſe, when to Leave off; and when any of the Fleet are too-far a-Head, to Retire, by ſparing ſome Sail; or when they are too-far a-Stern, to cauſe them to make all the Sail they can, to get up: Theſe *Signals* are particulariz'd and communicated in the Inſtructions ſent to the *Commander* of every Ship of the Fleet, before their putting out to Sea.

As, The Day that the *Admiral* deſigns to Set out to Sea, He uſually cauſes his Top-Sails to lie looſe upon the Cap, very early in the Morning: And if it be Hazy, Dark-Weather, and that the Fleet be large, then about two or three Hours before he begins to Weigh his Anchors, He Fires a Gun.

If the *Admiral* finds Occaſion to Tack about in the Day-time, He Fires a Gun, to cauſe the reſt of the Fleet to Look out, and do as He does. If it be in the Night-time, He, beſides his uſual Lights, puts one Light either in the Main-Top, or ſome remarkable Place as determin'd for ſuch a *Signal*, and Fires a Gun.

If he deſigns to lie a-Hull, then he puts Two Lights there.

If to lie a-Try, they put Three Lights out in the ſame Place.

If Occaſion requires a General Convention of the *Captains* on-Board the *Admiral*, He uſually hangs out a *Yellow* Flag in the uppermoſt part of his Main, or Miſen-Shrouds.

If a Council of War is Call'd, He hangs a *Blue* Flag in the ſame Place.

If a Fleet Meets an Enemy's Fleet, and that, after due Conſultation on-Board the *Admiral*, it be thought fit to Engage them; He takes in his Ordinary *Enſign*, and heaves out another all *Red*, which is call'd, by ſome, the *Bloody Colours*. By which *Signal*, the whole Fleet may Prepare and Order themſelves for a Sea-Engagement.

If any Ship of a Fleet Diſcovers any ſtrange Ship, and would acquaint the whole Fleet with it; he puts abroad ſome Flag on that Part of the Ship which points moſt to the Ship ſeen; and Fires a Gun, to give Notice of the *Signal*. If it be a Fleet that is Diſcover'd, he puts out two Flags.

If, in the Day-time, any Ship in the Fleet, finds her ſelf in Danger of Foundering in the Sea, by Springing of a Leak, or any other Miſchance, ſhe may give Three Guns, then ſhew a

Waft from the Main-Top. If it be in the Night-time, ſhew a Light, and continue Firing now and then a ſingle Gun; by that means they may be taken notice of, found out, and reliev'd.

If a Fleet, coming out of Sea, expects a Land-fall, the firſt Ship that diſcovers it, is to give preſent Notice thereof to the reſt of the Fleet: If it be in the Day-time, by ſhewing her Colours abroad, tho' it be (for a time) in the Main-Top it ſelf, inclining them to that Part whence the Land is ſeen: If it be in the Night-time, give Two Guns, ſhew a Light, Tack about and ſtand off, that the reſt of the Fleet may obſerve the ſame.

There are peculiar *Signals* made and known to every Ship of a Fleet, that having loſt Company, and coming afterwards in ſight, they may preſently be diſcover'd one to the other. There are alſo Variety of other *Signals*, either for the whole Fleet, ſome particular Squadron, or Diviſion, or for ſome particular Ship or Ships, all known to the Perſons concern'd, from their Inſtructions: So that what is here ſaid, ſerves only to give a general Notion of *Signals*, and their Uſe. But whatever they be, at Sea, a diligent looking after them is to be practis'd, and at all times, a careful Attention is to be given, by every particular Ship in the Fleet.

SINICAL-*Quadrant*, is made of Braſs, or Wood, with *Sines* drawn from each Side interſecting one another; and an *Index* divided by *Sines* alſo; with 90 *degeees* on the Limb, and two Sights to the Edge, to take the Altitude of the Sun But ſometimes, inſtead of *Sines*, 'tis divided all into equal Parts: It ſerves to Solve any Problem in *Plain-Sailing*, and is much in Uſe among the *French* Sermen.

SINNET, is a Line made of Rope-Yarn, commonly conſiſting of two, four, ſix or nine Strings, platted in three Parts over one another, and then beaten ſmooth and flat with a a Wooden Mallet: It's Uſe is to ſerve Ropes, *&c.* to keep keep them from Galling.

SKARFED, the ſame with *Scarfed.*

SKEGG, is that ſmall and ſlender Part of the Keel, which is cut ſlanting, and left a little without the Stern-Poſt; 'tis now much out of Uſe.

SKIFF, is the leaſt of Ship-Boats: It ſerves chiefly to go a-ſhore in, when the Ship is in Harbour.

SKUPPERS, or *Skupper-Holes*, the ſame with *Scoper-Holes.*

SLATCH, when any Rope or Cable hangs ſlack, the middle Part which hangs down, is called the *Slatch of the Cable*, or *Rope*.

SLATCH *of fair Weather*, ſo the Sea-men call a ſmall Interval

val of fair, that comes after foul Weather.

SLEEPERS, are Timbers ly- before and aft in the Bottom of a Ship on each Side of the Keelson; the lower-most of them is bolted to the Rung-heads, and the uppermost to the *Futtocks*, in order to strengthen and fasten the *Futtocks* and *Rungs*.

SLINGS for Hoising up *Casks* or any heavy Things; are made, commonly, of a Rope spliced at either End into it self, with one Eye at either End, so long as to be sufficient to receive the Cask, *&c.* The middle Part of the Rope is also seiz'd together, and so makes another Eye to hitch the Hook of the Tackle.

SLINGS for Hoising of *Ordnance*, are made much longer, and with a small Eye at each End, one of which is put over the Breech of a Piece *of Ordnance*, and the other Eye comes over the End of an Iron-Crow, which is put into the Mouth of the Piece, to weigh and Hoise the Gun, as they please.

SLINGS for *Yards*, are strong Ropes, or Iron-Chains, with which they are bound fast to the Cross-Trees a-loft, and to the Head of the Mast; that if the Tye should happen to break, or to be shot to pieces in a Fight, the *Yard* nevertheless may not fall down upon the Hatches.

To SLING *a Man over Board, to stop the Leaks*, is done thus; The Man is truss'd up about the Middle, in a piece of Canvass, and a Rope, to keep him from sinking; with his Arms at Liberty, a Mallet in one Hand, and a Plug wrapp'd in Okum, and well Tarr'd in a Tarpawlin-Clout in the other, to beat it quickly into the Hole.

SLOOPS, are Vessels attending our Men of War, and generally of about 66 Tuns, carrying about 35 Men, and commonly two Guns. See *Shallops*.

SMACKS, are small Vessels, with one Mast, which attend on Ships of War, either in carrying the Men or Provision aboard: They also serve near the Coasts for Fishing, and the like.

SMALL CRAFT. See *Craft*.

SMITING-*Line*, is a small Rope fasten'd to the Misen-Yard-Arm, below at the Deck, and is always furl'd up with the Misen-Sail, even to the upper End of the *Yard*, and from thence it comes down to the Poop. It serves to loose the Misen-Sail, without striking down the Yard; which is easily done, since the Misen-Sail is furl'd up only with Rope-Yarn; and therefore when this Rope is pull'd hard, the Rope-Yarn breaks, and so the Sail falls down of it self.

SMITE *the Misen!* that is, Hale by the *Smiting-Line*, that the Sail may fall down.

SNATCH-

SNATCH-*Block*, or SNAP-*Block*. See *Block*.

SOCKETS, in a Ship, are the Holes which the Pintles of the Murdering Pieces go into.

SOUND, in *Hydography*, is any great In-draught of the Sea, betwixt two Head-lands, where there is no Paſſage thro': As that of the *Baltic-Sea*, *Hudſon's Bay*, *White-Sea*, &c.

To SOUND, or *Sounding*, is to try the Depth of Water with a *Line* and Plummer, called,

SOUNDING-*Line*; that is, a Line, by which Sea-men find where the Ship may Sail by the Depth of the *Water*. 'Tis bigger than a *Deep-Sea-Line*, and is not much above 20 *Fathom* in Length. 'Tis mark'd at 2 Fathom next the Lead, with a piece of Black Leather; at 3 Fathom the like, but ſlit; at 5 Fathom with a piece of white Cloth; at 7 Fathom, with a piece of Red, *&c.* with ſome ſuch Diſtinction all along. This *Line* may be uſed, when the Ship is under *Sail*; Which cannot be done ſo well with the *Deep-Sea-Line*: in uſing of that, the Ship is uſually brought upon the *Back-Stays*.

SOUND *the Pump*, that is, to let a ſmall Line, with ſome Weight at the End, down into the Pump, to know what Depth of Water there is in it.

SOUNDING-*Lead*, is fix or ſeven pound weight of *Lead*, near a Foot long. He that Heaves this *Lead*, ſtands by the Horſe, or in the Chains, and there Sings, *Fathom by the Mark* 5, o *and a ſhaftment* leſs 4, o *&c.*

SOUND *the Pump*; that is, to let a ſmall Line, with ſome Weight at the End, down into the Pump, to know what Depth of Water there is in it.

SOUTH, one of the four Cardinal-Points of the Mariner's-Compaſs, 'tis directly oppoſite to the North Point.

SOUTHERN-*Hemiſphere*. See *Hemiſphere*.

SOUTHERN-*Tropic*. See *Tropic*.

SOUTHING, is the *Difference of Latitude* a Ship makes in Sailing to the *Southward*. See *Difference of Latitude*.

SOUTH-*Pole*, is a Point in the *Southern Hemiſphere* of the Heavens, 90 Degrees, every way, diſtant from the Equinoctial.

SOUTHWARD; that is, towards the *South*.

SOUTH *Declination*. See *Declination*.

SPEEKE, the ſame with *Spike*.

SPELL, ſignifies to let go the *Sheat* and *Bow-lines* of a *Sail*, chiefly the Miſen; and Braceing the Weather-Brace to the Wind, that the Sail may lie more looſe, in caſe it has ſo much Wind in it, that the Maſt is in danger of being wrong'd.

SPELL *the Miſen!* that is, take in the *Miſen*, and Peek it up.

To do a Spell, signifies doing any Work for a short time, and then leaving it.

To Give a Spell; that is, to Work instead of such a One.

A Fresh Spell; that is, when Fresh Men come to Work, or to Relieve another Gang.

SPENT: As, *The Ship has Spent her Mast, or Yard*; that is, Her Masts, or Yards, have bin Broken down by Foul-Weather, or any such Accident: But if a Ship lose her *Mast*, or *Yard*, in a Fight; then 'tis said, *Her Mast, or Yard, was Shot by the Board*.

SPHERE, or *Globe*, is an Artificial Representation of the Whole Surface of the *Terraqueous* Ball; for the more distinctly Viewing of its Constituent Parts, and the better Comprehending its Various Positions; which are either *Parallel, Right*, or *Oblique*.

PARALLEL-SPHERE, is that Position of the *Globe*, which has,

1. The *Poles* of the *World* in the *Zenith* and *Nadir*.

2. The *Equator* in the *Horizon*.

3. The *Parallels* of *Latitude* Parallel to the *Horizon*.

And the Inhabitants of this *Sphere*, are those, if any, who live under the Two *Poles* of the World.

RIGHT-SPHERE, is that Position of the *Globe*, which has,

1. The *Poles* of the *World* in the *Horizon*.

2. The *Equator* passing thro' the *Zenith* and *Nadir*.

3. The *Parallels of Latitude* Perpendicular to the *Horizon*.

And the Inhabitants of this *Sphere*, are those who live under the *Equinoctial-Line*.

OBLIQUE-SPHERE, is that Position of the *Globe*, which has,

1. One of the *Poles* of the *World* Above, and the other Below the *Horizon*.

2. The *Equator* partly Above, and partly Below the *Horizon*.

3. The *Parallels of Latitude* cutting the *Horizon* Obliquely.

The Inhabitants of this *Sphere*, are those who live on all Parts of the Earth, except those exactly under the *Poles* of the World, and *Equinoctial-Line*.

Ptolemaic Armilar SPHERE, is that where the Great Circles of the *Sphere*, *viz.* the *Horizon*, *Equator*, *Meridian*, *Ecliptic*, with the Two *Colures*, and the Lesser Circles, *viz.* the Two *Tropics*, and Two *Polar Circles*, being made in *Brass*, *Wood*, &c. are put together in their natural Order, and placed in a Frame, so as to represent the True Position and Motion of those Circles, and the *Cœlestial Bodies*, supposing the *Earth* the *Centre* of the *Universe*, according to the Notion

tion of *Ptolemy*, and his Followers.

Copernican SPHERE, is that which shews the *Phænomena*, supposing the *Earth* to *Move*, and the *Sun* placed in the *Centre* of its *System* according to the now-Receiv'd and Approv'd Opinion.

SPIKES, or *Speeks*, are large long Iron Nails, with Flat Heads; they are of divers Lengths, some a Foot or two long; and some are Jagged, so that they cannot be drawn out again: They are used to fasten the Planks and Timbers of a Ship.

MARLINE-SPIKE. See *Marline*.

SPIKING *up the Ordnance*; that is, fastening a *Coin* with *Spikes* to the Deck, close to the Breech of the Carriages of the Great-Guns, that they may keep Close and Firm to the Ship's-Sides, and not break-loose when the Ship Rolls, and by that means endanger the breaking-out the But-head of a Plank.

SPINDLE, is the Smallest Part of a Ship's *Capstan* which is betwixt the Decks. The *Spindle* of the *Jeer-Capstan* has *Whelps* to heave the *Viol*. See *Capstan*.

To SPLICE *a Rope*, is to untwist two Ends of *Ropes*, then twist them both together, and fasten them one into the other.

A Cut SPLICE, is when a Rope is let into another with as much Distance as one pleases, so as to have it undone at any time, and yet be strong enough.

A Round SPLICE, is when a Rope's-End is so let into another, that they shall be as Firm, as if they were but one Rope.

To SPLICE *a Cable*. See *Cable*.

SPLIT: As, *The Sail is Split*; that is, The Sail is blown to Pieces.

SPOOMING, or *Spooning*: When a Ship being under-Sail in a Storm at Sea, cannot bear it, but is forc'd to put right before the Wind, without any Sail; then they say, *The Ship Spooms*: So that if a Ship will neither *Try*, nor *Hull*, then *Spoom*; that is, Put her Right before the Wind.

SPOOMING *with the Fore-sail*; that is, when the Ship *Spooms*, and that there is danger lest she should bring her Masts by the Board, with her Rolling, or *Seel* under-Water, and so Founder; then 'tis usual to set up the Fore-sail, to make her go the steadier, especially if there be Sea-room enough.

SPOUTS, in the *West-Indies*, and other Parts of the World, are, as it were, Rivers falling intirely from the Clouds, as out of *Water-Spouts*; they make the Sea, where they fall, rebound in exceeding high Flashes.

SPRING *a Leak*, is said of a Ship that begins to *Leak*.

SPRING *a Mast*: When a Mast is only crack'd, and not quite broken, in any Part of it, as in the *Partners*, the *Hounds*, &c. then 'tis said, *The Mast is Sprung*.

SPRING-TIDE, is the Encreasing higher of a *Tide*, after a *Dead Neap*; this is about Three Days before the *Full*, or *Change* of the *Moon*: But the Highest *Spring-Tide* is Three Days after the *Full*, or *Change*; then the Water runs highest with the Flood, and lowest with the Ebb, and the Tides run more strong and swift than in the *Neaps*.

SPRIT-SAIL. See *Sail*.

SPUNGING *of a Great-Gun*, is Clearing of its Inside after 'tis Discharg'd, with a Wad of Sheep-Skin, or the like, roll'd about one end of the Rammer: And this prevents all Danger of any Fire being left in the Piece; which might prove of ill-Consequence to him who should Load, or Charge it again.

SPUN-YARN; is Rope-Yarn, whose Ends are beaten, or scrap'd thin, and so Spun one Piece to another, that it may be as long as is necessary: It serves for several Purposes on-board a Ship, as to make *Caburns*, and the like.

SPURKETS, are the Spaces betwixt the Timbers along the Ship's Side, in all Parts, betwixt the Upper and Lower Futtocks, or betwixt the Rungs fore and aft. Those in the Howl, below the Sleepers, are broad Boards, which are now and then taken up, to clear the *Spurkets*, when any thing has got between the Timbers.

SQUADRON, is a particular Detachment of *Ships of War*; or one of the Three Bodies, which, in Order of Battle, composes the *Van*, *Centre*, and *Rear*: Each of which is sometimes distributed into Three Divisions, and distinguish'd by their *Flags* and *Colours*. So that when a Fleet is Divided into *Three Squadrons*, each *Squadron* has its *Admiral*, and each *Admiral* has its *Flag*; by which the *Squadron* is named, and distinguish'd, as the *Flags* are, either *White*, *Blue*, or *Red*: The *Flag* of the *White Squdron* being *White*, with a Frank-Quarter, and a Cross *Gules*. That of the *Blue Squadron* is *Blue*, with a Frank-Quarter *Argent*, and Cross *Gules*, &c.

In *Sailing*, also, a great Fleet is usually Divided into *Three Squadrons*; The *Admiral*'s; the *Vice-Admiral*'s; and the *Rear-Admiral*'s Squadron: The which being distinguish'd by their *Flags* and *Pendants*, are to put themselves, and, as near as may be, to keep themselves in their Customary Places; *viz*.

The *Admiral* with his Squadron, to *Sail* in the *Van*; that so he may Lead the Way to all the rest, in the Day-time, by the Sight of his *Flag* in the Main-Top-Mast-Head; and in the Night-time, by his Lights or Lanterns.

The *Vice-Admiral* and his Squadron is to *Sail* in the Centre or Middle of the Fleet.

The *Rear-Admiral*, and the Ships of his Squadron, to bring up the Rear.

But

But ſometimes other Diviſions are made; and thoſe compos'd of the Lighter Ships, and Beſt Sailors, and are placed as Wings to the Van, Centre, and Rear.

SQUARE: As, *Square the Yards!* See *Yards*.

STAFF, *Back-Staff*; the ſame with *Quadrant*: Which ſee.

Flag-STAFF, is that long *Staff*, or Piece of Wood, whereto the *Flag* is made faſt, and along which 'tis hois'd up.

FORE-STAFF, or *Croſs-Staff*, is an Inſtrument uſed at *Sea*, chiefly for obſerving the *Sun* or *Star's Altitude*: It is call'd *Fore-Staff*, from the Poſition of the Obſerver in Uſing it, whoſe Face is towards the Thing Obſerved, generally: Tho', for the *Sun*, it's ſo Contriv'd (for Preſerving the Eye) to be uſed Backward.

It's call'd a *Croſs-Staff*, from its Form, being a Square *Staff* with three or four Pieces a-croſs it, which are call'd *Croſſes*.

The *Staff* is uſually 30 Inches, or 3 Feet long, and more than Half an Inch Square, having four Sides, each Graduated unequally, like the *Scale of Tangents*; to each Side there belongs a diſtinct *Croſs*: Tho' ſometimes, the Shorteſt *Croſs* is made to ſerve two Sides of the *Staff*; that is, the Breadth is for the *Ten-Croſs*, and Length for the *Thirty-Croſs*: Beſides this, it has two other *Croſſes*; the Longeſt is call'd, the *Ninety-Croſs*; the other, the *Sixty-Croſs*. And theſe Four *Croſſes* are thus to be underſtood:

The *Ten-Croſs*, belongs to that Side beginning at 3 *deg.* and ending at 10 *deg.*

The *Thirty-Croſs*, belongs to that Side beginning at 10 *deg.* and ending at 30 *deg.*

The *Sixty-Croſs*, belongs to that Side beginning at 20 *deg.* and ending at 60 *deg.*

The *Ninety-Croſs*, belongs to that Side beginning at 30 *deg.* and ending at 90 *deg.*

The Uſe of the FORE-STAFF.

It's commonly uſed at *Sea*, to take the *Meridian Altitude* of the *Sun* or *Stars* by; in order to find the *Latitude of the Place*: And that is perform'd in the following manner;

1. The *Ten*, *Thirty*, *Sixty*, and *Ninety Croſſes*, are to be uſed according as the *Meridian Altitude* is more or leſs; that is,

If leſs than 10 *deg.* uſe the *Ten-Croſs*.

If between 10 and 30 *deg.* uſe the *Thirty-Croſs*.

If between 30 and 60 *deg.* uſe the *Sixty-Croſs*.

If more than 60 *deg.* uſe the *Ninety-Croſs*.

2. Having conſider'd which *Croſs* is ſuitable, Put it on the *Staff* ſo, that the flat Side of the *Croſs* may be towards the flat End of the *Staff*.

3. Then

3. Then hold the flat End of the *Staff* to the Corner of your Eye; there let it rest upon your Eye-bone, as near the Corner of your Eye as you can, so it does not hinder the Sight.

4. And look to the Upper-end of the *Cross* for the *Sun*, or *Star*; and at the Lower-end, for the *Horizon*.

5. But if, at the Lower-end of the *Cross*, you see all Sky, and no Water; then draw the *Cross* a little nearer to your Eye.

6. If, on the contrary, you see all Water, and no Sky; then slide the *Cross* a little farther from you.

7. Then, if you see the Centre of the *Sun* or *Star* at the Upper-end of the *Cross*, and the *Horizon* at the Lower-end, the *Cross* stands as it ought.

8. Wait till the *Sun* or *Star* be on the *Meridian*: Upon frequent Tryals, as the *Sun* or *Star* Rises, draw the *Cross* a little nearer to your Eye.

9. If the *Sun* or *Star* be fallen, you'll presently have the *Horizon* hid from you by the Water; then stir not the *Cross* out of its Place, (for the Observation is over,) and see at what *Degree*, &c. it rests, (on that side of the *Staff* belonging to it;) and that will be the *Meridian Altitude*, or its *Complement*, sought, according to the word *Alt.* or *Compl.* on the *Staff*.

How to Use the FORE-STAFF *Backwards.*

This is only used with the *Sun*; and for this Purpose, the *Ten-Cross* has another Piece put cross it; so that the Lower-edge of this Cross-Piece lies even with the Middle of the Square-Hole in the *Ten-Cross*; which also answers to the Middle of the Thickness of the *Staff*.

And there is a Plate of Brass, with a Hole in it, so fitted, that it will slide off and on the Ends of the *Ninety*, *Sixty*, or *Thirty-Crosses*.

These Two Things added to the *Fore-Staff*, make it fit for a *Backward* Observation of the *Sun*; which is thus:

1. According as the *Meridian Altitude* of the *Sun* is more or less, so use the *Ninety*, *Sixty*, or *Thirty-Crosses*; putting it on the *Staff*, the Flat-Side of it even with the Flat-End of the *Staff*, there screw it fast; and at one End of the *Cross*, slip on the foresaid Brass-Plate, so as to leave a Slit-Sight thro' it near the Lower-end of the *Cross*.

2. Put the *Ten-Cross*, (having a Cross-Piece on it,) on the *Staff*, the Flat-side of it towards the other Cross, at the *Staff*'s End.

3. Turn your *Back* to the *Sun*; Look thro' the Slit-Sight, at the Lower-end of the Cross, for the Shadow of the Upper end, lying on the *Ten-Cross*, in the Line answering to the Middle of the *Staff*,

Staff, and on each Side of the *Staff*.

4. At the ſame time, the *Horizon* ſhould be ſeen (thro' the ſaid Slit) to lie even with the Shadow on the Middle-Line, in the *Ten-Croſs*, at each End of it, on both Sides of the *Croſs*.

5. In looking thro' the Slit-Sight, you ſee the Shadow on the Middle-Line; but inſtead of the *Horizon*, you ſee only Water there; then draw the *Ten-Croſs* nearer, till the Shadow and Horizon agree or meet in the ſaid Middle-Line.

6. On the contrary, Looking as before, inſtead of the *Horizon*, you ſee *Sky* meet the Shadow on the Middle-Line; then put the *Ten-Croſs* from you, till you ſee the Horizon and Shadow meet at the Line.

7. Continue Obſerving, till the *Sun* be at the Higheſt; and as the *Sun* Riſes, you muſt draw the *Ten-Croſs* nearer, in order to keep the *Horizon* and *Shadow* together on the Middle-Line of it.

8. If the *Sun* be fallen (after you have continued Obſerving, as before Directed) the *Horizon* will lie below the *Shadow*, on the Middle-Line; then is the *Obſervation* ended. Stir not the *Ten-Croſs* out of its Place; for where it now ſtands, on that ſide of the *Staff* belonging to the *Croſs*, at the End of it, is the *Sun*'s *Meridional Altitude*, or *Complement* thereof, which was required.

STANCHIONS, or *Stantions*, are thoſe Pieces of Timber, in a Ship, that are ſet up Pillar-wiſe, to ſupport and ſtrengthen the Waſt-Trees.

STANDARD-ROYAL. See *Flags*.

STANDING-*Coins*, or *Standing-Quoins*. See *Coins*.

STANDING-*Part of the Sheat*. See *Sheat*.

STANDING-*Ropes*, are thoſe which do not run in any Block, but are ſet *tau't*, or let ſlack, as Occaſion ſerves, as the Sheat-Stays, Back-Stays, and the like.

STANDING-*Part of a Tackle*, is the End of the Rope where the Block is ſeiz'd or faſten'd; as the other, which is hal'd, is call'd the *Fall*.

STARBOARD, is the Right-ſide of the Ship, or the Right-Hand; as *Larboard-ſide* is the Left.

STARBOARD *the Helm!* or, *Helm a Starboard!* in Conning a Ship, is to put the *Helm* to the Right-ſide of the Ship, or to put the Helm *a-Starboard*, to make the Ship go to the *Larboard*; for the Ship always Sails contrary to the *Helm*.

STAYS; are Ropes belonging to all Maſts, Top-Maſts, and Flag-Staffs, except the Spritſail-Top-Maſt: They ſerve to keep the Maſts from falling aftward, or too much forward.

Main STAY; that is, the *Stay* of the *Main-Maſt*: 'Tis made faſt by a Lannier to the Collar,

Collar, which comes down about a Knee, below the Head.

Main-Top-Mast-STAY, is made fast to the Head of the *Fore-Mast*, by a Strap and Dead-Man's-Eye.

Main-Top Gallant-Mast-STAY, is, in like manner, made fast to the Head of the *Fore-Top-Mast*.

The *Fore-Mast*, and *Masts* belonging to it, are likewise *Stay'd* to the *Bowsprit*, and *Spritsail-Topsail-Mast*: And those *Stays*, do likewise Stay the *Bowsprit* it self.

Misen-STAY comes to the Main-Mast; and the *Misen-Top-Mast-Stays*, to the Shrouds, with Crows-Feet.

All *Stays* have their Blocks; and their Length are generally, the same with that of the Mast they belong to.

Back-STAYS, are those which serve to keep the Masts from pitching forward, or over-board, because they go on either Side of the Ship; and therefore the Main-Mast, Fore-Mast, with the Masts belonging to them, have these *Back-Stays*.

STAYING *a Ship*, or, *To bring a Ship upon the Stays*, is in order to her Tacking, and is done thus:

At the same time,

Bear up the Helm!
Let Flie the Fore-sail-Sheat!
Let go the Fore-Bow-line!
Brace the Weather-Brace of the Fore-sail.

The same is also done to the *Topsail* and *Top-Gallant-sail*, only their Sheats must be kept fast.

If the *Spritsail* be out, then,

Let go the Spritsail-Sheat, with the *Fore-Sheat*,
And Brace the Weather-Brace!

But the *Tacks*, *Sheats*, *Bracings*, and *Bow lines* of the *Main-sail*, *Main-Top-sail*, and *Misen*, are not altered.

And when the Wind comes in at the *Bow*, which before was *Lee-Bow*, it drives all the Sails Backwards against the Shrouds and Masts; so that the Ship makes no Way forward, but drives with her Broad-side. And those are accounted the best Ships, which will *Stay* with the fewest Sail.

STEADY! is a Word of Command to the Man at *Helm*, to keep the Ship right upon that Point he Steers by, and not to make Yaws in or out.

STEER: *To Steer a Ship*; is to Guide, or Direct her Course by the Helm.

And there are Three ways to *Steer* by; *viz*.

1. By any *Mark* on the *Land*, so as to keep the Ship even by it.

2. By the *Compass*; which is, to keep the Ship's-Head upon such a *Rumb*, or *Point* of the *Compass*, as best leads to the design'd Port.

3. To *Steer* as one is bidden, or Conn'd; which is the Duty

(in

(in Great Ships) of him that is taking his Turn at the Helm.

And he is reckon'd the beſt *Steers-man*, that uſes the leaſt Motion in putting the Helm over, to and again, and that keeps the Ship beſt from making *Taws*, that is, from running in and out.

STEERAGE, is that Part of the Ship next below the Quarter-Deck, before the Bulk-head of the Great-Cabbin : 'Tis here, in moſt Ships of War, ſtands the *Steers-man*, or he that Guides the Helm or Rudder of the Ship.

STEEVE : The *Bowſprit*, or *Beak-head of the Ship, does Steeve* ; that is, it either ſtands too-upright, or not ſtrait enough forward.

STEEVING, is alſo a Word uſed by *Merchant-men*, when they ſtow Cotton, or Wool, which is forc'd in with Screws ; this they call *Steeving* their *Cotton*, or *Wool*.

STEM of a Ship, *Main-Stem*, is that great Piece of Timber which is wrought Compaſſing, and Scarf'd at one End into the Keel of a Ship, and comes bowing or bending right before the Fore-Caſtle. This *Stem* guides the Rake of the Ship ; and all the But-ends of the Planks forwards are fix'd into it.

Falſe-STEM, is fix'd before the *Right one*, where that is made too-flat for the Ship to keep the Wind well : Which Defect is remedied by a *Falſe-Stem* ; that will make her Rid more Way, and bear a better Sail.

STEP *of the Maſt* or *Capſtan*, in a Ship, is that Piece of Timber whereon the *Maſts* or *Capſtans* do ſtand at bottom.

STERN of a *Ship*, is, properly, the Outmoſt-part of a Ship, abaft : Tho', generally ſpeaking, 'tis all the Hindermoſt or Aftmoſt-part of a Ship.

STERN-*Chaſe*. See *Chaſe*.

STERN-*Faſt*, is ſome Faſtenings of Ropes, *&c.* behind the *Stern* of a Ship, to which a Cable, or Hawſer, may be brought or fix'd, in order to hold her *Stern* faſt to a Wharf, *&c.*

STERN-*Poſt*, is a great Timber let into the Keel, at the *Stern* of the Ship, ſomewhat ſloaping, into which are faſten'd the *After-Planks* : And on this *Poſt* hangs the Rudder, by its Pintles and Gudgeons.

STEWARD *of a Ship*, is he that receives all the Victuals from the *Purſer* ; and he is to ſee it well Stow'd in the Hold : In his Cuſtody are all Things of that nature belonging to the Ship's Uſe. He is to look after the Bread, and to diſtribute out the ſeveral Meſſes of Victuals in the Ship. He has an Appartment for himſelf in the Hold, call'd the *Steward-Room* : And he has uſually a *Mate* under him.

STIRRUP *of a Ship*, is a Piece of Timber put on the Ship's Keel, when ſome of it is

is beaten off, and the *Carpenter* cannot come to mend it, or put in a new Piece; then they patch on a Piece of Timber, and bind it with an Iron which goes under the Ship's Keel, and comes up on the other Side of the Ship, where 'tis nail'd fast with Spikes; and this they call a *Stirrup*.

STOAK'D: *The Ship is Stoak'd*, or *a-Stoak*; that is, when the Water in the Bottom of the Ship, cannot come to the Well of the Pump.

The Limber-Holes are Stoak'd; that is, when the Water cannot pass thro' them.

The Pump is Stoak'd; that is, when something is got into it, which choaks it up, so that it will not work.

STOCKS; is a Frame of Timber, and great Posts, made a-Shore, to Build *Pinnaces*, *Ketches*, *Boats*, and such Small-Crafts, and sometimes Small *Frigats*.

A Ship is on the Stocks; that is said of a Ship when she is a Building.

STOCK *of an Anchor*. See *Anchor*.

STOP *a Leak*. See *Leak*.

STOP: *To Stop the Ship*; that is, when she comes to an *Anchor*, and the Cable is veer'd out but by degrees, till the Ship Rides well.

STOP! is a Word also, used by him that holds the *Half-Minute-Glass*, in Heaving the *Log*: For as soon as the Glass is out, he calls, *Stop!* to them that let run the Line.

STOPPER, is a Piece of Rope, in a Ship, having a Wail-Knot at one end, with a Lannier spliced into it; and at the other end 'tis made fast in the Place where 'tis to be used: It serves to stop the Main-Halliards, or the Cable.

The *Stopper* for the Halliards, is fasten'd at the Main-Knight; and it serves, when they are Hoising the Main-Yard, to Stop it, while the Men that Hale, may stay and rest a little. But 'tis chiefly used for the Cable, to stop it, that it don't run out too fast.

Lay on the Stopper! That is, Bind the Wail-Knot about the Cable with the Lanniers, and that Stops it, so that it can't slip away. This *Stopper* is fasten'd to the bottom of the *Bits*, by the Decks.

The Ship Rides by the Stoppers; that is, when the Cable is fasten'd or Stay'd only by them, and not Bitted: But this is not safe Riding, in a Stress of Weather.

STOW: *To Stow*, is to put the Goods, in order, into the Hold of the Ship; the most Ponderous and Heavy, next the Ballast.

STOWAGE; the putting of Goods into the Hold of the Ship.

STORM, or *Tempest*, is well known to be, a violent Wind, which raises the Sea to a high degree

degree of Rage and Foam, and endangers all Ships that lie under it, by its exceeding Breaks. In this Case, a Ship that has Sea-Room enough, 'tis best to let her Top-Masts stand, for then she will make better way thro' the Sea.

STRAIT, is a Narrow Sea, shut up between Lands on either Side, affording a Passage from one Great Sea into another.

In EUROPE *there are*,

The *Straits* of *Dover*, joining the *German* Sea to the *English* Channel.

The *Straits* of the *Sound*, joining the *Danish* to the *Baltic* Sea.

The Straits of *Gibralter*, join the *Mediterranean* to the *Western* Ocean.

The *Straits* of *Caffa*, joining *Palus Meotis* to *Pontus Euxinus*, or the *Black Sea*.

The *Thracian Bosphorus*, joining the *Black Sea* to the *Propontis*.

The *Hellespont*, joining the *Propontis* to the *Archipelago*.

The Veer of *Messina*, joining one Part of the *Mediterranean* to another.

In ASIA, *there are*,

The *Straits* of *Sunda*, joining the *Eastern* and *Indian Ocean*.

The *Straits* of *Ormus*, joining the Gulf of *Persia* to the *Southern* Ocean.

In AFRICA, *only*,

The *Straits* of *Babelmandel*, joining the *Red-Sea* to the *Eastern* Ocean.

In AMERICA, *there are*,

The *Straits* of *Magellan*, joining the Vast *Eastern* and *Western* Ocean.

Davies's *Straits*, joining *Baffin*'s Bay to the *Eastern* Ocean.

Hudson's *Straits*, joining *Button*'s Bay to the *Eastern* Ocean.

STRAKE, is a Seam between two Planks.

Garboard-STRAKE, is the first Seam next the Keel.

The Ship heels a-Strake; that is, the Ship hangs or inclines to one Side the Quantity of an whole Plank's Breadth.

STRAP, in a Ship, is the Rope which is splic'd about any Block; and made with an Eye, to fasten it any where on Occasion.

STREAM-*Anchor* is only a small one made fast to a small Stream-Cable, for a Ship to Ride by in gentle Streams, and in Fair Weather, when they would only stop a Tide.

STREAMERS, the same with *Pendants*.

STRETCH; as, *Stretch forwards the Halliards, or Sheets!* that signifies, (in Hoising the

Yard, or Haling the Sheat,) that the Part which the Men are to *Hale* by, should be put into their Hands, in order to their *Hoising*, or *Haling*.

STRIKE, is a Word variously used at Sea; as

The Ship strikes; that is, when a Ship in a Fight, or on meeting with a Man of War, lets down or lowers her Top-Sails, at least half Mast high, which is called *Striking the Top-Sails upon the Bunt*; and signifies, that she yields or submits, or pays her Devoir to that *Man of War* she passes by.

STRIKE; it is also said, that *the Ship strikes*, when she touches Ground in Shoal Water.

STRIKE *the Top-Mast!* That is said, when any Top-Mast is to be taken down.

STRIKING *down*, of any thing, *into the Hold*; that is, to let or lower down any thing into the Hold.

STRIP; As, *the Chase Strips himself into Short or Fighting-Sails*; that is, puts out his Colours in the Poop; his Flag in the Main-Top; his Streamers or Pendants at the End of his Yard-Arm; Furls his Sprit-sail; Peeks his Misen, and Slings his Main-Yard. In which Case, the Chaser must provide himself to Fight.

STUDDING-*Sails*. See *Sails*.

SUN's *Altitude* or *Height*, is an Arc of a Vertical Circle intercepted between the *Sun* and the *Horizon*.

SUN's *Meridian Altitude*, is an Arc of the *Meridian* intercepted between the *Sun* at Noon, and the *Horizon*.

This is easily found by means of a *Quadrant*, *Cross-Staff*, or the like *Instrument*; when an *Observation* can be made, in order to get the *Latitude of the Place*. See *Observation*.

SUN's *Amplitude*, is an Arc of the *Horizon* intercepted between the *East* and *West* Point and the Place of the *Sun*'s Rising and Setting. 'Tis useful for finding the *Variation of the Com-* See *Variation*.

And the *Sun*'s *Amplitude* is readily found (having the Latitude and Declination given) by this

Proportion.

As the Cosine of the Latitude of the Place,
Is to the Radius,
So is the Sine of the Sun's Declination,
To the Sine of the Sun's Amplitude.

SUN's *Azimuth*, is an Arc of the *Horizon* intercepted between the Meridian and the Vertical Circle which passes thro' the Centre of the *Sun*. It serves chiefly for to find the *Variation of the Compass*. See *Variation*.

And having the *Latitude of the Place*, *Sun's Declination* and *Altitude* given; his *Azimuth* is found by this RULE.

Add

Add the Complements of your Data together, and find the Difference between their Half-Sum and the Co-Declination.

Then set down the Logarithm-Sines of the Half-Sum, and Difference found.

As also the Arithmetical Complement of the Logarithm-Sines of the Co-Latitude and *Co-Altitude:*

And twice the Complement of the Arc answering to ½ the Sum of these four Logarithms, (taken among the Sines) is the Azimuth sought.

SURFF *of the Sea*, is the great Breakings, or Rolling of the Sea against some Shores; making it dangerous Landing in such Places.

SURGE; that is, a Wave or Billow of the Sea.

To SURGE: As, *the Cable Surges*; that is, slips back a little, when there is Heaving at the *Capstan.*

SURGEON, or *Chirurgeon* of a Ship, is he that attends and administers Physick to the sick, and takes care of the wounded: Therefore his Chest ought to be well furnish'd with all Necessaries proper for the Climate the Ship is bound to; The Neglect of which has bin the Loss of of many Men's Lives: Besides 'tis well known, That many of the *Surgeons*, but more especially *their Mates* (who are their *Assistants*) which are employed in the Fleet, are not altogether so well Qualified as they ought to be; and yet the Poor Men are forc'd to depend on their Skill, not only in Surgery, but in Physick also, in the Absence of a Physician: And it may be question'd whether there are many of the Ablest of the *Sea-Surgeons*, Qualified to judge nicely of many Distempers incident to a Sailor, therefore they must, if not so Qualified, be greatly to seek for proper Remedies.

SWABBER, is an Inferior Officer a board a Man of War; whose Business is to see that the Ship be kept Clean and Neat; in order to which, he is to see her well wash'd once or twice a Week at least: And he ought to burn Pitch, or some such thing, now and then between the Decks, to prevent Infection; and to acquaint the Commanding Officers of such as are Nasty and Offensive.

SWEEP: as *the Sweep of the Ship*, or *Sweep of the Futtocks*; that is, the Mould of a Ship when she begins to Compass in at the Run, heads.

SWEEPING; that is, dredging along the Ground, at the Bottom of the Sea or Channel, with a Three-hook'd Grapnel, to find some *Hawser*, or *Cable*, which is slip'd from an Anchor

SWIFTERS, are Ropes belonging to the Main-Masts, and Fore-Masts; they serve to succour and strengthen the Shrouds, and to keep the Masts stiff: They have Pennants fasten'd

under the Shrouds at the Head of the Masts, with a double Block thro' which the Swifter is Reev'd, which at the standing Part has a single Block with a Hook hitch'd in a Ring at the Chain-Waile; and so the Fall being Haled up, helps to strengthen the Mast; and 'tis belay'd about the Timber-heads of the lower Rails aloft.

SWIFTING *of a Boat*, is Compassiing her Gun-Wall round with a good Rope, to strength-en her in stress of Weather, that she be not shatter'd by the Violence of the Sea

SWIFTING *the Capstan Bars*, is straining a Rope all round the outer Ends of the *Capstan-Bars*, in order to strengthen them, and make them bear all alike, and together when the Men heave or work there.

SWIFTING *of a Ship*, is either bringing her a-ground, or upon a Carreen.

To SWIFT *the Masts*: In Swifting of a Ship, to ease and strengthen the Masts, that all the Weight may not hang by the Head, and to keep them from Rising out of the Step: All the Tenuants of the *Swifters* and *Tackles* are laid close to the Mast with a Rope, and as near to the Blocks as can be; then the Tackles are carried forwards, and hal'd down, as tau't as possible.

T

TACK, is a great Rope with a Wale-Knot at one End, which seiz'd into the Clew of the Sail, so is Reev'd thro' tht Chesse-Trees, and then is brought thro' a Hole in the Ship's-Side. It serves to carry forward the Clew of the Sail, to make it stand close by a Wind. The *Tacks* of a Ship are usually belay'd to the *Bitts*, or else there is a *Chevel* on purpose to fasten them; they belong only to the *Main-Sail*, *Fire-Sail*, and *Misen*. When the Sails are to be Trimm'd so as to stand close to a Wind: Then

The *Main*, *Fore*, and *Misen-Tack* are brought close by the Board, and over-hal'd as forward as they can be;

The *Bowlings* are so also on the *Weather Side*;

The Lee-Sheats are Haled close Aft.

As also the *Lee-Braces* of all the Sails are likewise Braced Aft: And the Top-Sails are Braced and govern'd, as the Sails whereto they belong.

Hale aboard the Tacks! That is, bring the *Tack* down close to the Chess-Trees.

Ease the Tack! That is, slacken it, or let it go, or run out

Let rise the Tack! That is, Let it go all out.

The

The Ship Sails, or *stands close upon a Tack*; that is, close by the Wind.

To TACK *about*; is to bring the Ship's-Head about, so as to lie the contrary way: Which is perform'd thus;

First, Make the Ship *Stay*; and when she is *Stay'd*, they say she is *Pay'd*.

Then, *Let Rise, and Hale!* That is, Let the Lee-Tack Rise, and Hale aft the Sheats; and so Trim all the Sails, by a Wind as before: For they cast off that which was before the *Weather-Bow-line*, and set up the other Tau't.

And thus they do also by all *Sheats*, *Braces*, and *Tacks*, which a Ship that is Trimm'd by a Wind must have.

TACKLES, are Ropes running in three Parts; having a *Pennant* with a Block at one end; and a Block with a Hook at the other end, to hang any Goods upon, which is to be Heav'd in, or out of the Ship. And there are several sorts of *Tackles*: As,

Boat's and Bores TACKS, made fast to the *Main* and *Fore-Shrouds*, to Hoise the Boat in, or out: And to keep firm the *Masts* from from *straying*; as also, for many other Uses

Ground-TACKLE. See *Ground*.

Gunner's-TACKLES, serving for Haling the Ordnance in or out.

Winding-TACKLE; is a *Tackle* that serves as a Pennant, with a great Double-Block and Three Shivers in each, seiz'd fast to the end of a small Cable about the Head of the Mast: It has a Guy brought to it from the Fore-Mast. Into this Block there is reev'd a Hawser, which is also reev'd thro' another Double-Block, having a Strap at the end of it; which being put thro' the Eye of the Slings, is lock'd into it with a Fidd, in order to Hoise in Goods.

The Fall of this *Tackle* is reev'd into a Snatch-Block, and so is brought to the Capstan, whereby the Goods are Heav'd.

There is also another *Tackle*, call'd *Burton*; which see under the word *Burton*.

TAFFEREL, is the uppermost Part, Frame or Rail of a Ship, abaft over the Poop.

TALLY *the Sheats!* is a word of Command, at Sea, when the Sheats of the Main-sail or Fore-sail is to be Haled aft.

TAMPION, *Tampkin*, or *Tampin*, is a a Round Piece of Wood made fit for the Mouth of any Great Gun: It serves to stop it, so that no Sea-Water or Rain may get in to wet the Powder.

TAR PAWLING, is a Piece of Canvase well *Tar'd* over, to lay on or over any Place, to keep off Rain.

And a Downright Seaman, that has bin all his Life-time bred up to the Sea, is call'd, in a Burlesque Sence, *Tar-Pawling*; as when we say, *He is a True Tar-Pawling*.

TARTANE, is a sort of *Barque*, very common in the *Straits*, and upon the Coasts of *Spain*: They carry One Mast, and generally a Three-Corner'd Sail'd.

TAU'T, signifies Stiff, or Fast; As,

Hale Tau't a Rope! That is, Set it Stiff, or Strait.

Set Tau't the Shrouds, or *Stays!* That is, Let them be more Tight and Stiff, , that are too-Slack, and Loose.

Wind-TAU'T See *Wind*.

TAUNT: *The Ship is Taunt-Masted*, or her Masts are very *Taunt*; that is, Her Masts are too Tall for her.

TEMPEST. See *Storm*.

TENDERS, are Small Ships employ'd in the Service of *Men War*, for Carrying of Men, Provision, or whatever else is necessary.

THAUGHTS, or *Thoughts*, are the Benches on which the Rowers Sit in a Boat to Row; or the Rowers Seats, in a Ship's-Boat.

THOWLES, are those Pins in the Gunnel of a Boat, against which the Rowers bear the Oars, or between which they put their Oars, when they Row.

THWART *the Hawse*. See *Hawse*.

Ride a-THWART. See *Ride*.

TIDE, signifies as well the *Ebbing*, as the *Flowing* of the Sea.

TIDE of *Flood*, so is the *Flowing* of the Sea call'd.

TIDE *of Ebb*; that is, the *Ebbing* or *Reflux* of the Sea. See *Ebb*.

A Windward TIDE, is when the *Tide* runs against the *Wind*; this makes the Water, generally, Rough.

Leeward-TIDE, is when the *Wind* and *Tide* go both together the same Way.

Spring-TIDES, are the *Tides* at *New* and *Full Moon*: They *Flow* Highest, and *Ebb* runs Strongest. See *Spring*.

Neap-TIDES, are the *Tides*, when the *Moon* is in the Second and Last Quarter; being neither so High, nor so Swift as the *Spring-Tides*.

TIDE-*Gate*; that is, where the *Tide* runs very strong.

To TIDE *it over*, or *up* into any Place; is to go with the *Tide of Flood*, or *Ebb*, as long as that lasts, then to stay at an Anchor all the Time the contrary *Tide* lasts, and the next *Tide*, set in again, till the same *Tide* returns: Thus one may often-times work against the Wind, if it don't Over-blow.

It Flows Tide and Half-Tide; that is, it will be Half-Flood by the Shore, before it begins to Flow in the Offing: Or it will be High-Water sooner by Three Hours at the Shore, than at the Offing.

They brought the Tide along with them; that is, (speaking of a Ship that came into the Harbour, over a Bar, or Sand,) They came in with the *Flood*.

Eddy-

Eddy-TIDE. See *Eddy*.

And the *Flux* and *Reflux* of the Sea, is occasion'd by the Water, (if suppofed to cover all our Globe,) its putting on two Oblong Spheroïdal Figures, whofe Axes produced, would pafs thro' the Moon and Sun. This Spheroïdal Figure of the Ocean, being ftretch'd out like two Mountains, the one towards the Moon, and the other to the Part oppofite to her, is continually fhifting, according to the Daily Motion of the Moon, which it follows; Or rather, The Earth, from its Diurnal Motion, fhifts its felf away from thefe Mountains of Water, which keep, as it were, Immoveable under the Moon. Therefore, it follows, that the Waters twice Rife and Fall in 25 Hours; in which Time the Moon is fuppos'd to move from the Meridian of any Place, to the fame again.

And fince the Waters alfo will fwell in thofe Places to whom the Sun is in the Zenith or Nadir, tho much lefs than when the Moon is fo pofited; therefore, in the Conjunction and Oppofition of thefe Luminaries, the aforefaid Protuberances of the Water will be conjoin'd; and confequently, caufe then the Higheft *Spring-Tides*; for the Water is then Elevated by the conjoint Forces of both: So that the Loweft Ebbs will be when both thefe Luminaries are in the Horizon of any Place; for the Water is now Deprefs'd by the conjoint Force of both.

Alfo, in the Quadratures of the Luminaries, the Sun Elevates the Water where the Moon makes it Fall, and makes it Fall where the Moon Elevates it: So that the Elevation of the Water, depending only on the Difference of thefe Forces, will be the leaft of all; and fo for the Depreffion.

And when the Moon is in the Equinoctial, the two oppofite Protuberances of the Water, will be alfo in the Earth's Equator, and each of them defcribing that Great Circle of the Earth, by its Diurnal Revolution, it will move fwifter; and when it is thrown towards the Shores, will rife higher there. Alfo, fomething muft be allow'd for the Equatorial Diameter of the Earth, being its longeft, and confequently the Water there being fomething nearer to the Luminaries, will, by their Influence, be raifed higher, than in other Parts.

So that whenever the Luminaries are either in Conjunction, or Oppofition, in the Equator, their Forces will be conjoin'd to Raife or Elevate the Sea at the Equator; as happens at the *Syzygies* next the Equinoxes, or in or near *March*, or *September*, when there is always the greateft *Annual Tides*.

All thefe Things would regularly come to pafs, if the whole Earth were cover'd with Water very deep: But by reafon of the Shoalnefs of fome Places,

Narrowness of the Straits, of Islands, &c. there will be an infinite Variety in the *Phenomena* of the *Tides*, which are not to be accounted for, without an exact Knowlege of all the Circumstances of the Places; as of the Position of the Land, and of the Breadth and Depth of the Chanels by which the *Tide* Flows, &c.

TIES, are those Ropes by which the Yards do hang; and when the Halliards are strain'd to Hoise the Yards, these *Ties* carry them up.

The *Main-Yard* and *Fore-Yard-Ties* are first reev'd throw the Ram's-Head. then throw the Hounds, with a Turn in the Eye of the Slings made fast to the Yard

The *Misen-Yard* and *Top Yard* have but single *Ties*.

The *Spritsail-Yard* has none; being made fast to the *Bowsprit* with a Pair of Slings.

TILLER, the same with the *Helm* of a Ship: It is most properly us'd in a Boat, where that which would be the *Helm* of a Ship, is call'd the *Tiller*, and serves for the same Use.

Tilt-*Boats*, are those which have Bails or Hoops over the Stern, cover'd with a Tar-Pawling, or the like; and this so fit up, is call'd the *Tilt* of the Boat; as is common in *Wherries*.

TIMBERS: In a Ship, there are the

Floor, or *Ground*-TIMBERS, which form the *Floor* of the Ship: These *Timbers* are strait, except at the Ends, where they begin to Compass; they lie on the Keel, and are fasten'd to it with Bolts thro' the Keelson; and are call'd *Ground-Timbers*, because the Ship rests upon them when she is a-Ground.

Knee-TIMBERS. See *Knee*.

Rising TIMBERS, are the Hooks placed on the Keel: They are so call'd, because as these Rise in Proportion, so the Ship's *Rake* and *Run* Rise on her Flat-Floor by degrees.

TIRE, or *Teer of Guns*, is a Row of them placed along the Ship's-Side, either above upon Deck, or below.

Also, Three-Deck'd Ships have Three *Tire of Guns*, call'd the *Lower*, *Second*, and *Third* or *Upper Tire*: The *Lower-Tire* have, generally, the Greatest Guns; and *Upper-Tire*, the Smallest.

Half a TIRE: The Fore-Castle and the Half-Decks are also furnish'd with *Half a Tire of Guns*.

TITE, or *Tight*: As, *The Ship is Tite*; that is, She is so Staunch, as to let in but little Water: And this is known by the Smell of the Water Pumpt out; for if she let in but little Water, it will always Stink, otherwise not.

TON, or *Tun*, is a certain Weight by which the Burden of Ships are Estimated, and is *Twenty-Hundred Weight*, or 2240 *Pounds Averdupoise*.

TONNAGE,

TONNAGE, is a Custom, or Impost paid to the King, for Merchandise carry'd out, or brought in Ships, and such-like Vessels, according to a certain Rate upon every Ton.

TOP of a *Ship*, or *Round-Top*, is a Round Frame of Boards lying upon the Cross-Trees, near the Head of the Mast; where the Men may stand to Furl and Loose the Topsails, *&c.*

TOP-ARMINGS, are a sort of Cloths hung about the *Round-Tops* of the Masts, for Show; and also to hide the Men which are at the *Top*, in a Fight; who lie there to fling Stink-Pots, *&c.* or to fire Small-shot down on the Enemy, in case of Boarding.

Man the TOP! See *Man.*

TOP-GALLANT-MASTS of a Ship, are those two, *viz.* *Main-Top-Gallant-Mast*, and *Fore-Top-Gallant-Mast*: Which are small round Pieces of Timber, set on their respective *Top-Masts*; having on their *Top* the Flag-Staffs set, on which the *Colours*, as *Flags*, *Pendants*, &c. hang.

TOP-MASTS of a Ship, are these Four, *viz.*

Main-Top-Mast,
Fore-Top-Mast,
Misen-Top-Mast,
Spritsail-Top-Mast,

Which are made fast, and settled unto the Heads of the *Main-Mast*, *Fore-Mast*, *Misen-Mast*, and *Bowsprit* respectively.

TOP-MARTNETS. See *Martnets.*

TOP-ROPES, are those with which the *Top-Masts* are Set, and Struck.

They are reev'd thro' a great Block, which is seiz'd on one side under the Cap; and they are reev'd thro' the Heel of the *Top-Mast*, where there is a Brass-Shiver plac'd athwart Ships; after this, they are brought up, and fasten'd on either side of the Cap with a Ring: The other Part of them comes down by the *Ties*, and so is reev'd into the *Knight-head*; and when it is to be Heav'd, 'tis brought to the Capstan:

These *Top-Ropes* belong only to the *Main-Mast* and *Fore-Mast.*

TOP-SAILS, and *Top-Gallant-Sails.* See *Sails.*

TOP-SAIL-*Lifts.* See *Lifts.*

TOP *the Yard Arms!* That is, Make the Ends of the Yards hang Higher, or Lower. See *Yards.*

TOP *a-Starboard!* That is, Hale upon the Larbord-Lift.

TOPPING *the Lifts*; that is, Haling of the *Topsail-Lifts.*

TORNADO'S, the same with *Turnado's.*

TOUCH: As, *Touch the Wind!* That is, Keep the Ship near the Wind; as the Method is, in Sailing against the Wind.

TOW: *To Tow*, signifies, to Draw or Drag any Thing after the Ship.

To Tow a Ship, or *Boat*; That is, to Draw them with a Rope, after the Ship; then they are ſaid to be *Tow'd*, or to be in her *Tow*.

TRADE-*Winds*. See *Winds*.

TRAILE-*Board*, is a Carv'd Board on each ſide of the Ship's Beak, reaching from her *Main-Stem* to the Figure, or to the *Brackets*.

TRANSOM, is a Piece of Timber that lies a-thwart the Ship's Stern, between the two Faſhion-Pieces, directly under the Gun-Room-Port; and it lays out the Breadth of the Ship at the Buttocks: As when a Ship is built Broad or Narrow at her *Tranſom*, ſhe is ſaid to have a Broad or a Narrow Buttock.

TRANSOM, is alſo the Vane of a *Croſs-Staff*, which is made to ſlide along upon the *Staff*, by means of a Square Socket; and may be ſet to any of the Graduations thereon, in an *Obſervation*.

TRAVERSE, is the Way of a Ship, when ſhe makes Angles in and out, and cannot keep directly to her True Courſe; becauſe of the Shifting of Winds, *&c.*

And in Sailing by, or againſt the Wind, there ſhould be an Allowance made for Leeward-Way, and that according to the Mould of the Ship, the Sail ſhe Bears, the Growth of the Sea, *&c.*

Reſolving, or *Working a Traverſe*, is the Reducing of theſe *Courſes*, found by the *Compaſs*; and Diſtances found by the *Log*, into one *Courſe* and *Diſtance*: And that is done after this manner:

1. Make a *Table*, conſiſting of *Six Columns*; in the *Firſt Column* ſet the ſeveral different *Courſes* of your *Day's Sailing*.

In the *Second Column*, over-againſt the reſpective *Courſes*, ſet the *Diſtances* in *Miles*; and let the other Columns be titled at the Top, with *South*, *North*, *Eaſt*, *Weſt*.

2. By the *Table* of *Difference of Latitude* and *Departure*, found in moſt Books of *Navigation*, ſeek the *Difference of Latitude* and *Departure*, for each *Courſe* and *Diſtance*, in your *Traverſe*; and ſet each *Difference of Latitude*, as the *Courſe* is *South* or *North*, under *Southing* or *Northing*: Alſo, ſet each *Departure*, as the *Courſe* is *Eaſt* or *Weſt*, under *Eaſting* or *Weſting*, reſpectively.

3. Add up the *Northings* and *Southings*; their Difference will be the *Difference of Latitude*.

And add up the *Eaſtings* and *Weſtings*; their Difference alſo, will be the *Departure* ſought.

4. With this *Difference of Latitude* and *Departure* found: The *direct Courſe* and *Diſtance* is

is readily had, either by the *Traverse-Table*, *Gunter's Scales*, or *Logarithms*.

TRAVERSE-*Table*, is the same with the *Table of Difference of Latitude* and *Departure*, in Books of *Navigation*; being only the *Difference of Latitude and Departure* ready Calculated to every *Degree*, *Point*, *Half-Point*, and *Quarter-Point* of the *Quadrant*, and for any *Distance* under 100 Miles, tho' it may conveniently serve for more.

This *Table* is one of the most necessary Instruments, that a *Navigator* has occasion for; since by this he may readily reduce all his *Courses* and *Distances*; run in the space of 24 *Hours*, into one *Course* and *Distance*; whence the *Latitude* he is in, and his *Departure* from the *Meridian* is found; and not only that, but the *Difference of Longitude* is gain'd also by Inspection, and that near enough for a Sailor's Purpose, in common and daily Practice. See *Longitude*.

TRAVERSE-*Board*, is a little round Board hanging up in the Steerage of the Ship, or elsewhere, and bored full of Holes upon Lines, shewing the Points of the Compass: And upon this Board, the Steers-man keeps an Account how many Glasses (that is, Half-Hour) the Ship Steers upon any Point, by moving a little Peg from Hole to Hole.

TRAVERSE, *To Cast a Point of Traverse*. See *Cast*.

TRAVERSE, also signifies to turn or remove a Piece of *Ordnance*, this, or that way, in order to bring it to Bear: and this is called *Traversing the Piece*.

TRAVERSE *the Yard!* that is, Brace it Aft.

TREE-NAILS, or *Trennels*, are long Pins, or Nails of Wood, made out of the Heart of *Oak*, to fasten the Planks to the Timbers; and these have always Oakum driven with them, to prevent any Leak.

TREES, *in a Ship*: Of *Timbers* so called, there are several, as

CHESS-TREES, are *Timbers* on each Side of the Ship, for the Main-Tack to run thro' and to Hale it down to. See *Chess*.

CROSS-TREES. See *Cross*.

ROUFE-TREES, are small Timbers that bear up the Gratings from the Half-Deck to the Fore-Castle; supported by Stantions that rest upon the Half-Deck.

WASTE-TREES, are those Timbers of a Ship that lie in the *Waste*.

TRESSEL-TREES, are those Timbers of the *Cross-Trees*, that stand along Ships, or Fore and Aft at the Top of the Mast. See *Cross-Trees*.

TRICE, or *Trise*, signifies, haling up any thing by a Dead Rope, or one that does not run in a Block; But is done by hand, or by main Strength: So if any Chest, Cask, or other Goods, has only a Rope made fast to it, and is pulled

up by hand into the Ship, without a Tackle, then they say, 'tis *Tris'd up*.

TRIM *of a Ship*, is her best Posture, Proportion of Ballast, and hanging of her Masts, &c. for Sailing.

To find the Trim of a Ship; that is, to find the best Way of making any Ship Sail swiftly, or how she will Sail best.

This is done, either by trying her Sailing with another Ship so many Glasses *Trimm'd* a Head, so many a-Stern, and so many upon an even Keel: Or else by easing of her Masts, and Shrouds; some Ships Sailing much better when Slack, than when they are Tau't.

But this depends very much upon Experience and Judgment, and the several Trials and Observations which the Commander and other Officers may make a-board.

TRIP; *as the Ship goes with her Top-Sails a-Trip*; that is, when she carries her Top-Sails hoisted up to the highest, and when it blows not too hard, but a gentle or Loom-Gale.

TROPICS, or *Tropicks*, are two Circles supposed to be drawn parallel to the *Equinoctial*, at 23 degrees and 30 minutes distant from it.

These Circles are term'd the Limits of the Sun's Progress; for between them the Sun is supposed to have his continual Course, and never exceeds beyond any of them.

TROPIC *of Cancer*, is that Circle which is suppos'd to be drawn parallel to the *Equinoctial Northward*, 23 degrees 30 minutes distant from it; and therefore is called the *Northern Tropic*; it passes thro' the Beginning of the Sign *Cancer*: And when the Sun comes to this Circle, that is, about the 12, or 13th of *June*, it makes the *Summer Solstice*, and the Longest Day (and *Shortest Night*) in the *Northern Hemisphere*; and therefore, the contrary in the *Southern* one.

TROPIC *of Capricorn*, is that Circle which is suppos'd to be drawn parallel to the *Equinoctial Southward*, about 23 degrees, 30 minutes distant from it, and therefore is called the *Southern-Tropic*; it passes thro' the Beginning of the Sign *Capricorn*: And when the Sun is said to come to this Circle, that is, about the 12th or 13th of *December*, it makes the *Winter Solstice*, and the *Longest Day* and *Shortest Night* in the *Southern Hemisphere*, and therefore the contrary in the *Northern one*.

TROUGH *of the Sea*, is the Hollow or Cavity made between two Waves or Billows, in a Rolling Sea.

The Ship lies in the Trough of the Sea; that is, She lies down between two Waves.

TRUCK,

TRUCK, is a ſquare Piece of Wood at the Top, wherein the *Flag-Staff* is put.

TRUMPETER, a *Sea-Trumpet*, whoſe Office is always to attend the Captain's Command; to be ready at the Entertainment of Strangers: Alſo when a Ship is Haled; and an Enemy's Ship Charg'd, Boarded or Enter'd. The Poop is his proper Station.

TRUCKS, belonging to the Carriage of a Piece of Ordnance; are the Wheels which are on the Axle-Tree, to move the Piece.

TRUNDLE-*Shot*. See *Shot*.

TRUNNIONS, are Pieces of well ſeaſon'd Wood, to make faſt the Ship's Timber with.

TRUNNIONS, of a Piece of Ordnance, are thoſe Nobs or Bunches of the Gun's Mettal, which bear it up upon the Cheeks of the Carriage.

TRUSSES; are Ropes made faſt to the Parrels of a Yard in a Ship; they ſerve either to bind the Yards faſt to the Maſts, when the Ship rolls, lying either a-Hull, or at Anchor; or to Hale down the Yards in a Storm, or Guſt of Wind.

These *Truſſes* belong to the *Main-Yard*, *Fore-Yard*, and *Miſen*; and are all brought to upon Occaſion.

TRY; as *the Ship lies a-Try under a (Main-Sail, or) Main-Courſe*; that is, when ſhe has no Sails abroad, but her Main-Courſe; the Tacks Hal'd cloſe aboard; the Bow Lines ſet up; the Sheats haled cloſe Aft; and the Helm tyed down cloſe by the Board; and ſo ſhe is let lie in the Sea: But at this time, if ſhe be upon a Wind, ſhe uſually makes her Way good, near Three *Points* before the Beam.

To Try under a Miſen-Sail: This is done when it blows ſo hard, that they cannot maintain the Main-Sail; that is, cannot bear the Main-Sail out: And then, if ſhe lie cloſe by the Wind, ſhe'll make her Way about Two *Points* before the Beam

But if ſhe *Try under a Main-Sail and Miſen*, ſhe'll make her Way good about Four Points before the Beam.

If the Ship will neither Try nor Hale, then Spoom; that is, put her Right before the Wind.

TUCK *of a Ship*, is the gathering up of the Works upon her Quarter under Water; which if it lie too low or two deep, makes her have a broad or a flat Quarter, and hinders her Steering, by keeping the Water from paſſing ſwiftly to the Rudder: But if it lie too high above the Water ſhe muſt be laid out in that Part, otherwiſe ſhe will want Bearing for her After-Works.

TURN, as *Land-Turn*. See *Land*.

TURNA-

TURNADO'S; are violent, sudden Gusts of Wind from all Points of the Compass; frequent upon the Coast of *Guinea*.

V

VALE *of a Pump*, is the Trough by which the Water runs from the Pump along the Ship's Sides, that it may go out at the Scupper-Holes.

VAN *of a Fleet*. See *Squadron*.

VANES; are those Sights which are made to move and slide upon *Cross-Staffs*. See *Quadrants*, &c.

VARIATION *of the Magnetic Needle* or *Compass*, is the Deflection of the *Magnetic Needle* from the *True Meridian*; or that Arc of the Horizon, either *Eastward* or *Westward*, intercepted between the *True* and *Magnetic Meridian*. See *Compass*.

The *Variation* is found either by the *Sun's Amplitude*, or *Azimuth*.

1. *To find the Variation of the Compass by the Sun's Amplitude, and Magnetic Amplitude.*

RULE. If the Amplitudes be of one Kind, that is, both North, or both South, their *Difference* is the *Variation*.

But if the *Amplitudes* be of different Kind, that is, one North, the other South, their *Sum* is the *Variation*.

And in counting both *Amplitudes* from the North, Observe that,

If the *Magnetic Amplitude* be to the Right of the *True Amplitude*, the *Variation* is *West*; but if to the Left, the *Variation* is *East*.

How the *Sun's Amplitude* is found, see *Sun*.

But the *Magnetical Amplitude* is found, by Observing, with a *Compass* for that Purpose, on what Point thereof the *Sun* Rises, or Sets: And the *Observation* should be made when the *Sun's* under-Limb is about $\frac{2}{3}$ of his Diameter above the *Horizon*; because of the Refraction, and the Height of the Observer's Eye above the Surface of the Water.

2. *To find the Variation of the Compass by the Sun's Azimuth, and the Magnetic Azimuth.*

RULE. In counting both *Azimuths* from the *North* in *North Latitude*, or *South* in *South Latitude*, and their *Difference* is the *Variation*.

And if the *True Azimuth* be to the Right of the *Magnetic Azimuth*, the *Variation* is *East*; But if to the Left, 'tis *West*.

The chief Use of Knowing the *Variation* at Sea, is to Correct the *Courses* of Ships. For if the *Variation* of the *Compass* be not allow'd,

allow'd, all Reckonings must be Erroneous.

And if the *Variation of the Compass*, and *Course* Steer'd upon, be given; the True *Course* is found by this RULE:

> *If the* Course *and* Variation *be of the same Kind, their* Sum *is the* True Course: *But if* Different *Kind, their* Difference *is the* True Course.

And this *Correction of the Course*, is in no Case so necessary, as in running on a *Parallel East* or *West* to hit a *Port*: For if being in your *Latitude*, at the Distance of 70 or 80 Leagues, you allow not the *Variation*, but Steer *East* or *West* by *Compass*, you shall fall to the *Northward* or *Southward* of your *Port*, on each side 19 Leagues of Distance, *one Mile* for each *Degree of Variation*, which may produce very dangerous Errors, where the *Variation* is considerable.

As for Instance; Having a good Observation in *Latitude* 49 *deg.* 40 *min.* about 80 *Leagues* without *Scilly*, and not considering that there is 8 *deg. West Variation*, I Steer away *East* by *Compass* for the *Chanel*; but making my Way truly *East* 8 deg. *North*, when I come up with *Scilly*, instead of being 3 or 4 Leagues to the *South* thereof, I shall find my self as much to the *Northward*: And this Evil will be more, or less, according to the Distance you sail in the *Parallel*. And therefore, to keep your *Parallel* truly,

> *You must go so many Degrees to the Southward of the East, and Northward of the West, as is the West Variation: But contrariwise, so many Degrees to the Northward of the East, and Southward of the West, as there is East Variation.*

VARIATION-CHART, Design'd by Capt. *Halley*: The Projection thereof is according to *Mercator*'s; and the Situation and Form of the Surface of the *Terraqueous Globe*, as to its Principal Parts, and the Dimensions of the several *Oceans*, are therein ascertain'd with the utmost Accuracy, as well from *Astronomical* Observations, as from *Journals*. And the *Curve-Lines* drawn over the several Seas shew the Degrees of the *Variation* of the *Magnetical-Needle*, or *Sea-Compass*. The Double Line passing near *Bermudas*, the *Cape Verde Isles*, and St. *Helena*, every where Divides the *East* and *West Variation* in this *Ocean*; and that on the whole Coast of *Europe* and *Africa* the *Variation* is *Westerly*, as on the more *Northerly* Coasts of *America*: but on the more *Southerly* Parts of *America* 'tis *Easterly*.

The *Degrees* of the *Variation*, or how much the *Compass* declines from

from the True *North* on either Side, is reckon'd by the Number of the Lines on each Side of the Double Curve, which is called the *Line of No Variation*; and each Fifth and Tenth is distinguish'd in its Stroak, and numbred accordingly: So that in what Place soever your Ship is, you find the *Variation* by Inspection.

Also in the *Indian Ocean*, where the Variation is altogether *Westerly*, encreasing till you come to the Meridian of the *East* Part of *Madagascar*, (where 'tis of about two Points) and from thence it decreases, till you arrive on the East-Coast of *China*, or at the *Philippine Isles*.

Here another Double Line on which there is *No Variation* divides again the *West* from the *East Variation*, that in all probability is to be met with almost all over that Immence Ocean called the *Southern-Sea*, where the *Curves of the Variation* are wanting, for want of Accounts and Journals to ascertain them.

In this *Chart*, 'tis plain, that the Degrees of *Westerly Variation*, in the *Mediterranean*, are all mark'd and numbred from 5° at *Malaga*, to 12° at *Cyprus*. In the *Baltic*, from 8° to 10°. In the *Red-Sea*, from 12° to 16°. And in the Gulf of *Persia*, from 13° to 14°, all *Westerly*.

And at the *Madera*, the Variation is 3° ½ *West*; at *Barbadoes* 5° ½ *East*; at *Annobon* 7° *West*; at *Diego Roiz* 19° *West*; at C. *Raze* in *Newfoundland* 14° *West*; at the Mouth of *Rio de Plata* 18° *East*; at *Java-head* 2° ½ *West*, &c.

So that this *Chart*, shewing the *Variation of the Compass* in any Place at Sight, which is unavoidably necessary for correcting the *Ship's Course* at Sea; Therefore, in a continued Cloudy Weather, or where the Mariner is not provided to observe this *Variation* duly, the Chart will readily shew him what Allowances he must make for this Default of the Compass, and thereby rectify his Journal. And besides you may also in many Cases estimate the *Longitude* at Sea; for where the Curves run near by *North* and *South*, and are thick together, as about C. *Bonne Esperance*, it gives a very good Indication of the Distance of the Land, to Ships come from far; for there the *Variation* alters a Degree to each two Degrees of *Longitude* nearly: But in the *Western Ocean*, between *Europe* and *North America*, the Curves lying nearly East and West, cannot be serviceable for this purpose.

Note, That there is a perpetual, tho' slow Change in the *Variation* almost every where; As that about C. *Bonne Esperance* the *West Variation* encreases at the rate of about a Degree in Nine Years. In the *English*

Channel,

Channel, it encreases a Degree in Seven Years; but slower the nearer the *Equinoctial Line*, as on the Coast of *Guinea*, a Degree in a 11 or 12 Years. On the *America-Side*, the *West Variation* alters but little; and the *East Variation* on the *Southern America* decreases, the more Southerly the faster; the *Line of no Variation*, moving gradually towards it. In the *Indian Seas*, after you pass St. *Lawrence*, the *West Variation* is on the Decrease, the faster the more Westerly and Southerly; and is in a manner at a Stand, when you come to the Length of *Java*. Therefore 'twill be necessary in time to alter the whole System from what it now stands in. Where this *Chart* is to be had, see under the Word *Chart*.

VEER; As, *Veer out a Rope!* That is, let it go by hand, or let it run out of it self.

VEER *more Cable!* That is, let more *Cable* run out.

VEER *more Sheat!* That is, let more of it out: And we don't use this Word for letting out of any Running Ropes, except the *Sheat*.

VEER, is also used in reference to the Wind: As,

The Wind Veers; that is, when it changes often, and suddenly, or shifts from Point to Point.

VEERING, as, *the Ship goes Veering*; that is, she goes at large; neither by a Wind, nor directly before it, but between both; which is also called *Quartering*.

VICE-ADMIRAL. See *Flag-Officers*.

VICTUALLING *the Navy*: This Affair was formerly performed by Contractors, but is now under the Care of Commissioners, who keep their general Office on *Tower-Hill*, and have not only their proper Agents at the several *Victualling* Ports in this Kingdom, but also their Correspondence abroad: And indeed it requires more than ordinary Care in supplying the Fleet with good and wholsome Provisions, the Want whereof subjects the Men to so many Distempers. This Care ought to extend it self as well to Quantity as Quality; for as nothing does more discourage a Sailor than his being wrong'd in the First, so is there not any thing subjects them to Diseases so much as a Defect in the Latter

VIOL, a kind of Hawser made use of to purchase in the Cable, when the Main Capstan cannot do it; because the Ground in which the Anchor is let fall, is too stiff, or else the Sea runs too high, so that they cannot weigh it: in which Case they take a Hawser, opening one End of it, they put therein Nippers, of about 8 Fathoms from each other, wherewith they bind this Hawser fast to the Cable, and then bring it

to the *Jear Capstan*, to heave upon it; And this will purchase far more than the *Main-Capstan* can.

W

WAD-HOOK, is a Rod, or Wire of Iron, turn'd Spiral-wise; and its End is put upon a Handle or Staff, to draw out *Wads* or *Oakum*, in unloading a Piece of Ordnance.

WAFT; as, *To Waft a Ship*, is to Convoy her safe, as Men of War do by Merchant-Ships.

To make a Waft, is to hang out some Coat, Sea-Gown, or the like, in the Main-Shrouds of the Ship, as a Sign for Men to come on Board, *&c.* as signifying oftentimes, that the Ship is in great Danger by a Leak, *&c.* and therefore wants Help from the Shore, or some other Ship.

WAIL, the same with *Wale*.

WAKE *of a Ship*, is the Smooth Water a-Stern, when she is under Sail: This shews the Way she has gone in the Sea, whereby the Mariners judge what Way she makes. For if the *Wake* be right a-Stern, they conclude she makes her Way forwards; But if the Wake be to Lee-ward a Point or two, then they conclude she falls to the Lee-ward of her Course.

The Ship stays to the Weather of her Wake; that is, when in her *staying* she is so Quick, that she does not fall to Lee-ward upon a Tack, but that when she is Tack'd, her *Wake* is to the *Lee-ward*; then 'tis a Sign that she feels her Helm very well, and is quick of Steerage.

WALE, or *Wail*, the same with *Bends*, which See.

Chain-WALE. See *Chain*.

Gun-WALE, See *Gun*.

WALE or *Waile-Knot*. See *Knot*.

WALE, or *Wail-Rear'd*; when a Ship is built upright, after she comes to her Bearing, she is said to be *Wale-Rear'd*; the which tho' it be unsightly, and as the Sea-men term it, not *Ship-Shapen*; yet causes a Ship to be much more Roomy, (that is, larger) within Board, and withal, makes her a wholsome Ship in the Sea, especially, if her Bearing be well Lay'd out.

WALT; *a Ship is Walt*, when she has not her due Ballast, that is, not enough to enable her to bear her Sails, and to keep her stiff

WAP, in a Ship, is that Rope wherewith the Shrouds are set Taught with *Wale-Knots*; one End is made fast to the Shrouds, and to the other are brought the *Lanniers*.

WAR *no more!* is only to bid him at the Helm, to keep the Ship as near the Wind as possible.

WARPING, is to Hale a Ship up by a Hawser, or any other Rope, sufficient for that Purpose, with an Anchor bent to it.

This

This is uſed, when a Wind is wanting to carry her into, or out of a Harbour.

WARP; ſo is the Hawſer, or Rope us'd to hale the Ship up, call'd.

WAST-BOARDS, are Boards ſometimes ſet upon the Sides of a Boat, or other Veſſel, to keep the Sea from breaking into her.

WAST-CLOTHS, are *Cloths* hung upon the uppermoſt Work of a Ship's Hull, to ſhadow the Men from an Enemy in a Fight; and therefore, by ſome, are call'd the *Fights*.

WASTE *of a Ship*, is that Part of her between the two Maſts, that is, between the *Main-Maſt* and *Fore-Maſt*.

WAST-TREES. See *Trees*.

WATCH, at *Sea*, ſignifies the Space of Four Hours; for Half the Ship's Company, or Crew, (when ſhe is at Sea,) *Watch*, and do Duty in their Turns, ſo long a Time.

Alſo, a Ship's Company is is Divided into Two Parts, call'd the *Larboard-Watch*, and the *Starboard-Watch*.

Quarter-WATCH; that is, when but a *Quarter* of the Ship's Company *Watch* at a time; as when they are in Harbour, having but little to do, or look after.

WATCH-GLASS, being Four Hours, is uſed at *Sea*, to Shift or Change their *Watches*.

There are alſo *Half-Watch-Glaſſes*; *Minute*, and *Half-Minute-Glaſſes*; whereby they count the *Knots*, when they Heave the *Log*, in order to find the Ship's Way.

WATER-*Born*, is ſaid of a Ship, when ſhe is where there's no more *Water* than will juſt *bear* her from the Ground.

Dead-WATER, is the *Eddy-Water* that follows the Stern of a Ship, not paſſing away ſo faſt as that which ſlides by her Sides.

WATER - *Draught*. See *Draught*.

Eddy-WATER. See *Eddy*.

Foul-WATER; As, *The Ship makes Foul-Water*. See *Foul*.

High-WATER; that is, when the *Tide* is at the *Higheſt*.

WATER-*Line*, is that which diſtinguiſhes that Part of a Ship which is *under Water*, from that *above*, when ſhe is duly Laden.

Low-WATER; that is, when the *Tide* is at the *Loweſt*.

WATER-*Shot*, is a kind of Riding at Anchor, when the Ship is Moor'd neither Croſs the Tide, nor Right up and down, but Quartering betwixt both.

WATER-*Way*, in a Ship, is a ſmall Piece of Timber lying *fore* and *aft* on her Deck, cloſe by her Sides, to prevent the *Water* running down there.

To find the Time of High-Water, or Full-Sea, at any Place.

The RULE.

To the *Moon's Southing*, add the Point of the *Compass* making *Full Sea* (on *Full* and *Change-Day*) for the Place propos'd; and the Sum is the Time sought.

Note, The Point of the *Compass* making *Full-Sea* on the *Full* and *Change-Day*, is found in the *Tide-Table*, in *the Seaman's Kalender*, and other Books of *Navigation*.

The Time of the Moon's Southing, is found by this

RULE.

Multiply the *Age of the Moon* by 4, and Divide the Product by 5, the Quotient is *Hours*; and the Remainder is so many times 12 *Minutes* of an *Hour*, and both is her *Southing* sought.

The Moon's Age is found by this

RULE.

To the *Epact* Add the *Day* of the *Month*, and the *Number* of the *Month*; the Sum, if it exceeds not 30, is her *Age*. But if it does, Subtract 30 as oft as you can, and the Remainder is her *Age*.

The *Numbers* of the Months are these:

Jan.	*Feb.*	*Mar.*	*Apr.*
0	2	1	2
May	*June*	*July*	*Aug.*
3	4	5	6
Sept.	*Oct.*	*Nov.*	*Dec.*
8	8	10	10

The Epact *is found by this*

RULE.

1. Divide the *Golden Number* by 3; and Note the Remainder.

2. Multiply that Remainder by 10, and Note the Product.

3. To that Product Add the *Golden Number*: The Sum (if it exceed not 30) is the *Epact*; but if it does, Subtract 30 from it, and the Remainder is the *Epact* sought.

The Golden Number *is found by this*

RULE.

Add 1 to the Year of Our Lord, and Divide the Sum by 19; the Remainder will be the *Golden Number* sought.

WAVES, are the Billows made by the rolling of the Sea: Their Breadth is estimated to be the distance between the top Edges of any two *Waves*; or

or a Right Line drawn from the Middle of one Trough or Hollow, to the Middle of the other.

And the Velocity of *Waves*, is always in a *Subduplicate Ratio* of their Breadth. Therefore, those *Waves* whose Breadth is $3\frac{1}{15}$ *Parisian* Feet, will pass over that Distance in a Second of Time; and consequently, in One Minute, will run $183\frac{3}{4}$ Feet; and in an Hour's Time, will have gone about 11000 such Feet.

WAY *of a Ship*: As, *To keep an Account of the Ship's Way*, is one of the most necessary Things in the Practice of *Navigation*; and the most-Approv'd Method of doing this, is by the *Log* and *Log-Line*; which see.

Lee-WAY, or *Leeward-Way*, is the Angle made by the *Point Steer'd* upon, and the *Way of the Ship*. And since most Ships are apt to fall a little to *Leeward* of their Course, 'tis necessary, at all time, in Casting up the *Log-Board*, to Allow something for the *Lee Way*. As,

1. If the Ship be upon a Wind, allow One *Point* for *Lee-Way*.

2. If the Wind blow hard, and cause one *Top-sail* to be taken in, allow Two *Points* for *Lee-Way*.

3. If the Wind blow so hard, that both *Top-sails* are taken in, and the Sea run high; allow Three *Points* for *Lee-Way*.

4. If the *Fore-sail* be Furl'd, and she *Try* under a *Main-sail* and *Misen*; she'll make her *Way* Four *Points* before the Beam.

5. If she *Try* under a *Main-sail* only, she will make her *Way* near Three *Points* before the Beam.

6. But if she *Try* under a *Misen* only, she'll make her *Way* about Two *Points* before the Beam.

7. If she lie *a-Hull*, with all her Sails Furl'd; she will make her *Way* One *Point* before the Beam.

As may be seen more at large, in Mr. *Jones*'s *Compendium* of *Navigation*.

WAY *Afrward on*, or *Forward on*; that is, the *Rake* or *Run* of a Ship: Which see.

WEATHER: As, *To Weather a Point*; that is, to sail by it with a Fair Wind.

To WEATHER *a Ship*; that is, to go to Windward of her.

WEATHER-*Coyle* or *Quoil*; that is, when a Ship has her Head brought about, so as to lie that way, which her Stern did before, without loosing of

any Sail, but only by bearing up the Helm.

WEATHER-*Gage*. See *Gage*.

To WEIGH *Anchor*. See *Anchor*.

WENDING, ſignifies, bringing the Ship's Head about; and ſeems to be the ſame as *Winding*.

How Wends the Ship? the ſame with, *How Winds the Ship?*

WEST, is one of the Four *Cardinal-Points* of the World: And is that Point of the *Horizon* where the *Sun* is ſeen to *Set*, when 'tis in the *Equinoctial*.

WESTERN-*Amplitude*, is an Arc of the *Horizon* intercepted between the Point of the *Sun*'s Riſing, and the *Weſt-Point* of the *Horizon*.

WESTERN-*Hemiſphere*. See *Hemiſphere*.

WESTING. See *Departure*.

WHELP of the *Capſtan*. See *Capſtan*.

WHERREYS, are Small Boats employ'd in Rivers; they are too weak and tender for any Service at Sea.

WHIP, or *Whip-Staff*, is a Piece of Timber like a ſtrong Staff, faſten'd into the Helm, for him that Steers, in ſmall Ships, to hold in his Hand, thereby to move the Helm, and Steer the Ship: It goes thro' the Rowle, and is made faſt to the Tiller with a Ring.

WHIRLWIND, a ſort of Wind that turns round, and blows divers Ways at once.

WHOLSOM *Ship*; that is, a Ship that will *Try*, *Hull*, and *Ride* well, without Rolling or Labouring in the Sea. A Long Ship that draws much Water, may *Try*, *Hull*, and *Ride* well; But if ſhe draw's little Water, ſhe may *Try* and *Ride* well, yet never *Hull* well: And a Short Ship, that draws much Water, may *Hull* well, but neither *Ride* nor *Try* well; and ſuch is call'd *an Unwholſom Ship*.

WHOODINGS, are thoſe Planks which are join'd and made faſt along the Ship's Side into the Stern.

WIND, is defin'd to be the Stream, or Current of the Air; which blows from ſome One of the Two and Thirty Points of the *Horizon*, or *Compaſs*: Hence they are call'd the Two and Thirty *Winds*; and therefore the Four *Cardinal-Points*, *North*, *South*, *Eaſt*, and *Weſt*, are call'd *Cardinal-Winds*.

WINDLASS, is an Inſtrument, in ſmall Ships, placed upon the Deck, juſt abaft the Fore-Maſt. 'Tis made of a Piece of Timber having Six or Eight Squares, and is in the Form of an Axle-Tree, whoſe Length is placed Horizontally upon Two Pieces of Wood at the Ends thereof, and upon which it's turn'd about by the Help of Handſpikes, put into Holes made for that Purpoſe: This

This Instrument, serves for Weighing of Anchors, or Hoising of any Weight in or out of the Ship; and will Purchase much more than any Capstan, and that without any Danger to those that Heave: For if in heaving the *Windlass* about, any of the Hanspikes should happen to break, yet the *Windlass* would paul of it self.

Trade-WINDS; the *General Trade Winds*, are those which blow perpetually from the *East*, between the Latitudes of 30 Degrees North and South, in the *Atlantic*, *Ethiopic*, and *Pacific Oceans*; But that to the Northward of the *Equator*, inclines to the Northward of the *East*; and that to the Southward of the *Equator*, to the Southward of the *East*:

Which is thus Accounted for by Mr. *Halley*, from the Laws of *Statics*: The *Air* which is less Rarified or Expanded by Heat, and consequently, more Ponderous, must have a Motion towards those Parts thereof, which are more Rarified and less Ponderous, to bring it to an *Æquilibrium*: And the Presence of the *Sun* continually shifting to the Westward, that Part towards which the Air tends, by reason of the Rarification made by his greatest Meridian Heat, is with him carried Westward, and consequently the Tendency of the whole Body of the Lower Air is that way. Thus a general *Easterly Wind* is form'd, which being impress'd upon all the Air of a Vast Ocean, the Parts impel one the other, and so keep moving till the next Return of the Sun; whereby so much of the Motion as was lost, is again restor'd: Thus the *Easterly Wind* is made perpetual.

And since near the *Line*, the Air is much more Rarified, than at a greater Distance from it; because the Sun is twice in a Year Vertical there, and at no time distant above 23 Degrees and a half; at which Distance the Heat, being as the Sine of the *Angle of Incidence*, is but little short of that of the Perpendicular Ray. Whereas, under the Tropics, tho' the *Sun* stay long Vertical, yet he is as long 47 Degrees off; which is a kind of Winter, wherein the Air so Cools, as that the Summer-Heat cannot warm it to the same Degree with that under the *Equator*. Wherefore the Air to the Northwards and Southwards, being less Rarified than that in the Middle, it follows, that from both Sides it ought to tend towards the *Equator*: This *Motion* compounded with the former *Easterly Wind*, answers all the *Phænomena* of the *General Trade-Winds*; which, if the whole Surface of the Globe were Sea, would undoubtedly bow all round the World, as they are found to

do in the *Atlantic* and *Ethiopic Oceans.*

COASTING-TRADE-WINDS, are such as the *Southerly* and *South-Westerly Winds*, which blow perpetually all along the Coast of *Africa* on the Southward of the *Equator* within the *Trade-Wind* Limits: and the *Northerly* and *North-Westerly Winds* on the North Side of the *Equator*, blowing on the same Continent.

SHIFTING-TRADE-WINDS, the same with *Monsoons*; which See.

Note, That from the Original View of these *Winds* given by Mr. *Halley*; they have bin lately represented on a Large and Correct *Mercator*'s Chart of the *World*; as also on a New *Terrestrial-Globe*, both being carefully done by *Char. Price*, and *Jer. Sellers*, *Hydrographers*, at the *Hermitage-Stairs.*

To WIND *a Ship*, or *Boat*; that is, to bring the *Ship*, or *Boat*'s Head about.

How WINDS *Ship?* that is, Upon what Point of the Compass does she lie with her Head.

To WIND *up*; that is when the Ship comes to Ride at an Anchor.

Eddy-WIND. See *Eddy*.

Side-Wind, is that which blows Sideward.

WINDWARD *of Ship*; that is, to the Wind of a Ship.

To have the WIND *of a Ship*; that is, to be to *Windward* of her.

WINDWARD-*Tide.* See *Tide.*

Large-WIND: As, *to Sail with a Large Wind*; that is, with a Fair Wind.

Quarter-WIND. See *Quarter.*

Touch the WIND. See *Touch.*

WIND-*Taught*: As, too much Rigging, and high Masts, *&c.* is said to hold the Ship *Wind-taught*, and this makes her stoop too much in her Sailing in a Stiff Gale of Wind.

And when a Ship Rides in in any Main-Stress of Wind and Weather, the Yards are brought alongst Ship, and the Top-Masts are struck down, which else would hold much *Wind*, or be *Wind-Taught.*

WINDING-*Tackle.* See *Tackle.*

To Ride WIND-*Road.* See *Ride.*

The WIND *Veers.* See *Veers.*

WOOD *and* WOOD; that is said, when Two Pieces of Timber; being so let into each other, that the Wood of the one joins close to the other. See *Scarfed.*

WORKING-PART *of Navigation.* See *Navigation.*

WORMING, is laying a small Line or Rope all along betwixt the Strouds of any Cable or Hawser, in order to strengthen or succour it.

To

To WOULD, or to Bind; as to *would a Mast*, or *Yard*; is the winding of Ropes fast about a Mast, or Yard that is fish'd, in order to make it hold the better.

WRACK, *Ship-wrack*, is a Sea-Term, when a Ship either splits or sinks; and those that escape, are said *to have suffer'd Shipwrack*.

WRIGHT's, or *Mercator's Sailing*, is the Method of finding on a Plane, the Place of a Ship upon any assign'd *Course*, true in *Longitude*, *Latitude*, and *Distance*; the *Meridians* being supposed Parallel, and the *Parallels of Latitude* strait Lines.

CASE 1.

Given, *the Latitude and Longitude of two Places*;

Required? *Course*, *Distance*, *and Departure*.

1. To find the *Course*, say,

As the Meridional Difference of Latitude,
To the Radius:
So is the Difference of Longitude,
To the Tangent of the Course.

2. Find the *Distance* and *Departure* by Case 3. of *Plain-Sailing*.

CASE 2.

Given, *the Latitudes* and *Course*;

Required? *Difference of Longitude, Distance, and Departure.*

1. To find the *Difference of Longitude*, say,

As the Radius,
To Meridional Difference of Latitude:
So is the Tangent of the Course,
To the Difference of Longitude.

2. Find the *Distance* and *Departure* by Case 3. of *Plain Sailing*.

CASE 3.

Given, *the Latitudes and Distance*;

Required? *Course, Difference of Longitude, and Departure.*

1. Find the *Course* and *Departure* of *Case* 5. of *Plain Sailing*.

2. Find the *Difference of Longitude*, by *Case* 2. of this: Or by saying,

As the Difference of Latitude,
To the Departure:

So

So is the Meridional Difference of Latitude,
To the Tangent of the Course.

CASE 4.

Given, *Latitudes and Departure.*
Requir'd? *Difference of Longitude, Course, and Distance.*

1. Find the *Difference of Longitude*, by *Case* 3. of this.
2. Find *Course* and *Distance*, by *Case* 6. of *Plain-Sailing.*

CASE 5.

Given, *One Latitude, Course, and Distance.*
Requir'd? *Difference of Latitude, and Difference of Longitude.*

1. Find the *Difference of Latitude*, by *Case* 1. of *Plain-Sailing.*
2. Find the *Difference of Longitude*, by *Case* 2. of this.

CASE 6.

Given, *One Latitude, Course, and Departure.*
Requir'd? *Distance, Difference of Latitude, and Difference of Longitude.*

1. Find the *Distance* and *Difference of Latitude*, by *Case* 2. of *Plain-Sailing.*
2. To find the *Difference of Longitude*, say,

As the Difference of Latitude,
To Meridional Difference of Latitude,
So is the Departure,
To the Difference of Longitude.

CASE 7.

Given, *One Latitude, Distance, and Departure.*
Requir'd? *Course, Difference of Latitude, and Difference of Longitude.*

1. Find the *Course* and *Difference of Latitude*, by *Case* 4. of *Plain-Sailing.*
2. Find the *Difference of Longitude*, by *Case* 2. or 6. of this.

CASE 8.

Given, *One Latitude, Course, and Difference of Longitude.*
Requir'd? *Difference of Latitude, Distance, and Departure.*

1. To find the *Difference of Latitude*, say,

As the Radius
To the Difference of Longitude:
So is the Co-Tangent of the Course,
To the Meridional Difference of Latitude,

2. Find the *Distance* and *Departure*, by *Case* 3 of *Plain-Sailing.*

Y

YARDS *of a Ship*, are thoſe long Pieces of Timber which are made a little Tapering at each End, and are fitted each a-thwart its proper Maſt, with the Sails made faſt to them; ſo as to be Hois'd up, or Lower'd down, as Occaſion ſerves. They have their Names from the Maſts unto which they belong.

As for the *Length* of the *Main-Yard*, 'tis uſually $\frac{5}{6}$ of the Length of the *Keel*, or $\frac{6}{7}$ of the Length of the *Main-Maſt*.

Their *Thickneſs* is commonly $\frac{3}{4}$ of an Inch for every Yard in Length.

The Length of the *Main-Top-Yard* is $\frac{2}{5}$ of the *Main-Yard*: And the Length of the *Fore-Yard*, according to ſome, is $\frac{4}{5}$ of the *Main-Yard*; but by others, $\frac{}{9}$ thereof.

The *Spritſail-Yard*, and *Croſs-Jack-Yard*, are of the ſame Length; that is to ſay, $\frac{1}{2}$ the *Miſen-Yard*.

And the *Thickneſs* of the *Miſen-Yard* and *Spritſail-Yard* is $\frac{1}{2}$ an Inch for every Yard in Length.

All Small *Yards* are Half the Great *Yards* from *Cleat* to *Cleat*.

When a *Yard* is *down a Portlaſt*, it gives the Length of all *Topſail-Sheats*, *Lifts*, *Ties*, and *Bunt-Lines*; as alſo, of the *Leech-Lines* and *Halliards*, meaſuring from the *Hounds* to the *Deck*: And when it is Hois'd, it gives the Length of *Clew-Lines*, *Clew-Garnets*, *Braces*, *Tackles*, *Sheats*, and *Bow-Lines*.

And if all Ships were Built after one and the ſame Mould; and all their *Maſts*, *Yards*, *Cables*, *Cordage* and *Sails*, were of like Goodneſs, a Methodical Rule of their Proportion might be given: But ſince their Length, Breadth, Depths, Rakes and Burthens are ſo variable and different; it is hardly poſſible to give any certain Rules, which may be taken as Standard: Tho', in Time, we may expect more Regularity in the Method of Building Ships, that is, when the *Theory* thereof is *better known* and minded.

YARD-ARM, is that half of the *Yard* that is on either ſide of the Maſt, when they lie a-thwart the Ship.

Brace the Yard! That is, Traverſe aſt the *Yard-Arm*, whoſe *Brace* is haled.

Square the Yards! That is, See that they hang right a-croſs the Ship, and one *Yard-Arm* not Travers'd more than another.

Top the Yards! That is, Make the *Yards* ſtand Even: For to *Top* the *Main* and *Fore-Yards*, the Clew-Lines are the moſt

most proper; but when the Top-sails are Stow'd, then the Top-sail-Sheats will *Top* them.

YARE: As, *Be Yare at the Helm!* That is, Be Quick, Ready, and Expeditious at the Helm.

YARE, is sometimes used by Seamen for Bright: As, *To keep his Arms Yare*; that is, to keep them Clean, and Bright.

YARN, as *Rope-Yarn*; by this is meant the *Yarn* of any Rope untwisted: 'Tis usually made of Cables-Ends which are worn out. It's Use is, to serve small Ropes, to make Sinners, Mats, Plats, and Caburns. It serves also to fasten the Sails to the Yard-Arms, with several other Uses.

Spun-YARN, is nothing but *Rope-Yarn* made small at the Ends, and so spun one to another with a Winch, as long as One would have it.

YATCHES, are One-Deck'd Vessels, carrying Four, Eight, or Twelve *Guns*; with Thirty or 40 *Men*; and from Thirty, to an Hundred and sixty *Tuns*: They have the Gun-Deck from Thirty to Sixty three *Feet* Long, and from Thirteen to Twenty-one *Feet* Broad. They Draw little Water; and are Excellent for running of small Boards, making short Trips. They generally serve for carrying of Persons of Quality, &c.

YAWS: As, *The Ship makes Yaws*; that is, does not Steer steady, but goes in and out when there is a stiff Gale. And he that keeps the Ship most from *Yawing*, does commonly use the least Motion with the Helm, and those Steer best.

YOKE, *Sea-Yoke*: When the *Sea* is so Rough, that the Men cannot govern the Helm with their Hands, then they seize two Blocks to the End of the Helm, one on each side, and reeving two small Ropes thro' them, which they call Falls, and which are made fast to the Sides of the Ship: By having some Men at each Tackle, they govern the Helm according to Direction: This they call a *Yoke* to Steer by.

And sometimes, Seamen make a *Yoke*, by taking a double Turn about the End of the Helm by a single Rope, the Ends being laid to the Ship's Sides; and by this means they guide the Helm.

YOUNKERS, are those Young-Men on-board a Ship, that Take in the Top-sails, and are for Furling the Sails, Slinging the Yards, Bowsing, Tricing, &c.

ZENITH,

Z

ZENITH, is the Point in the Heavens, right-over One's Head; or is the Upper *Pole* of the *Horizon*; and therefore is 90 Degrees every way distant from it.

ZENITH-*Distance*, is the Complement of the *Sun* or *Star's Meridian Altitude*; that is, what the *Meridian Altitude* wants of 90 *Degrees*.

ZONE: The Surface of the *Terraqueous Globe*, was by the Ancients divided into Five *Zones*; that is to say, One *Torrid* or *Burning Zone*, Two *Temperate*, and Two *Frigid* or *Frozen Zones*.

Torrid or *Burning* ZONE, contains all that Space that lies between the Two *Tropics*.

Temperate ZONES, are, one on the *South* side of the *Equator*, between the *Antarctic Circle* and the *Tropic of Capricorn*; the other on the *North* side of the *Equator*, between the *Arctic Circle* and the *Tropic of Cancer*.

Frozen ZONES, are comprehended between the *Polar Circles* and the *Pole*: And therefore the

Northern Frozen ZONE, lying between the *Arctic Circle*, or the *Parallel* of 67 *deg.* 30 *min. North Latitude*, and the *Arctic Pole*, contains *Nova-Zembla*, and Part of *Greenland*, *Samojeda*, *Finmark*, *Lapland*, *Norway*, *Island*, and some Parts of *North-America*.

Southern Frozen ZONE, lying between the *Antarctic Circle* and the *Antarctic* or *South Pole*, is not yet known whether it contains Land or Sea.

FINIS.

☞ For EASTERN *Amplitude*, read *Eastern Magnetic Amplitude*.

PROPOSALS

For Printing

A Compleat HISTORY *of* ENGLAND, *from the Earliest Account of Time, to the Death of His Late Majesty* King WILLIAM the Third: *Containing a Faithful Relation of all Affairs of State, Ecclesiastical and Civil; with the* Effigies *of all the* Kings *and* Queens, *taken from the Orginals, and Curiously Engrav'd by the best Masters.*

MANY have been the Attempts, by Great Men, to make a Compleat History of the Lives of the Kings of *England*; but they have all been Disappointed, either by unavoidable Avocations, Sickness, Death, or many times want of Resolution, being Discourag'd with the Prospect of so vast a Work; even the Learned Mr. *Camden* confesses the Task too weighty, and so confin'd himself to the History of Queen *Elizabeth*, which Reign alone cost him Sixteen Years Labour, tho' he had all the Materials before him, and was an Eye-witness of most Transactions. There being little hopes therefore ever to see a Design of this Nature compleated by One Man, we have taken another Method, which is, by uniting the Labours of Several, who being Men of great Reputation, and living in or near the Times of those Princes whose Reigns they writ, had better Opportunities of throughly informing themselves. This is the Way which Sir *William Temple*, and several other Learned Men, approv'd and recommended, and which has been above even Years Carrying on with great Application, and is now Ready for the Press.

The Method and Authors made use of, are as follow:

I. *Milton*'s History of *England*, to the Conquest.

II. From the Conquest, to the End of the Reign of K. *Edward* III. by Mr. *Daniel*.

III. The Reigns of K. *Richard* the IId. K. *Henry* IVth, Vth, and VIth, are all new writ in Mr. *Daniel*'s Method; Mr. *Trussel*'s Performance not having met with such Approbation as Mr. *Milton*'s and Mr. *Daniel*'s, we chose therefore to be at the Charge of writing those Lives a-new, to make the Work all of a Piece.

IV. The

have a Seventh *gratis*; which will reduce it to about 2 *l.* 15 *s.* 4 *d.* *per* Book. The Price to any but *Subscribers* not under 3 *l.* 15 *s.* in Quires.

III. A few will be Printed on Royal Paper, at Five Guinea's each, Two to be Paid in Hand, the other Three on Delivery.

IV. *Subscriptions* will be taken in till the First of *June* next, and the Book Finish'd in *Michaelmas-Term* following.

V. The *Subscribers* shall have their Names and Titles Printed before the Book, as Encouragers of so Useful and Chargeable an Undertaking.

SUBSCRIPTIONS are Taken in by the UNDERTAKERS,

Brab. Aylmer,
Henry Bonwicke,
Sam. Smith and *Benj. Walford*,
Will. Freeman,
Tim. Goodwin,
Tho. Bennet,
Matth. Wotton,
Jo. Walthoe,
S. Manship,
Tho Newborough,
John Nicholson,
Richard Parker, and
Benj. Tooke,

BOOKSELLERS, In *Fleet-street*, St. *Paul's Church yard*, *Little-Britain*, and near the *Royal-Exchange* in *Cornhil*.

IV. The Reign of K. *Edward* the IVth, is excellently well writ by Mr. *Habington*.

V. The Lives of K. *Edward* the Vth, and K. *Richard* the III. by Sir *Thomas Moore*; and that also of the last by Mr. *Buck*, have met with a general Esteem, answerable to the Characters of the Authors.

VI. The Reign of K. *Henry* the VIIth. by the Lord *Bacon*; as also the Annals of the said King, writ by Sir *Ja. Ware* (chiefly relating to *Ireland*, and which he publish'd to supply some Defects in the Lord *Bacon*'s History) we have translated, keeping each Author by himself.

VII. The Reign of K. *Henry* the VIIIth. by the Lord *Herbert*.

VIII. The Reign of King *Edward* the VIth, and Queen *Mary*, by Bishop *Godwin*, Sir *John Heyward*, and Others.

IX. The Reign of Queen *Elizabeth*, by Mr. *Camden*; but this being originally writ in Latin by the Author, we have it new translated by a good Hand, it having suffer'd very much by the two former Translators.

X. The Reign of King *James* the 1st. by Mr. *Wilson*; which Author being suspected of some Partiality, we have had him carefully examin'd throughout; and tho' we have kept his Text entire as he writ it, have added several considerable Observations, which which will set all the Matters of Fact in a true Light.

> *N.B.* This Reign was intended to be writ by Mr. *Camden*, but he only drew a Plan of his Design, and left some short Memorials of it, which shall be here Printed.

XI. The Reigns of K. *Charles* the 1st. and IId. King *James* the IId. King *William* and Queen *Mary*, are new writ by a Learned and Impartial Hand.

All the Authors that are Reprinted are accurately Collated, with the Ancient Historians of the respective Times, and Improv'd from them with useful Additions, Observations, and Corrections.

The PROPOSALS *are as follow:*

I. THE Book will be Printed in Three Volumes, containing about 600 Sheets; which with the Charge of Cutts, will make the Expence near one Fourth-part more.

II. That the Price to *Subscribers* shall be Three Guinea's in Quires, One to be Paid down, and the other Two at the Delivery; and for the Encouragement of those who Subscribe for Six, shall have